Topics in Mining, Metallurgy and Materials Engineering

Series Editor

Carlos P. Bergmann, Federal University of Rio Grande do Sul, Porto Alegre, Rio Grande do Sul, Brazil

"Topics in Mining, Metallurgy and Materials Engineering" welcomes manuscripts in these three main focus areas: Extractive Metallurgy/Mineral Technology; Manufacturing Processes, and Materials Science and Technology. Manuscripts should present scientific solutions for technological problems. The three focus areas have a vertically lined multidisciplinarity, starting from mineral assets, their extraction and processing, their transformation into materials useful for the society, and their interaction with the environment.

More information about this series at http://www.springer.com/series/11054

José Ignacio Verdeja González •
Daniel Fernández González •
Luis Felipe Verdeja González

Operations and Basic Processes in Ironmaking

José Ignacio Verdeja González
Escuela de Minas, Energía y Materiales
University of Oviedo
Oviedo, Asturias, Spain

Daniel Fernández González ⓘ
Escuela de Minas, Energía y Materiales
University of Oviedo
Oviedo, Asturias, Spain

Luis Felipe Verdeja González ⓘ
Escuela de Minas, Energía y Materiales
University of Oviedo
Oviedo, Asturias, Spain

ISSN 2364-3293   ISSN 2364-3307   (electronic)
Topics in Mining, Metallurgy and Materials Engineering
ISBN 978-3-030-54608-3   ISBN 978-3-030-54606-9   (eBook)
https://doi.org/10.1007/978-3-030-54606-9

This Springer imprint is published by the registered company Springer Nature Switzerland AG
The registered company address is: Gewerbestrasse 11, 6330 Cham, Switzerland

This book is dedicated to the memory of José Antonio Pero-Sanz Elorz (1934–2012), who founded in the middle of the seventies of the last century the research group in iron and steelmaking, metallurgy and materials at the University of Oviedo (Asturias, Spain). As a result of this long period of research and teaching, several books have been published, including Physical Metallurgy of Cast Irons *(Springer, 2018) and* Structural Materials: Properties and Selection *(Springer, 2019).*

Prologue

Iron, together with the copper, is the only element of the periodic table related to the human's ages. According to the history, the Hittites Kingdom (located in the Anatolia, former Turkey) marked the beginning of the Iron Age in the Twelfth century BC.

Even when the iron obtaining sinks its roots in the history of the humans, the Iron Science and Technology has only had a marked development in the second half of the Nineteenth century. Along the seven chapters of this manuscript, we describe the operations and industrial processes connected with the production of iron with carbon contents within 1.50 and 5.00%. These seven chapters constitute the first part of the iron and steelmaking process, called ironmaking, as focuses on the stages of the process until obtaining the pig iron in the blast furnace.

The book tries, above all, to be a textbook for the university teaching of undergraduate, master, and Ph.D. students. At the same time, this textbook might be a useful instrument in the basic formation of those people who are going to work in the iron and steelmaking industry and need to refresh–update the basic knowledge acquired in the university studies.

The manuscript shows for the first time in a book of these characteristics, together with the most significative operating variables of the processes and basic operations of the iron obtaining numerous solved exercises. These solved exercises are conceptually supported on the thermodynamic and kinetic fundamentals of the sintering, pelletizing, coking, reduction, and smelting reduction of iron ores, as well as on those subjects related to the desulfurizing and dephosphorizing of the iron melts with high carbon contents.

It is possible that, for professionals and experts from the iron and steelmaking process, the contents of these seven chapters of the manuscript, raw materials, agglomeration, ironmaking coke, or technologies related to the reduction with gas or coke, might be limited. However, in the enclosed references and bibliography, we have included specific manuals and reviews that lead the reader to both the reinforcement of the concepts developed in the text and the increase of the specialization on the matters, developments, and technological changes presented in the text.

Finally, we want to express our gratitude to the Prof. José Antonio Pero-Sanz Elorz because he was the Head of the Research Group in Iron and Steelmaking,

Metallurgy and Materials that started, in the seventies of the Twentieth century, the researches in the field of materials science and metallurgy at the University of Oviedo. We also want to express our gratitude to the Professors José Pedro Sancho Martínez and Ángel Alfonso Fernández from the University of Oviedo and Roberto González Ojeda from the Universidad Panamericana de México for their contributions both in the thermodynamic fundamentals and in the simulation and modeling of processes.

Oviedo, Asturias, Spain   José Ignacio Verdeja González
May 2020   Daniel Fernández González
                                              Luis Felipe Verdeja González

Contents

Raw Materials

1

1.1 Introduction: Iron and Steel in the History

The metallurgy can be defined as the science and technique of the extraction, concentration, and production of the different types of metals, including two of the most closely related to the development and technological progress of the man, the iron and the steel.

The Iron Age began when the man, probably by chance, discovered the iron sponge when he was heating iron ores that were easily reducible. According to the historians, the Hittites probably had the monopoly of the iron production (1500 BC) in that period. After the disappearance of this kingdom (approximately in the 1200 BC), the experts in the manufacture (i.e., transformation) of the metal moved geographically and their skills and knowledge were transferred to other civilizations and nations. On the one hand, humans were able to give different shapes to a resistant and deformable material as the steel. On the other hand, humans were able to improve and technologically develop the steel production process. These are the reasons for the maintenance of the ironmaking as one of the pillars of both the economic activity and the creation of infrastructures without interruption over the years.

Nowadays, the production of steel is mainly carried out following one of the next routes:

A. **Integrated iron and steel route**, which uses as raw material iron ore and follows the sequence blast furnace–converter–secondary metallurgy to produce steel by means of continuous casting or conventional casting (solidification in metallic or ceramic ingot). Approximately 65% of the world steel production (according to *International Iron and Steel Institute*) is obtained by the integrated route. The Corex process is the single alternative to the smelting reduction performed in the blast furnace. In this process, a molten metal with similar characteristics to that of the pig iron obtained in the blast furnace is

J. I. Verdeja González et al., *Operations and Basic Processes in Ironmaking*, Topics in Mining, Metallurgy and Materials Engineering, https://doi.org/10.1007/978-3-030-54606-9_1

produced (Corex pig iron, Chap. 6, Sect. 6.6). The capacity of production (Corex) installed in China, India, and South Africa adds up to around 10 million tons per year.

B. **Electric furnace steel plant**, which is based on the utilization of either scrap or prereduced iron ore (synthetic scrap). The electric furnace together with the processes of the secondary metallurgy is the main stage for the production of the steel. 30% of the world steel production is obtained using this method, which is an instrument available in the iron and steelmaking to recycle the steel.

C. **DRI (Direct Reduction Iron)**, which uses iron ore to obtain, by means of a reductant gas, a porous metallic product contaminated with the mineral gangue. However, the following stages (as happens in the electric furnace steel plants) are necessaries to obtain the final product: smelting, adjustment of the composition, and casting. Approximately 5% of the world steel production is obtained using this route.

The Martin–Siemens process with configuration of Basic Open Hearth, BOH (Sancho et al. 2000, page 16) was considered at the end of the twentieth century as an alternative for the production of steel in Poland, Romania, Turkey, India, China, and some republics of the ancient Soviet Union. However, this process has fallen in disuse to produce steel nowadays. Nevertheless, the Martin–Siemens furnaces are still in use in the glass industry, and especially in the last years, most of the installations to produce flat or hollow glass have as principal equipment metallic oxides smelting furnaces (technologically developed, with high energetic and economic efficiencies).

In Table 1.1, we present the world steel production since the middle of the twentieth century. The fast growth in the decade of the fifties stopped in the sixties, and the next two decades showed a production standstill in around 750 million tons. However, an increase in the production of steel took place during the last years of the twentieth century, and it reached 850 Mt in 2000.

However, in the last decades of the twentieth century, researchers focused on the development of new structural materials and put the technologies of traditional materials production aside. In this context, some documents forecasted (against the majority opinion) a new period of growth in the production of steel (Verdeja et al. 2002, pages 88–89; Sancho et al. 2000, page 16). Equally, it was possible to appreciate during the last years of the twentieth century a noticeable increase of the steel production in the Asiatic countries (China, India, and South Korea), and in a lesser extent in Latin America. In that period (Sancho et al. 2000, page 16), the value of the steel consumption per capita (kg man^{-1} $year^{-1}$) and its importance to foresee the growth of the sector in the different regions were relevant. The steel consumptions per capita in countries of the European Union and in the countries associated with the North America Free Trade Agreement (NAFTA) were around 350 kg in periods of weak economic growth or recession and 450 kg in periods of economic growth. Current data indicates that the mean world steel consumption per capita is of 225 kg. Consequently, the possibility of growth in the world steel market, considering the current production (Table 1.1) would be within 925 Mt in recession periods and 1665 Mt in periods of economic growth.

Table 1.1 World steel
production from 1950 to 2018

Year	Production (Mt)
1950	189
1960	347
1970	595
1980	717
1985	719
1990	770
1995	753
2000	850
2005	1148
2006	1250
2008	1343
2010	1433
2012	1559
2014	1662
2015	1620
2016	1627
2017	1691
2018	1809

Data from the World Steel Association

Considering all above indicated, the expectations that the world of the steel offers (structural material *par excellence*) for the twenty-first century are very positives, and the production standstill occurred in the eighties can be attributed to the excessive growth of the installed capacity in the previous decades. The adaptation of the production to the market and the emergence of deep technological transformations in the production process have made up the most solid guarantees of the sector and can ensure a sustained growth in the next years.

1.2   Steel and Materials Science

Interest in Materials Science has grown in the last decades of the twentieth century, both as an academic discipline and as a topic that often receives funding via R&D programs. The review, coordination, and development of a common language and a working model have been enriching both for engineers and for scientists (physicists, chemists, geologists, and mathematicians). However, there is a problem that has yet to be resolved satisfactorily: how to predict which materials will be used extensively in the future and how the materials currently in use will evolve. Experts (for instance, Gordon 2012, page 313) indicate that the objective is to identify the materials that will be in great demand in the future, although categorical assertions are very questionable. Ashby 2005 (page 4) analyzed the "relative importance" of the different families of materials in the first two decades of the twenty-first century, although his forecast differs slightly with the current usage of metals, ceramics, polymers, and composites. Ashby (2005) bases the "relative importance" on the work of several authors and from 1960 onward also on the number of teaching hours dedicated to each family of materials at universities throughout the UK and the USA. Moreover, the predictions made for 2020 in Ashby (2005) assume which materials will be used currently or foregoing in the automotive and aircraft industries.

If we calculate the "relative importance" of each family of materials as the quotient between the estimated consumption for each group of materials and the total production of materials in the considered year, we obtain the data shown in brackets in Table 1.2 and the global consumption of materials for the years 2000, 2005, and 2010.

Table 1.2 indicates that *Metals* (steels, cast irons, and non-ferrous materials) have a relative importance of 23%. The predominant families are the *Ceramics and Glasses* (with cement, which is the raw material for the manufacture of concrete and steel-reinforced concrete/rebar concrete, being the most important), *Natural Polymers* (wood) and *Metals*. If we compare this data with the Ashby's predictions for 2020 (metals, 26%; ceramics and glasses, 26%; and polymers, 30%), the results are similar. However, Ashby predicts that *Artificial Composites* (Kevlar, CFRP or GFRP) will represent 18% in 2020 although their yearly production in tons is small. However, if the production of paper (approximately 400 Mt) is considered, the relative importance of composite materials would rise (up to almost 6%), and the relative importance of polymers would decrease (down to 14%), as 2.5 tons of wood are required to produce one ton of paper. Moreover, the same author includes in his analysis the "relative importance" of the materials in 10,000 BC: 3% for metals, 43% for polymers, 12% for composites, and 41% for ceramics and glasses. Based upon the foregoing, we can conclude that:

- The importance assigned to the metals in the first few decades of the twenty-first century, far from showing a decreasing tendency, is stable at 23%, with the potential to increase slightly in the short to medium term ($\simeq$1% for sustainable development).

Table 1.2 "Relative importance" by percentage and world production in millions of tons

	Year 2000	Year 2005	Year 2010
Metals	990 (21%)	1328 (23%)	1653 (23%)
Polymers total	1891 (41%)	2000 (34%)	1988 (27%)
Natural (wood)	1703	1776	1718
Artificial (synthetic)	188	224	270
Ceramics and glasses[a]	1786 (38%)	2550 (43%)	3584 (50%)
Composites	7 (–)	9 (–)	12 (–)
Total	4674	5887	7237

[a]Cement is included in this family of materials

- The "relative importance" of metals has continued to grow throughout history, with very fast growth in some specific decades, such as the second half of the nineteenth century (Industrial Revolution). However, since the beginning of the twenty-first century, its relative importance has been fairly stable, with a slight increase due to emerging economies beginning to adopt the consumption habits of developed countries.
- The "relative importance" of synthetic composites, regardless of the nature of their matrix, is not significant enough to warrant being considered as an independent group of materials. Even though they can demand a high price, they can be incorporated to the family of metals (those with a metallic matrix), polymers (those with a polymeric matrix), or ceramics (those with a ceramic matrix).

The "relative importance" of the three families can also be analyzed on the basis of the market price for the materials. While the percentages obtained by this method are different, they do not differ significantly from those previously indicated, as the unit price of a ceramic semiconductor is not the same as that of structural steel. The *Rare Earth Elements (REE)*, which command a very high price, compensate for the cost of the *CFRPs* or *Graphene* in the polymers: The market for rare earth elements is around 200,000 t per year, 80% of which is used in the production of functional materials, although they are also used as micro-alloy elements in steels and non-ferrous metals.

1.3 Coking Coal

For the integrated ironmaking, it is basic ensuring the coal supply to produce coke. The current world production of pig iron, including that obtained in the foundries with cupola furnace (around 11% of the world steel production, Table 1.1), reports around 1800 Mt (see Table 1.1), approximately. As raw material, the blast furnace coke provides the reductant gas and the energy required to perform the smelting reduction in the blast furnace.

Coal reserves are much bigger than those of the rest of the fossil fuels, and the world production of coal, for thermal and coking uses, is estimated in 6500 Mt (85% thermal coal and 15% coking coal, Schwartz 2018). Nevertheless, it is matter of concern the geographical distribution of the coking coal, as it is only available in few regions. For that reason, great efforts have been done with the purpose of limiting this dependence by broadening the types of coal to be transformed in blast furnace coke. For the non-coking coal, procedures have been developed by which a mixture of these coals can reach the specifications of the blast furnace coke. On this matter, technologies that preheat the coal to increase the range of raw materials susceptible of coking are used, and, simultaneously, the quality of the blast furnace coke is improved.

Table 1.3 Evolution of the CR (*Coke Rate*) and the *RAR* (*Reduction Agent Rate*) in the production of pig iron (Naito et al. 2015)

Year	CR (kg/t pig iron)	RAR (kg/t pig iron)
1950	1100	1100
1975	500	510
1985	400	510
2000	350	510
2014	330	500

In the future, it is feasible the appearance of new procedures based on the technology of nut coke: a series of systems, in experimental phases, aimed at taking advantage of the different coal qualities in the obtaining of products with performance like that of the blast furnace coke. Recently, see Table 1.3, it is possible to notice a drastic reduction in the coke consumption per ton of pig iron (*Coke Rate, CR*) due to, principally, the replacement of its reductant and energetic function with other fuels as the natural gas, the *fuel-oil,* or the pulverized coal, and also, to a lesser extent, due to the technological improvements in the control of the process. However, the *Reduction Agent Rate, RAR*, has been stable for the last 35 years (Sect. 6.4).

Table 1.4 indicates the physical–chemical properties required to the coals used in the coke batteries of the integrated steel plant. Only ≃15% of the world coal production is used in the manufacture of blast furnace coke.

Nowadays, the world coke production is around 700 Mt. Several coking plants were closed in Europe during the last years of the twentieth century, and as a result, the coke availability (difference between the coke produced and the coke consumed) was negative. Today, however, the equilibrium between the coke produced and the coke consumed was stablished again with the reactivation of several construction and remodeling projects.

Lately, the production of coke, in the world in general, and in Europe in particular, has been subjected to strong environmental restrictions. The relocation of the production in other countries or regions of the world with less environmental restrictions has influenced in the coke availability. Nevertheless, the necessity of ensuring a supply of high-quality coke to the blast furnace, and the availability of raw materials (distilled volatiles from the coking process, Sects. 4.1 and 4.2), with

Table 1.4 Most relevant physical–chemical properties of the coking coals (Álvarez et al. 2007; Babich et al. 2008, page 95)

Property	Value
Mean reflectance	1.05–1.20
Fluidity (log)	2.6–3.0
Grain size	
>3 mm	<18%
<0.5 mm	<38%
Moisture	<10%
Ash	<7.5%
Volatile matter	25%

reasonable price, to facilitate the development of the industry of the composites reinforced with carbon fiber, has promoted different projects for the opening of new coke ovens or coke batteries adapted to the environmental requirements of the European Union.

1.4 Iron Ore

Iron oxides are compounds that can be found in the nature in almost all the layers of the earth's crust. The weathering of the iron (which is initially formed as Fe^{3+}, in ground and marine environments) takes place as result of the mechanical transportation by the wind and/or the water, followed by the re-precipitation of the oxides. Thus, the formation of the different minerals and iron oxides is an example of weathering processes.

The world iron ore trade (mainly made up by hematite, Fe_2O_3, and magnetite, Fe_3O_4) has a slightly higher growth than the ironmaking and steelmaking productions (Table 1.1). This trend might be increased in the future because of the bigger share of the developing countries, apart from the constant increase of the quality of the imported ores. The average iron content varied from the 50% in 1940 to the 59% in 1985 and to the 60–65% in this moment. The consumption of high-quality ores with low level of impurities (chiefly alkalis, sulfur, and phosphorus) has positive expectations for the future due to the requirement of feeding the blast furnace with the raw materials with the suitable characteristics to reach the maximum operating performance of the installation. Consequently, it is possible that a devaluation of the local deposits, whose composition (iron content) or volume of reserves was far from the profitability ratio defined by the market, might take place.

The evolution of the prices and the production of iron ore have been always subjected to the production of steel, and that last one has increased from 1984 after a long stage of standstill. This might make attractive investing in new mining projects. Anyway, the analysis of the iron and steelmaking situation advises a constant vigilance because, despite the abundance of reserves, the reactivation of the iron and steelmaking market in this twenty-first century would require an increase of the investment in mining exploration and operations to adjust the supply and the demand, which will be translated into more financial requirements for the sector.

1.4.1 Iron Ore to Produce Materials with Structural Applications

The ferric products transported via maritime are classified into three categories: ore fines (concentrated), coarse ore (as a result of the gravimetric concentration treatment using spirals), and pellets. From the 410 Mt of iron oxides transported in the

Table 1.5 Chemical
characteristics of the sinter
and the pellets in percentage
and the size in mm

Element/metallic oxide	Pellet	Sinter
Fe_2O_3	94	64
Fe	–	10
FeO	–	5.0
SiO_2	3.3	6.5
CaO	1.0	10.5
MnO	0.20	1.0
MgO	0.50	1.0
Al_2O_3	1.0	2.0
Size (mm)	10–20	5–30

1995, 66% corresponded to the fines (<10 mm), 18% to the pellets, and 16% to the coarse fraction (within 10 and 30 mm). The production of iron ore in the same year exceeded the 1000 Mt (1014×10^6 t) which, expressed as a function of the iron content, represented a total of 580×10^6 t of iron. This way, it would be possible to estimate, by comparison, the quantity of fines, coarse fraction, and pellets commercialized nowadays because the world iron ore production is close to 3000 Mt.

Within 85 and 90% of the load that feeds the blast furnaces and the direct reduction reactors (*Midrex* and *HYL*, Sects. 5.2 and 5.3) comprise pretreated-agglomerated iron ore (sintered or pelletized). The chemical, physical, and mechanical characteristics of the sinter or the self-fluxing *pellets* are usually adapted to the standards defined by the reduction unities, although, in Table 1.5, we provide details about the most noticeable chemical and physical (size) properties.

1.4.2   Iron Oxides to Produce Materials with Functional Applications

The man is a consumer of iron in the form of metal, but he also uses iron oxides in several different industrial processes or applications (pigment, photocatalyst, magnetic recording, etc.). Iron oxides are also an important raw material researched for many years in different disciplines. There are six types of iron oxides that can be oxides (O^{2-}), hydroxides (OH^-), or hydroxide oxides ($O^{2-} + OH^-$).

Iron oxides and hydroxides (compounds of iron with oxygen or OH^-) generally appear, structurally talking, in the hexagonal (hcp) or cubic (ccp) systems, where the octahedral sites, and in some cases tetrahedral sites, are occupied by divalent or trivalent iron. Iron oxides differ in how the basic structural unities are disposed in the space, and many times other types of anions can be found in the structure.

The hydroxide oxides get relatively easily dehydrated to have a low quantity of OH^- and facilitate a transformation process. Other feature of these compounds is their low solubility, that is to say, high stability of the Fe^{3+} and their specific surface area. The most typical compounds are:

- Goethite, α-FeOOH.
- Lepidocrocite, γ-FeOOH.
- Akaganeite, β-FeOOH.
- δ-FeOOH (synthetic).
- Ferrihydrite.
- Bernalite, $Fe(OH)_3$.
- $Fe(OH)_2$.
- Hematite, Fe_2O_3.
- Magnetite, Fe_3O_4.
- Maghemite, γ-Fe_2O_3.
- β-Fe_2O_3 and ε-Fe_2O_3.
- Wustite, FeO.

1.5 Slag-Forming Agents and Fluxes

The chemical specifications of the slag-forming elements have taken great importance recently because of the low sulfur and phosphorus contents that are required in the steels. This implies that the raw materials used in the process should have the least quantities of the above-mentioned elements.

One of the raw materials affected by this requirement is the *lime*, CaO, which is a typical slag-forming element used in the ironmaking, both in the sintering process, in the blast furnace, in the conversion process, in the electric furnace, and in the operations of the secondary metallurgy. Table 1.6 collects the most relevant chemical characteristics of the lime used in the ironmaking processes.

The fluxes are used to facilitate the dissolution of the CaO and MgO in the conversion process and in the electric furnace, the fluorite/fluorspar CaF_2 is the most widely used. Despite how the fluorite acts in the process is not well understood, which is well known is that the fluorite provides liquid to the slag because it is a powerful flux (in the cement industry it is known as mineralizer). It is supposed that it attacks the calcium compounds ($CaO \cdot SiO_2$; $3CaO \cdot SiO_2$ and $CaO \cdot Fe_2O_3$) that wrap the lime particles and facilitates the contact between this oxide and the liquid. The specification for the fluorspar used in the ironmaking (Table 1.7) stablishes that the percentage of effective CaF_2 must be always >72%, where:

Table 1.6 Chemical characteristics (percentage) of the lime used in the ironmaking processes

Element/compound	wt%
CaO	95.6
SiO_2	0.85
MgO	0.8
CO_2	1.5
S	0.09
Losses at 950 °C	2.4

Table 1.7 Chemical analysis of the fluorspar used in the ironmaking

Element/compound	wt%
CaF_2	84.93
SiO_2	4.76
Fe_2O_3	1.96
$CaCO_3$	1.30
S	0.15

Table 1.8 Chemical analyses, wt%, of the ilmenite colemanite, bauxite, and dunite used in the ironmaking process

Metallic oxide	Ilmenite	Colemanite	Bauxite	Dunite
SiO_2	1.4	1.25	1.80	38
Fe_2O_3	28.4	0.38	9.11	8.8
FeO	9.7	–	–	–
TiO_2	60.3	–	10.11	–
P_2O_5	0.17	0.10	–	–
CaO	–	28.8	–	1.1
MgO	–	–	–	39
B_2O_3	–	42.5	–	–
Al_2O_3	–	–	54.28	2.0
Calcination losses (950 °C)	–	26.5	24.0	11.1

$$\%CaF_2(\text{effective}) = \%CaF_2 - 2.5(\%SiO_2) \qquad (1.1)$$

The attack of the CaF_2 to the refractory and other problems related to the environmental pollution have promoted the research of a replacement of the CaF_2 with other raw materials as the *ilmenite*, the *colemanite*, the *bauxite,* and the *dunite* (Table 1.8), also aimed at solving other operating problems that concern the ironmaking process: accumulation of volatile elements or integrity of the refractory.

1.6 Scrap. Recycled Steel

The *scrap* (or recycled steel) is one of the main raw materials in the countries with largest industrial tradition because the availability of iron-based materials is so high that can ensure the continuous growth of the supplies. Nevertheless, the prices of the scrap suffer from big fluctuations, being low in periods of economic recession (building and automobile sectors) and high when the economic situation improves.

In these countries, initiatives to reactivate the projects of direct reduction with gas (Chap. 5) appear when the price grows, as an alternative to the utilization of scrap as raw material. Developing countries do not usually have great quantities of scrap, but they usually have reserves of natural gas, petroleum, or low-quality coals, and they are interested in developing processes to replace the scrap with prereduced materials.

We can consider the steel and cast iron parts that cannot be used in any production process as scrap. Scraps come from ships, bridges, automobiles, electrical appliances, and machines scrapyards/junkyards, but also from built up edges, chips, and residues from the metalworking industry. It is also considered as scrap, the material rejected because of not meeting with the specifications of quality for the metallic product and result of the low metallurgical performance of the ironmaking process.

The quality of the scrap depends on three factors: the easiness to be transported and fed into the furnace, behavior in the smelting process and, chemical composition (mainly residual copper, tin, nickel, and cobalt contents). Thus, depending on the origin, it is possible to distinguish three types of scrap:

1. *Back/Recycled scrap*, which is produced as a consequence of the chips, trimmings, cuttings, and rejections of the own ironmaking and steelmaking plant.
2. *Metalworking/Transformation scrap*, which comes from the chips and burr generated in the manufacture of tools, parts, and components of steel.
3. *Obsolescence scrap*, which comes from equipment, machines, and installations that have finished their product life cycle. In this category, we include the most part of the scrap used in the steel works.

Logically, the world scrap consumption (recycled steel) relates to the production of steel, and it is estimated that the annual requirements of recycled steel are around 40% of the steel production (Table 1.1). For example, at the beginning of the twenty-first century, the requirements of scrap were around 360–400 Mt per year. Nevertheless, these values can vary depending on the examined country or region: In those zones whose steel production is mainly based on the electric steel plants, the percentage of recycled steel referred to the total production is very high; on the contrary, in those whose steel production is mainly based on the blast furnace, the scrap requirements would be below the 40% of the steel production.

As it was expected, the increase of the energetic and economic efficiencies of the process in the industrialized/developed countries has implied a significant reduction in the generation of both back/recycled and metalworking/transformation scraps, and it is the obsolescence scrap, the main supply for the production of recycled steel.

1.7   Ferroalloys

Ferroalloys are alloys with iron content equal or higher than the 4%, and with one or several elements in percentage higher than that indicated in Table 1.9. They are normally used as additions in the manufacture of other alloys or as deoxidizing or desulfurizing (or similar applications) agents, because they do not have ability for the plastic deformation (and as a result, they are not used as structural materials).

Despite the production of ferroalloys was mainly carried out in developed countries a few decades ago, the developing countries have joined to the advanced countries in the production of ferroalloys. The future requirements in the field of the ferroalloys will be conditioned by two self-opposing factors:

1. On the one hand, the tendency toward the reduction of the production costs and, for that reason, reaching the optimal performance in the addition of high added value raw materials will be the focus, as it is the case of the ferroalloys, during the operations and processes of the primary and secondary steel metallurgy.
2. On the other hand, an increase in both the production and the cost (higher restriction in the carbon and nitrogen contents) of the ferroalloys will take place in the future as a consequence of the evolution of the iron and steelmaking products demand toward micro-alloyed qualities of high added value (ULC and ELC steels, ultra-low-carbon and extra low-carbon steels, respectively), micro-alloyed weldable C-Mn steels, HSLA steels, and stainless steels.

From these two options, the second one seems that could be the most favored in the future, and as a result, the market of the ferroalloys will be more aimed at increasing the prices that at enlarging the capacity of producing ferroalloys.

1.8   Prereduced Materials

The problems associated with the reserves of coking coals, the alternative fuels to be used in the blast furnace, and the supply of scraps can be solved if the utilization of the prereduced materials (those materials whose metallic iron content is >92%, Table 1.10) in the market of the ironmaking raw materials was increased.

Table 1.9 Minimum percentages of the metals in a ferroalloy

Metallic element	wt%
Chromium, Cr	10
Manganese, Mn	30
Phosphorus, P	3
Nickel, Ni	15
Silicon, Si	8
Others (excluding carbon), Ti, V, Nb	10

Table 1.10 Chemical and physical properties of the *DRI* (*Direct Reduction Iron, Sponge Iron*) and *HBI* (*Hot Briquetted Iron*)

Parameter	DRI	HBI
Total Fe (%)	92–95	92–95
Metallic Fe (%)	85–90	85–90
Degree of metallization	92–95	92–95
Carbon, C (%)	1.0–1.5	1.0–1.5
Sulfur, S (%)	0.005–0.015	0.005–0.018
Phosphorus, P (%)	0.02–0.09	0.02–0.09
SiO_2	1.0–2.0	1.0–2.0
Mass density (kg m^{-3})[a]	1600	2500
Apparent density (kg m^{-3})[b]	3200	5000
Normal size (mm)	6–13	110 × 60 × 30

[a]Includes the volumes of open and closed pores and the crystallographic volume
[b]Includes the crystallographic volume and the volume of closed pores

The production of prereduced materials is carried out by reduction, in solid state, of iron ore with reductant gases such as the carbon monoxide, the hydrogen or the natural gas (*Direct Reduction Iron, DRI*). At the beginning, the commercialization of prereduced materials was difficult due to the high reactivity of the sponge iron. Physical–chemical variations of the charge/load were produced during the transportation, and this caused handling problems. Nowadays, technologies to stabilize the sponge iron, obtained by reduction with gas, were developed and, despite most of the prereduced material is used as raw material (cold load/charge) in electric furnaces located next to the direct reduction installations, there are two qualities *DRI* and *HBI* (*Hot Briquetted Iron*), which are successfully commercialized. In 1995, 4 Mt of *DRI* and 8 Mt of *HBI* were used.

Table 1.11 Production of prereduced material in million tons in the period 1970–2017 (www.midrex.com, 2017, World DRI Statistics)

Year	Production (Mt)
1970	0.79
1975	2.81
1980	7.14
1985	11.17
1990	17.68
1995	30.67
2000	43.78
2005	56.87
2010	70.28
2015	72.64
2017	87.10

Nearly 85% of the prereduced material production is controlled by the *Midrex* and *HYL* processes. Table 1.11 shows the production of prereduced material in the period 1970–2017. The quantities of carbon, sulfur, phosphorus, and silicon oxide show noticeable differences due to both the market requirements (regarding the quality of the products) and the geographical factor (because of the obtaining of the material in different places).

References

Álvarez R, Diez MA, Barricanal C, Díaz-Faes E, Cimadevilla JLG (2007) An approach to blast furnace coke quality prediction. Fuel 26:2159–2166

Ashby M (2005) Materials selection in mechanical design, 3rd edn. Butterworth-Heinemann, London, United Kingdom

Babich A, Senk D, Gudenau HW, Mavrommatis TTh (2008) Ironmaking. RWTH Aachen University, Department of Ferrous Metallurgy, Aachen, Germany

Gordon J (2012) Structures: or why things don't fall, down edn. Plenum Press, New York, EEUU

Naito M, Takeda K, Matsui Y (2015) Ironmaking technology for the last 100 years: development to advanced technologies from introduction of technological know-how, and evolution to next-generation process. ISIJ Int 55(1):7–35

Sancho J, Verdeja LF, Ballester A (2000) Metalurgia Extractiva. Volumen II: Procesos de Obtención. Síntesis, Madrid, Spain

Schwartz M (2018) Coking coal still needed. Mater World Mag 26(10):52–54

Verdeja LF, Alfonso-Fernández A (2002) Prácticas y Problemas de, Siderurgia edn. Fundación Luis Fernández Velasco, Oviedo, Asturias, Spain

Sintering

2

2.1 Introduction

Sintering is a process that consists in the thermal agglomeration of a mixture of fines that comprise iron ore, recycled iron and steelmaking products, slag-forming elements, and solid combustible (coke). The process has as objective obtaining a charge with the suitable physical–chemical and mechanical properties to be fed into the blast furnace. The sintering process was widely studied in:

- Fernández-González et al. (2017a): *Iron ore sintering: Raw materials and granulation*, where the concentration techniques used in the case of the iron minerals are described, as well as the current demand of iron ore in the world. The raw materials that are charged in the Dwight–Lloyd machine to be sintered are also described (iron ore, by-products, fluxes, and slag-forming products and coke). In this manuscript, the granulation process, widely used before the sintering process, is presented. It consists in the homogenization of iron ore mixture in a rotating drum with 7–8% water for a few minutes with the finality of obtaining a preagglomerated product that is then fed to sintering machine, ensuring a suitable sinter bed permeability and thus improving the productivity of the sinter plant.
- Fernández-González et al. (2017b): *Iron ore sintering: Process*, where the origins of the current sintering process are presented with the patent from 1909 of A. S. Dwight and R. L. Lloyd that consisted in a moving-bed of fine ore particles and additives supported on a metallic chain type strand with exposure to high temperatures. The different zones of the sintering process and the combustibles used in the sintering process are also reviewed in this manuscript. Additionally, different alternatives to the sintering and pelletizing techniques are described: HPS process, hybrid pelletized sinter; MEBIOS, mosaic embedding iron ore sintering; and, CAP, composite agglomeration process. Different parameters controlled in the sintering process are also studied in the manuscript, as the flame

J. I. Verdeja González et al., *Operations and Basic Processes in Ironmaking*,
Topics in Mining, Metallurgy and Materials Engineering,
https://doi.org/10.1007/978-3-030-54606-9_2

front and how the temperature of the flame front has influence in the sinter bed structure. The reactions that take place in the sintering process are also described as well as the influence of the different compounds (Fe_2O_3, Fe_3O_4, FeO, CaO, Al_2O_3, SiO_2, and MgO) on the final sinter structure.

- Fernández-González et al. (2017c): *Iron ore sintering: Quality indices*, where the indices used to define the quality of the sintered product to ensure blast furnace stable, homogeneous, and regular operation are described. The sinter structure (hematite, Fe_2O_3; magnetite, Fe_3O_4; wustite, FeO; calcium ferrites; and, silicates) and the conditions for the obtaining of one or other structure are also described as they have significant influence in the quality indices. The different quality indices are described and analyzed considering the influence of Fe_2O_3, Fe_3O_4, FeO, CaO, Al_2O_3, SiO_2, and MgO (as well as the alkalis): softening-melting test, as it is used to simulate the behavior of iron oxides in the cohesive zone of the blast furnace; Tumbler index, as it is used to describe the cold strength of the sinter; low-temperature degradation test, used to evaluate the sinter degradation in the blast furnace during the reduction in the low-temperature zone; reducibility test, as it gives a measure of the ability of the sintered products for transferring oxygen during the indirect reduction in the blast furnace stack; sinter porosity, as it is related to the sinter reduction behavior; Reduction Degradation Index (RDI), as it gives a measure of the sinter strength after the partial reduction of the material and serves information about the degradation behavior in the lower part of the blast furnace stack.
- Fernández-González et al. (2017d): *Iron ore sintering: Environment, automatic, and control techniques*, where the pollutants generated in the sintering process are described, including the source, the quantity, and the techniques used to reduce the emissions of each pollutant. The alternatives for emission control are also described, where primary, secondary, and tertiary measures are described. The use of biocombustibles in the sintering process is also reviewed. Finally, the developed automatization and control systems for the iron and steelmaking industry to improve the labor productivity, getting sinter with high yield and optimal quality (mechanical and chemical), as well as saving energy, are described.

A short description of the sintering process is also presented in Fernández-González et al. (2018).

Figure 2.1 shows the main parts of an iron ore sintering plant (Dwight–Lloyd machine). Despite at first instance the agglomeration of ores has imposed as a process required in the iron extractive metallurgy, this process has changed with the time until becoming in the main instrument to recycle in both the ironmaking and the steelmaking. Nevertheless, the main objections about the sintering plants are related to the environmental problems due to the emissions of polycyclic aromatic hydrocarbons, especially the polychlorinated dibenzodioxins (PCDDs), which are the compounds that receive the largest attention for their control and elimination from the gases (Cores et al. 2015, pages 231–235; Fernández-González et al. 2017d).

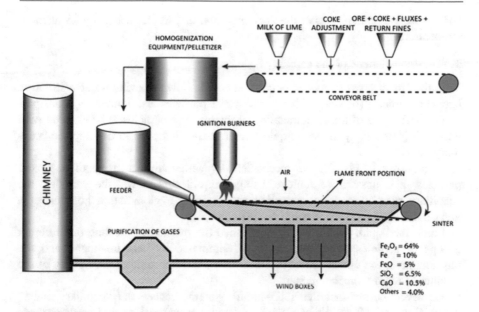

Fig. 2.1 Scheme of a sintering plant

The mixture of different iron ores, coke, fluxes, and return fines prepared in the homogenization beds is adjusted, regarding its composition, with coke and milk of lime to obtain a self-fluxing sinter with the specified basicity. The mixture, once adjusted the chemical composition and the basicity, goes into the blender of the plant where the impact forces between particles work against the liquid/solid interface tension forces and provide the largest surface of contact liquid/solid and the minimal distance between them.

As in the situation of the surface tension in liquids (liquid/air interphase), the liquid/solid surface tension also diminishes linearly with the temperature. Empirically, it is possible to confirm that an increase in the temperature of the water favors a better process efficiency because the quantity of recycled sinter that does not accomplish the particle size requirements diminishes.

If the Young–Laplace equation is applied in the water/particle interface, it is possible to deduce that the force, F, responsible of the union between solid particles can be expressed as follows:

$$F = \frac{2 \cdot \gamma_{l-s}}{x^2} \cdot V \tag{2.1}$$

where γ_{l-s} is the liquid/solid interfacial tension coefficient, x is the distance between particles, and V is the volume of liquid, on the condition that the solid/liquid ensemble was in the capillary state.

Brief note 1 Fundamentals of the agglomeration of particles using liquids at room temperature.

1. Thermodynamics of the capillary state.

With the aim of providing weak bonds that could allow achieving a stable structure between particles (between different groups of particles also known as "quasiparticles" in the case of the agglomeration of the iron ore particles to be sintered with water), it is necessary to use a liquid (water in the sintering) to develop the "capillary state."

It is shown in Fig. 2.2 the mechanism by which from a group of dispersed particles, it is possible to achieve, using a liquid able to wet the particles, the "capillary state" that allows obtaining an emerging agglomeration bonding of a great number of particles at room temperature.

Later, the liquid, once it has accomplished the mission of creating the different groups of particles ("quasiparticles"), is eliminated by heat treatment, and the physical processes and chemical reactions that are responsible of the rigidity of the product at room temperature start.

However, before that the solid–liquid system reaches the "capillary state," Stage III in Fig. 2.2, the effect of the liquid addition on the dispersed particles leads to a situation of minimum surface free energy that is known as "pendular state," Stage II in Fig. 2.2.

In the "pendular state," the balance of surface free energies involved in the adhesion reaches the equilibrium ($\Delta G_{ad} = 0$) when the wet surface area is equal to $\Delta A^=$, that is to say:

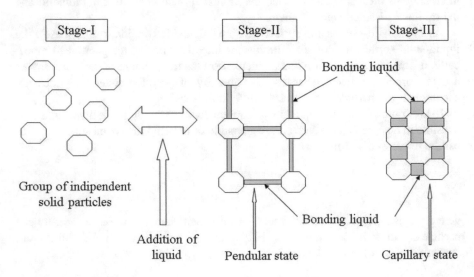

Fig. 2.2 Mechanism of capillary bonds between independent solid particles of the same size

$$\Delta G_{ad} = \left(\gamma_{l-g} \cdot \cos\theta + \gamma_{l-s} - \gamma_{s-g}\right) \cdot \Delta A^{\neq} = 0 \tag{2.2}$$

where γ_{l-s} is the liquid–solid interfacial energy, γ_{l-g} is the liquid–gas interfacial energy, γ_{s-g} is the solid–gas interfacial energy, and θ is the contact angle.

The process of adhesion of the liquid to the granular solid is possible, pass from the Stage I to the Stage II in Fig. 2.2, due to the free energy of the process during the first stages of the wetting is negative, $\Delta G_{ad} < 0$. This circumstance is promoted by the intervention of the following variables:

a. the value of the solid interfacial energy, γ_{s-g}, is very great.
b. the addition of the solid–liquid or liquid–gas interfacial variables, γ_{l-s} or γ_{l-g}, is small.
c. the solid is efficiently wet by the liquid: low values of the contact angle, θ.

Once reached the "pendular state" in the system, to progress toward the "capillary state," Fig. 2.2, it is necessary the participation of small compression stimulus caused by the impact of some particles with others. This impact is provided, for instance, by the mixing drum of the sintering plant or the pelletizing disk in the agglomeration of fine granulometries. The bond between solid particles depends on the interfacial energy of the water, γ_{l-g}, 72.0×10^{-3} N m^{-1} that is smaller than that of the liquid–solid interfacial energy, $\gamma_{l\ s}$, 250.0×10^{-3} N m^{-1}. That is to say, the part of liquid that joins each particle is the weakest and breaks before than those that are connected with the air.

When the volume of liquid grows, the "capillary state" progresses toward the "plastic state," where the system can plastically deform even without an external stimulus, that is, by the action of its own weight. In the "plastic state," the holes or those zones without bonding liquid that leads to the "capillary state" can be occupied by a liquid layer.

2. Stresses and forces in the capillary state.

The tension between the particles that are developed under the "capillary state," and that put up resistance to the work of the capillary forces that impede a greater approach between particles (the reduction of the curvature radius of the liquid between the particles), is identified by $\sigma_{t-capillary}$ and is expressed in Pa. The relation between $\sigma_{t-capillary}$, the size of the particles, D_p, in meters, and the liquid–gas interfacial energy, γ_{l-g}, in J m^{-2} can be expressed by means of the following equation (Verdeja et al. 1993, pages 109–110):

$$\sigma_{t-capillary} = \frac{9.58 \cdot \gamma_{l-g}}{D_p} \tag{2.3}$$

On the other hand, the capillary forces, $F_{\text{capillary}}$, in N, are related to the volume of liquid, V_1, in m^3, the distance between particles to which they can be brought closer, x, in meters, and the solid–liquid interfacial energy, γ_{l-g}, in J m^{-2}, by means of the following equation (Verdeja et al. 1993):

$$F_{\text{capillary}} = \frac{2 \cdot V_1 \cdot \gamma_{l-g}}{x^2} \tag{2.4}$$

Both $F_{\text{capillary}}$ and $\sigma_{t-\text{capillary}}$ decrease with the interfacial energy of the water, γ_{l-g}. Habitually, the interfacial energy of the liquids decreases with the temperature. In the particular case of the water, $\gamma(H_2O)_{s-g}$ in J m^{-2}, which is the liquid habitually used in the sintering, its variation with the temperature, T, in Kelvin, is represented by the following equation (TAPP Database 1994; Adamson 1982):

$$\gamma(H_2O)_{s-g} = (88.425 + 0.227 \cdot T) \cdot \left(1 - \frac{T}{647.1}\right)^{1.256} \tag{2.5}$$

An increase of the water temperature from 25 °C (71.99×10^{-3} J m^{-2}) to the 70 °C (64.50×10^{-3} J m^{-2}) makes both $F_{\text{capillary}}$ and $\sigma_{t-\text{capillary}}$ to decrease a 10.4%. This same effect can be increased with surfactants (alcohols or sugars) that reduce the interfacial energy of the water. For instance, a concentration of 0.05 mol/l of n-hexanol in water decreases a 44% the value of the water interfacial energy at 25 °C. This effect is very interesting to improve the performance of the sintering plants, as it would be possible to reduce even more the distance between particles in the system under the "capillary state."

On the other hand, the derivative of Eq. (2.4), which relates the capillary forces between particles, the volume of liquid to reach the capillary state and the distance between particles, can be obtained starting from the Young–Laplace equation. The creation by condensation, clustering, and growth of a liquid droplet (for instance, water) with radius size, r, in the air leads to an equilibrium between the pressure and surface forces (interfacial energy of the water), which leads to the Young–Laplace equation (see Adamson 1982, pages 8–11; Verdeja et al. 2014, page 249):

$$\Delta P = \frac{2 \cdot \gamma_{l-g}}{r} \tag{2.6}$$

On the other hand, the pressure stresses, ΔP, and the capillary forces, F, are related by the value of an area, A. If we identify the radius size of the water droplets of the Young–Laplace equation with the distance between particles, x, Eq. (2.6) is transformed in:

$$F = \frac{2 \cdot \gamma_{l-g} \cdot A}{x} \tag{2.7}$$

Finally, if the numerator and the denominator are multiplied by the distance between particles, x, Eq. (2.4) is obtained, where:

$$A(x) = V_1 \qquad (2.8)$$

3. Volume of water in the capillary state.

The force associated with the capillary state of a group of particles can be obtained starting from Eq. (2.3). In this case, to carry out the transformation between the stress and the applied force, we assume a mass of particles to be sintered with a cylindrical symmetry of 50 mm in diameter and 50 mm in length. This mass group of particles to be sintered, of cylindrical symmetry, is characteristic of the normalized ceramic specimens tested in compression. The force that maintains bonded this group of particles would be identifiable with the resistance to compression of the mentioned specimen of 50 mm in diameter that would contain a sample of the product to be sintered, that is to say:

$$F_{\text{capillary}} = \frac{9.58 \cdot \gamma_{l-g}}{D_p} \cdot \left(\pi \cdot \left(\frac{50.0 \times 10^{-3}}{2} \right)^2 \right) = \frac{2 \cdot V_1 \cdot \gamma_{l-g}}{x^2} \qquad (2.9)$$

Consequently, the calculation of the liquid volume that is required to reach the capillary state within a group of particles, when x is the distance between particles ("quasiparticles" in the case of the material to be sintered) and D_p is the average size in the granulometric distribution, is equal to:

$$V_1 = \frac{9.58 \cdot x^2}{2 \cdot D_p} \cdot \left[\pi \cdot \left(\frac{50.0 \times 10^{-3}}{2} \right)^2 \right] \qquad (2.10)$$

In the situation of the "capillary state" for the group of particles to be sintered, with a size comprised within 0.5 and 5.00 mm, if the distance between the "quasiparticles," x, is 2.25 mm and the average size of the granulometric distribution, D_p, is 2.75 mm, the volume of water-liquid associated with the "capillary state," according to Eq. (2.10), is 17.30 cm^3 or 17.31 g (assuming that the density of the water is 1.00 g cm^{-3}). If we consider that the real density of the material to be sintered is 5500 kg m^{-3}, and the global density is 2699.8 kg m^{-3}, the quantity of matter associated with a cylindric specimen of 50 mm in cross section and 50 mm in height, with the global density above indicated, is of 26,998 g, and consequently, the percentage of water in the specimen, which would correspond with the material that is available in the sintering grate, would be 6.03%.
End of the brief note.

The main constituents of the charge that feeds the sintering machine are (Sancho et al. 2000):

- Iron ores with $\geq 60\%$ of iron and particle size in the range 0.5–8 mm. Iron ores with particle size >8 mm can be directly fed into the blast furnace or any other reduction furnace reactor, and iron ores with particle size <0.5 mm must be processed in pelletizing plants.
- Recycled iron and steelmaking by-products as powders from the flue-gas dry and wet cleaning systems of the blast furnace, mill scale (mixture of iron and its oxides) and all those residues, metallic or not, generated in the iron and steel-making factory.
- Slag-forming elements such as the limestone, dolostone, or the milk of lime ($CaO \cdot H_2O$).
- Powder of coke with a grain size <5 mm, which is produced during the milling operations in the coke batteries.

So, it is possible to distinguish three constituents in the sinter from the structural point of view:

- A disperse constituent, which mainly comprises hematite (Fe_2O_3) and, in lower proportion, magnetite (Fe_3O_4) and secondary hematite.
- A matrix constituent, which mainly comprises chemical and eutectic compounds from the Fe_2O_3-CaO, FeO-CaO, and FeO-SiO_2 binary systems, such as: $CaO \cdot Fe_2O_3$ (monocalcium ferrite, CF), $2CaO \cdot Fe_2O_3$ (dicalcium ferrite, 2CF), and $2FeO \cdot SiO_2$ (fayalite).
- The porosity available in the mixture.

Figure 2.3 displays a micrograph of an iron ore sinter.

Commercial software such as MATLAB, APL, HSC, or APL-J can be useful for the calculations of the sintering plant mass balance. Tables 2.1 and 2.2 collect the results when the APL software is used under the operating conditions indicated by the variables. The restrictions imposed in the balance are:

$$\frac{m_{water}}{m_{ferriccharge}} \cdot 100 \leq 6\% \tag{2.11}$$

$$\frac{m_{coke}}{m_{ferriccharge}} \cdot 100 \geq 8\% \tag{2.12}$$

$$\frac{m_{air}}{m_{sinter}} \cdot 100 \geq 8\% \tag{2.13}$$

where m_i is the mass expressed in kilograms.

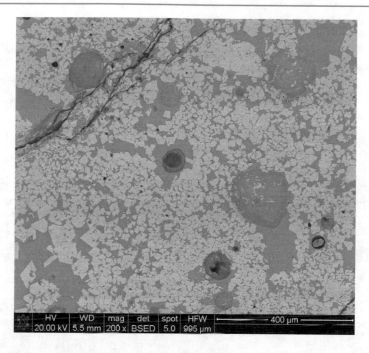

Fig. 2.3 Micrograph of an iron ore sinter. White/light gray: Iron oxides; dark gray: chemical and eutectic compounds from the Fe_2O_3-CaO, FeO-CaO, and FeO-SiO_2 binary systems; pores: round-shape holes in the figure

Table 2.1 Mass balance of the sintering process

Phase/compound	Ferric charge	Coke	Water	Air	Sinter	Gas
%N_2				76.7		
%O_2				23.3		
%C						
%P						
%S						
%Fe_2O_3					82.00	
%CaO					5.89	
%MnO					0.01	
%FeO					7.00	
%CO_2						
%CaS					0.09	
%H_2O			100			
%SiO_2					4.91	
%SO_2						
%P_2O_5					0.10	
Mass flow (kg)					1000	

Data from the entrance

Table 2.2 Mass balance of the sintering process

Phase/compound	Ferric charge	Coke	Water	Air	Sinter	Gas
%N$_2$						75.35
%O$_2$						20.50
%C						
%P	0.042					
%S	0.025					
%Fe$_2$O$_3$					82.00	
%CaO					5.90	
%MnO					0.013	
%FeO					7.00	
%CO$_2$						3.40
%CaS					0.09	
%H$_2$O						0.74
%SiO$_2$					4.91	
%SO$_2$						0.01
%P$_2$O$_5$					0.087	
Mass flow, kg	1003	80	60	8000		8143

Result of the balance (using the software APL)

$$m_{s,gas} = \frac{32}{64} \cdot m_{SO_2} \qquad (2.14)$$

30% of the sulfur is volatilized during the process as SO$_2$:

$$m_{s,gas} = \frac{30}{100} \cdot \left(m_{s,charge} + m_{s,coke} \right) \qquad (2.15)$$

Considering that the sintered product constitutes a raw material for those reduction furnaces that produce solid iron (sponge iron) or carburized liquid iron (pig iron), it is necessary to know the physical and chemical variations that the sintered product undergoes under parameters that reflect, as accurate as possible, the real working conditions for the material. This way, the following questions should be taken into account:

- Test to determine the cold mechanical strength of the sinter.
- Test to know the degradation of the sinter after a partial reduction process.
- Test to know the sinter reducibility.

The cold mechanical strength of the sinter is calculated by means of the mechanical self-degradation (autogenous crushing) test according to the scheme of Fig. 2.4. Quantitatively, the cold degradation indexes are used, MICUM (*Mission Interalliée de Contrôle des Usines et des Mines*), IRSID (*Institut de Recherches de la Sidérurgie Française*), or TUMBLER (TI), which standardize the quantity of material that passes through a certain sieve after a defined time of contact between agglomerated particles in a rotating drum. The cold mechanical strength can be similarly evaluated as well for all those materials involved in the ironmaking process: pellets, iron ores (coarse fraction), and coke. Habitually, the most used cold mechanical degradation index for the ironmaking sinter is the *Tumbler index*, TI (% > 6.3 mm), which indicates the percentage of particles with a size greater than 6.3 mm at the end of the abrasion test.

The sinter degradation after a partial reduction of the material is expressed by means of the RDI (*Reduction Degradation Index*). Figure 2.5 indicates the steps of the test. The same as in the cold degradation test, the other materials (apart from the sinter) that are used in the ironmaking (pellets, iron ore, and coke) can be analyzed and classified using this parameter. For that reason, the RDI evaluates the mechanical strength that the material has after a partial reduction because this partial reduction could have modified both the physical (density, porosity, size) and chemical properties of the material. Habitually, the hot degradation (or degradation after partial reduction) of the sintered product is expressed by means of the RDI (% < 3.15 mm), which indicates the percentage of sintered product with a size smaller than this value after the mechanical reduction degradation test. Thus, the quality of the sinter will be better, the lower the value of the RDI (% < 3.15 mm).

The improvements in the RDI in the particular case of the iron ore sinter are related to the following variables:

- Increase of the SiO_2 (and also CaO that keeps constant the basicity) and the MgO (dunite or dolostone) contents in the feeding ferric charge.
- Production of a sinter with greater FeO contents.
- Granulometry of the slag-forming elements.

Fig. 2.4 Cold mechanical strength test for the ironmaking raw materials, whose main characteristics are the cold degradation (MICUM, IRSID, or TUMBLER), abrasion, and SHATTER indices

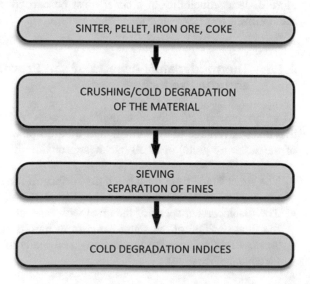

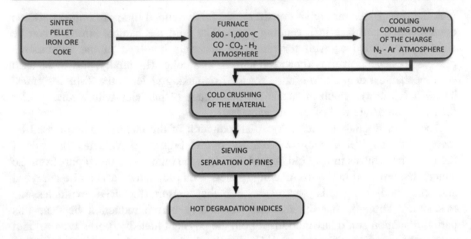

Fig. 2.5 RDI test (*Reduction Degradation Index*). Hot degradation of the ironmaking raw materials

- Permeability of the charge.
- Increase of the sinter cooling rate (because of the formation of few ordered structures with low thermal expansion coefficient).
- Diminishing the consumption of ores with high Al_2O_3, TiO_2, Na_2O, and K_2O.

The greater or smaller easiness for the iron ore reduction is calculated by means of the reducibility test schematized in Fig. 2.6. From the conceptual point of view, the reducibility of a sinter depends on the thermodynamic and kinetic variables, and thus, the most suitable phase for the reducibility might be defined. The valuation of more or less reducibility of a sinter must be carried out using the corresponding standardized test (ISO 4695).

2.1.1 Thermodynamic Aspects of the Process and the Product

At a commercial scale, the sintering is a process of ferric materials agglomeration that takes advantage of the formation of a liquid phase located along a band (width of the sintering grate) of small thickness, considering that:

- The sinter flame front must reach the surface of the grate before the discharge of the product in the crushing machine (Fig. 2.1).
- The displacement speed of the grate with respect to the wind boxes determines the final position of the sinter flame front before the discharge of the product.
- A slow rate reduces the productivity and leads to a re-oxidation of the sinter (secondary hematite).

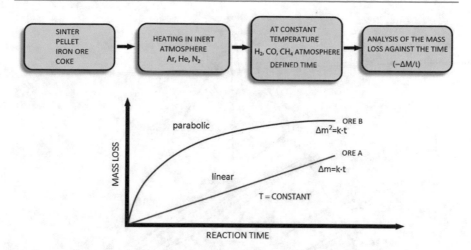

Fig. 2.6 Reducibility test for the ironmaking materials. Curves of mass loss against time according to JIS (*Japanese Industrial Standard*, M8713-1987)

On the other hand, a fast operation will produce a sinter with low mechanical properties and will also increase the proportion of recycled product that does not accomplish the granulometric requirements.

The reactions in the sinter flame front, where the temperatures exceed the 1400 °C, are:

$$C(coke, s) + O_2(g) \rightleftharpoons CO_2(g) \tag{2.16}$$

$$C(coke, s) + CO_2(g) \rightleftharpoons 2CO(g) \text{ (Boudouard equilibrium)} \tag{2.17}$$

$$Fe_2O_3(s) + CO(g) \rightleftharpoons 2FeO(s) + CO_2(g) \tag{2.18}$$

$$3Fe_2O_3(s) + CO(g) \rightleftharpoons 2Fe_3O_4(s) + CO_2(g) \tag{2.19}$$

$$2Fe_2O_3(s) + CaO(s) \rightleftharpoons 2Fe_2O_3 \cdot CaO(l) \text{ (calcium diferrite)} \tag{2.20}$$

$$Fe_2O_3(s) + CaO(s) \rightleftharpoons Fe_2O_3 \cdot CaO(l) \text{ (calcium ferrite)} \tag{2.21}$$

The sinter heating process in a reduction furnace can be simulated using the $CaO\text{-}Fe_2O_3\text{-}SiO_2$ ternary diagram of Fig. 2.7.

A pseudoternary composition of a sinter inside of the $CaO\text{-}Fe_2O_3\text{-}SiO_2$ ternary diagram can be considered (Ballester et al. 2000, pages 277–279) (determined the phases using the Roozeboom method): 78% Fe_2O_3, 9% SiO_2, and 13% CaO.

Even when there are several possible compatibility triangles for this global composition, the most suitable triangle is chosen considering the physical

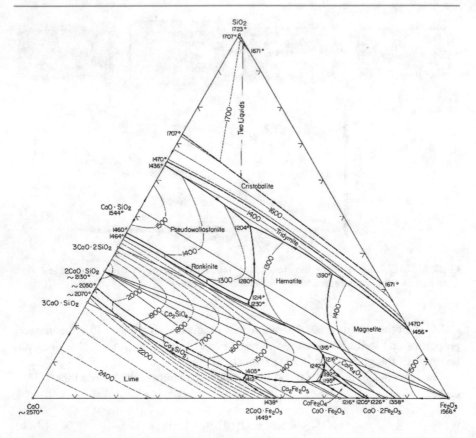

Fig. 2.7 CaO-Fe₂O₃-SiO₂ ternary diagram for the analysis of the raw materials with high Fe₂O₃ content (Levin et al. 1969)

characteristics of the sinter, that is to say, that triangle formed by the wollastonite (CaO·SiO₂), the lime (CaO), and the hematite (Fe₂O₃). See Fig. 2.8.

The existence of free lime in the sinter ($\sim 5\%$ according to the global composition of the sinter above indicated, see the CaO-Fe₂O₃-SiO₂ ternary diagram in Fig. 2.8) and its high ability for the chemical reaction with the ambient moisture can cause a mechanical degradation of the sinter due to swelling.

The first liquid phase during the heating of the sinter will appear at 1214 °C (eutectic point, point with the lowest temperature, inside of the compatibility triangle, see Fig. 2.9) as a consequence of the ternary eutectic affinity of three phases (wollastonite, rankinite, and hematite), according to the following equilibrium (see Fig. 2.9):

$$SiO_2 \cdot CaO(\text{wollastonite}) + 2SiO_2 \cdot 3CaO(\text{rankinite}) + Fe_2O_3(\text{hematite}) \rightleftharpoons \text{liquid}$$

$$(2.22)$$

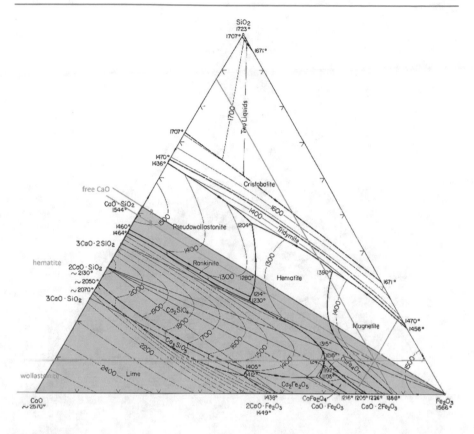

Fig. 2.8 Representation of the sinter global composition inside of the compatibility triangle. Determination of the proportion of the different phases at room temperature

Even though the wollastonite can be a compatible phase at room temperature in the sintered product (compatibility triangle), the rankinite is formed as a consequence of the reaction between ferrites and acid materials. The quantity of liquid phase in the sinter at 1214 °C is approximately 18%; see Fig. 2.9.

During the heating of the sinter, ternary diagram of Fig. 2.9, once exceeded the eutectic temperature, the solid wollastonite disappears (as it is possible to see in Fig. 2.9 in brown color, the wollastonite is the solid minority compound in the eutectic), and an equilibrium between the liquid and the solid rankinite and hematite is stablished. When the 1230 °C is reached, a new solid phase, formed as a consequence of a ternary peritectic reaction, appears (Fig. 2.10):

$$\text{liquid} + 2SiO_2 \cdot 3CaO(\text{rankinite}) + Fe_2O_3(\text{hematite}) \rightleftharpoons 2CaO \cdot SiO_2(\text{larnite})$$

$$(2.23)$$

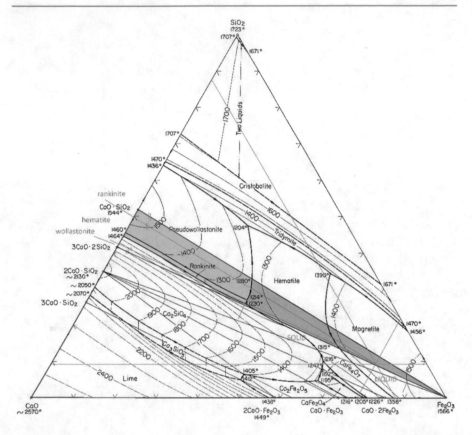

Fig. 2.9 Representation of the eutectic point (1214 °C), and scheme of how to calculate the quantity of liquid at this temperature

At 1230 °C, the quantity of liquid in the sintered product is increased up to 33% (Fig. 2.10), and above this temperature there are only: hematite (Fe_2O_3), larnite ($3CaO·SiO_2$), and liquid.

Between 1230 °C and around 1235 °C, the larnite ($2CaO·SiO_2$) is dissolved in the liquid, and above 1235 °C, there are only liquid and hematite (Fe_2O_3), being the hematite the last solid available in the system (see Fig. 2.11).

At 1300 °C, the quantity of liquid phase in equilibrium with the hematite is 42% (see Fig. 2.11).

The thermal decomposition of the hematite (Fe_2O_3) into magnetite (Fe_3O_4) takes place at a temperature slightly below 1395 °C (Fig. 2.12):

$$Fe_2O_3(s)(hematite) \rightleftharpoons Fe_3O_4(s)(magnetite) + O_2(g) \qquad (2.24)$$

Thus, above this temperature, the phase thermodynamically stable is the magnetite (primary field of crystallization of the sinter).

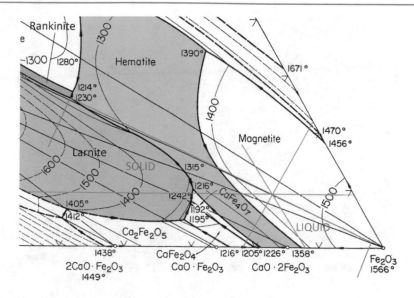

Fig. 2.10 Situation in the sinter at 1230 °C

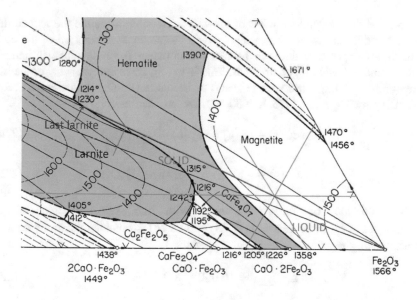

Fig. 2.11 Situation in the sinter between 1230 and 1300 °C

Finally, at approximately 1440 °C (which corresponds with the point defined by the sinter global composition, intersection of the blue lines), the liquidus temperature of the system is reached, which corresponds to a softening interval of 226 °C.

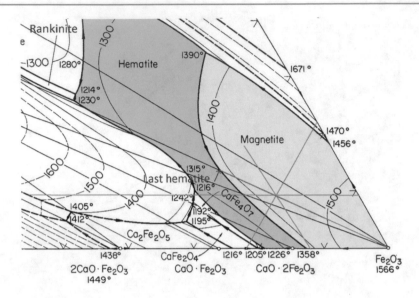

Fig. 2.12 Situation in the sinter from 1395 to 1440 °C

From the thermodynamic point of view, it is possible to define the reducibility criteria for different ferric phases involved in the sinter as a function of the equilibrium partial pressure that can be reached between the reductant gas and the oxidized and reduced iron compounds. That is to say, if we analyze what of the iron compounds, stoichiometric wustite (FeO) or hematite (Fe_2O_3), is the easiest reducible using hydrogen, at 1173 K (900 °C), we proceed as follows:

$$FeO(s) + H_2(g) \rightleftharpoons Fe(s) + H_2O(g) \tag{2.25}$$

$$\frac{1}{2}Fe_2O_3(s) + \frac{3}{2}H_2(g) \rightleftharpoons Fe(s) + \frac{3}{2}H_2O(g) \tag{2.26}$$

The variation of the temperature with the standard free energy, $\Delta_r G^0 \left(J \, mol \, Fe^{-1} \right)$, in the reactions (2.25) and (2.26) is obtained using the data for the ironmaking reactions collected in Appendix 1:

$$\Delta_r G^0_{(2.25)} = 16{,}000 - 8.0 \cdot T \text{ (K)} \tag{2.27}$$

$$\Delta_r G^0_{(2.26)} = 34{,}950 - 46.5 \cdot T \text{ (K)} \tag{2.28}$$

Thus:

$$\Delta_r G^0_{(2.25)}(T = 1173 \text{ K}) = 6616 \text{ J/mol Fe} \tag{2.29}$$

$$\Delta_r G^0_{(2.26)}(T = 1173 \text{ K}) = -19,595 \text{ J/mol Fe} \tag{2.30}$$

From these values, the relation of partial pressures between the hydrogen and the water vapor in equilibrium with the condensed phases, for the reactions (2.25) and (2.26), is

$$\left(\frac{P_{H_2}}{P_{H_2O}}\right)^=_{(2.25)} = 1.97 \text{(wustite)} \tag{2.31}$$

$$\left(\frac{P_{H_2}}{P_{H_2O}}\right)^=_{(2.26)} = 0.26 \text{ (hematite)} \tag{2.32}$$

As these values are the minimum required to reach the equilibrium, at 1173 K and with the thermodynamic data of Appendix 1, we see that more reductant conditions are required to obtain sponge iron using wustite than in the case of using hematite (the hematite is easily reducible). As well, it is possible to deduce that a sinter with high proportion of FeO is more difficultly reduced than other with lower percentage of FeO. It is possible to apply this thermodynamic criterion of reducibility both to the iron compounds and to the different reductant gases.

Asserting that the hydrogen is worse reductant than the carbon monoxide at low temperature or stating that the hydrogen is the best reductant at high temperatures is supported by means of the reducibility criterion of minimum equilibrium partial pressure for a defined temperature. However, it is important to consider that the wustite is non-stoichiometric. The relation Fe/O can be lower than one and, from the thermodynamic point of view, there are different values for the standard Gibbs free energy of formation, $\Delta_f G^0$. In Appendix 1, we present, together with the Gibbs free energy of formation of the stoichiometric wustite, the value for the Fe_xO ($x < 1$) according to Gaye and Welfringer (1984). As it is possible to see, the thermodynamic stability of the stoichiometric wustite is noticeable greater. On the other hand, the wustite is a compound available from the hematite or magnetite reduction processes. The nature of the wustite resulting from the iron ore reduction process clearly reflects the better or worse reducibility of the ferric charge.

2.2   Solid–Gas Kinetics in Porous Particles

Considering the sinter particles (also those of the coke) as porous solids, it is possible to determine the variables involved in the transformation (reduction) of the particles using a reductant gas. The mass global balance that considers the concentration of reductant gas around the porous particle, and also takes into account the terms of diffusional transfer, accumulation of matter, and the rate of chemical reaction, can be expressed by means of a differential equation:

$$\nabla^2 \left(C_i - C_i^= \right) - \frac{1}{h^2} \cdot \left(C_i - C_i^= \right) = 0 \tag{2.33}$$

where C_i is the concentration of reductant gas i inside of the porous solid, $C_i^=$ is the equilibrium concentration of gas i calculated using the thermodynamic data, and h is obtained using the following equation:

$$h^2 = \frac{D_i}{k_q \cdot S_e} \tag{2.34}$$

where D_i is the diffusion coefficient of the gas i, k_q is the reaction rate constant that appears in the generalized equation of the rate of solid/gas reactions of first order and molecularity equal to one (are those where the reaction rate is proportional to the concentration of the only reagent), and S_e is the specific surface area in $m^2 \ m^{-3}$. In the case of porous particles with spherical shape, such a sinter particle can be considered, Eq. (2.33) can be expressed in spherical coordinates as follows:

$$\frac{d^2 \left(C_i - C_i^= \right)}{dR^2} + \frac{2}{R} \cdot \frac{d \left(C_i - C_i^= \right)}{dR} - \frac{\left(C_i - C_i^= \right)}{h^2} = 0 \tag{2.35}$$

where R is the radius of the spherical particles.

In the case of flat particles, the differential equation that is obtained for the mass balance is:

$$\frac{d^2 \left(C_i - C_i^= \right)}{dx^2} - \frac{\left(C_i - C_i^= \right)}{h^2} = 0 \tag{2.36}$$

where x is the thickness of the particles (characteristic linear dimension).

Brief note 2 Transformation of a porous particle by a gas: General equation for the concentration gradient of a gas reacting with a porous solid (Eq. 2.33).

When a dense solid reacts with a gas, the reaction interface is simple and well defined, and the two phenomena that control the overall velocity of reaction (chemical or interface reaction and diffusion) are separated in space. When the solid is porous, the gas penetrates the pores and reacts throughout the mass of the particle and the reaction interface has no simple geometry. The phenomena of reaction and diffusion are dispersed throughout the solid in such small volume that they can be considered as occurring in the same space (Coudurier et al. 1985).

The progress of reaction cannot be observed, except in certain cases, by movement of the interface because it is not well defined. The kinetic expression which describes the process is different from that for a dense solid.

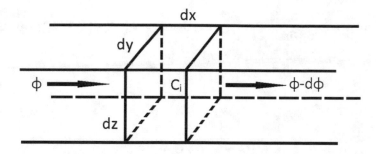

Fig. 2.13 Variation of the reactant gas flow in an elemental volume of porous solid

The elemental volume dx, dy, dz of Fig. 2.13 is porous, and the gaseous reactant flows through it in a direction parallel to dx.

The flux ϕ of reactant gas enters at one end, and $\phi - (d\phi/dx) \cdot dx$ comes out at the other. The change of flux $d\phi$ results from the reaction that consumes the gaseous reactant within the volume $dx \cdot dy \cdot dz$, since the amount of gas that accumulates in the pores in unit time is negligible. Consequently:

$$-\frac{d\phi}{dx} \cdot dx = r_r \tag{2.37}$$

The flow of gas is given by Fick's first law:

$$\phi = -D_i \cdot S \cdot \frac{dC}{dx} = -D_i \cdot \frac{dC}{dx} \cdot dz \cdot dy \tag{2.38}$$

and if D_i is assumed constant:

$$-\frac{d\phi}{dx} = D_i \cdot \frac{d^2C}{dx^2} \cdot dz \cdot dy = D_i \cdot \frac{d^2(C_i - C_i^=)}{dx^2} \cdot dz \cdot dy \tag{2.39}$$

The reaction velocity, r_r, has a form in which the reaction interface, S_i, is proportional to the volume $dx \cdot dy \cdot dz$. The coefficient of proportionality, S_e, is the specific surface of the porous solid, assumed constant:

$$r_r = k_q \cdot S_i \cdot \left(C_i - C_i^=\right) = k_q \cdot S_e \cdot \left(C_i - C_i^=\right) \cdot dx \cdot dy \cdot dz \tag{2.40}$$

Replacing $d\phi/dx$ and r_r by their values in Eq. (2.37), gives:

$$-\frac{d\phi}{dx} \cdot dx = r_r \rightarrow -\frac{\partial\phi}{\partial x} = D_i \cdot \frac{\partial^2 C}{\partial x^2} \cdot dx \cdot dy \cdot dz = k_q \cdot S_e \cdot \left(C_i - C_i^=\right) \cdot dx \cdot dy \cdot dz$$

$$\rightarrow D_i \cdot \frac{d^2\left(C_i - C_i^=\right)}{dx^2} = k_q \cdot S_e \cdot \left(C_i - C_i^=\right)$$

$$(2.41)$$

or:

$$\frac{d^2\left(C_i - C_i^=\right)}{dx^2} - \frac{C_i - C_i^=}{h^2} = 0 \tag{2.42}$$

with:

$$h^2 = \frac{D_i}{k_q \cdot S_e} = \frac{D_i}{k \cdot R \cdot T \cdot (1 + 1/K) \cdot S_e} \tag{2.43}$$

where k is the perfect gas constant. For a sphere, in which the flow is radial, we have Eq. (2.33). For flat particles, we have Eq. (2.35) or (2.42). And the general equation is (2.33).

End of the brief note.

The solution to Eqs. (2.35) and (2.36) gives the profiles of concentration/penetration of reductant gas inside of a porous solid. In the case of spherical particles:

$$\frac{C_i - C_i^=}{C_i^0 - C_i^=} = \frac{R_0 \cdot \sinh(R/h)}{R \cdot \sinh(R_0/h)} \tag{2.44}$$

where R_0 is the size of the non-transformed solid particles, and C_i^0 is the concentration of reductant gas in the gaseous phase.

In the case of flat particles of initial thickness e_0 ($x = e_0$ for $t = 0$):

$$\frac{C_i - C_i^=}{C_i^0 - C_i^=} = \frac{\cosh(x/h)}{\cosh(e_0/h)} \tag{2.45}$$

In Fig. 2.14, we show the profiles of reactive gas concentration/penetration inside of a porous particle, spherical or flat, as a function of the different values that the parameter h could take.

The diffusion coefficient of the reductant gas, D_i, must be corrected if the transfer inside of the sinter pores is included (*Knudsen* diffusion):

$$D_i^K = \frac{D_i \cdot P_A}{S} \tag{2.46}$$

where P_A is the open porosity (per unit) of material and S is the sinuosity of the pore (distance covered by the gas molecules from one point to other distant an unit

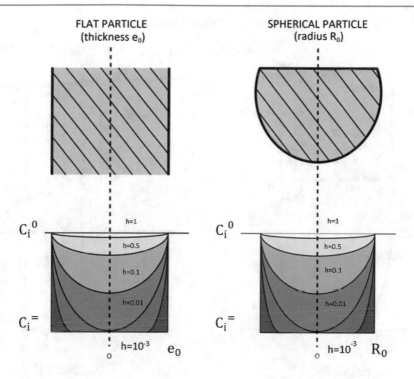

Fig. 2.14 Profiles of concentration/penetration of reductant gas in porous solids

of length from the other). Normally, the value assigned to the sinuosity of the pores is square root of two, and thus:

$$D_i^K = \frac{D_i \cdot P_A}{\sqrt{2}} \tag{2.47}$$

The fraction of solid reduced, X, is equal to:

$$X = \frac{V_0 - V_i}{V_0} \tag{2.48}$$

where V_0 is the initial volume of solid before the reduction beginning and V_i is the volume of solid that was not reduced in each instant of the process (after a time t).

It is necessary to know the variation with the time in each point of the particle and integrate the resulting function for the time of reaction to calculate X. However, the calculation of X can be simplified if the following points are considered:

1. A uniform transformation along the porous solid where all points of the sinter are uniformly reduced. In this case, the kinetical regime of the process is the chemical one (the slow stage or the controlling step is the interface controlled

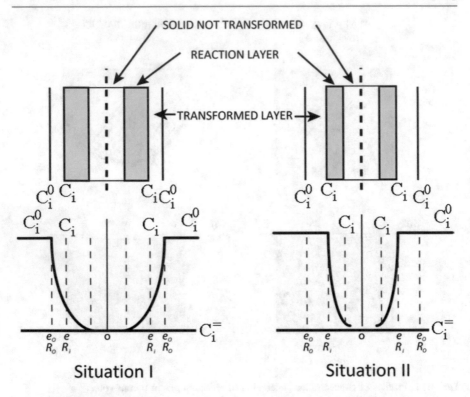

Fig. 2.15 Transformations localized along a reaction front of small thickness and profile of concentration/penetration. Situation I: the same porosity in the reactive solid and in the reaction product; Situation II: the porosity of the solid product of reaction is much bigger than that of the reactive solid

chemical reaction) with high values of h (Fig. 2.15), and the following equation is verified:

$$V_g = V_q = k_q \cdot A \cdot \left(C_i^0 - C_i^= \right) \tag{2.49}$$

And, besides:

$$X = k_q \cdot S_e \cdot \left(C_i^0 - C_i^= \right) \cdot t \tag{2.50}$$

where V_g is the global rate of the process, which coincides with the chemical reaction rate, V_q (controlling step), A is the area of the interface of reaction solid/gas and t is the time.

2. That the reaction concentrates around a small portion of porous solid.

This can give origin to two situations:

- *Situation I*: The porosity of the reactive solid and that of the product of reaction are equivalent. In this situation, $V_g = V_d$ (diffusional transfer rate), which means that the controlling step of the process is the diffusion. The gradient of concentration of the reactive is not changed when passing from the layer of transformed solid to the reaction layer because the thickness of the zone where the reduction takes place is small. In this situation, the kinetical problem is equivalent to that of a dense solid under the diffusional or mixed regime (Ballester et al. 2000, pages 232–235).
- *Situation II*: The porosity of the reaction product is much bigger than that of the reactive solid. The controlling step is, in this case, the diffusional transfer rate of the reductant gas along the sinter layer that is being transformed. This way, two equations are obtained:

$$V_g = V_d = D_i \cdot A \cdot \frac{C_i^0 - C_i^=}{h} \cdot \tanh\left(\frac{e}{h}\right) \tag{2.51}$$

in the case of flat particles, where e refers to the position of the reaction front (for $t = 0$, $e = e_0$), and:

$$V_g = V_d = D_i \cdot A \cdot \frac{C_i^0 - C_i^=}{h} \cdot \left[\frac{1}{\tanh\left(\frac{R_i}{h}\right)} - \frac{h}{R_i}\right] \tag{2.52}$$

in the case of spherical particles where R_i is the initial radius.

For flat particles, the value of the fraction of transformed solid, X, is:

$$X = \frac{D_i \cdot \left(C_i^0 - C_i^=\right) \cdot t}{\rho \cdot e_0 \cdot h} \tag{2.53}$$

where ρ is the density of the solid and e_0 is the initial thickness.

However, if e_0/h is greater than 2 ($\tanh 2 \simeq 0.96$), Eq. (2.51) is transformed into:

$$V_g = D_i \cdot A \cdot \frac{C_i^0 - C_i^=}{h} \tag{2.54}$$

On the other hand, in the case of spherical particles, if h is small in comparison with R_i,

$$\tanh\frac{R_i}{h} \tag{2.55}$$

tends to one, and Eq. (2.52) will be equal to:

$$V_g = D_i \cdot A \cdot \frac{C_i^0 - C_i^=}{h} \tag{2.56}$$

Besides, as:

$$\frac{dX}{dt} = \frac{V_g}{\rho \cdot V_0} \tag{2.57}$$

We obtain:

$$\frac{dX}{dt} = \frac{D_i \cdot A \cdot \left(C_i^0 - C_i^=\right)}{\rho \cdot V_0 \cdot h} \tag{2.58}$$

Equation that integrated leads to:

$$X = \frac{D_i \cdot A \cdot \left(C_i^0 - C_i^=\right) \cdot t}{\rho \cdot V_0 \cdot h} \tag{2.59}$$

2.3 Solved Exercises

Exercise 2.1 Estimate the capillary forces that are produced at the exit of the homogenization drum in a sintering plant (Fig. 2.1), when the materials that are going to be sintered are mixed, if the interfacial tension of the water is 74.0×10^{-3} N m^{-1} (74 mJ m^{-2}), the size of the agglomerates is 5 mm and the mean distance between particles of the agglomerate is 0.10 mm (100 μm).

In the resolution of this exercise, the Brief note 1 should be considered.

If the mean size of the agglomerates that are disposed in the sintering grate is 5 mm, the volume, V, is (if the agglomerates are considered spherical shaped):

$$V = \frac{4}{3} \cdot \pi \cdot r^3 = \frac{4}{3} \cdot \pi \cdot \left(\frac{5 \times 10^{-3} \text{ m}}{2}\right)^3 = 6.545 \times 10^{-8} \text{ m}^3 \tag{2.60}$$

If only 20% of the volume of the agglomerate is considered occupied by water, the volume of water in the agglomerate is:

$$V_{\text{water}} = 6.545 \times 10^{-8} \text{ m}^3 \cdot 0.2 = 1.309 \times 10^{-8} \text{ m}^3 \tag{2.61}$$

Replacing in Eq. (2.1):

$$F = \frac{2 \cdot \gamma_{l-s}}{x^2} \cdot V = \frac{2.74 \times 10^{-3}\,\mathrm{N\,m^{-1}}}{\left(0.1\,\mathrm{mm} \cdot \frac{1\,\mathrm{m}}{1000\,\mathrm{mm}}\right)^2} \cdot 1.309 \times 10^{-8}\,\mathrm{m^3} = 0.194\,\mathrm{N} \quad (2.62)$$

This value is similar to that obtained for the mechanical strength in compression of pellets that were not thermally treated (*green pellets*), which varies within 0.10 and 16 N, depending on their specific surface area.

Commentaries: The size of the raw materials to be sintered is in the interval 0.15–1 mm. The product obtained at the end of the sinter grate, before being crushed and classified, has a size in the range 10–100 mm. Moreover, the size of the sinter particles that are sent to the blast furnace is in the range 5–50 mm. Finally, the order of magnitude of the suction forces, which should sweep along the particles toward the wind boxes (lower zone of the sinter grate, Fig. 2.1), has similar values to the capillary forces:

- Pressure drop in the sinter bed, around 1 atm (1.01×10^5 Pa or N m^{-2}).
- Surface of each agglomerate (quasiparticle): 7.85×10^{-5} m^2.
- Suction force (F_{suction}):

$$F_{\text{suction}} = 1.01 \times 10^5\,\mathrm{N\,m^{-2}} \cdot 7.85 \times 10^{-5}\,\mathrm{m^2} = 7.95\,\mathrm{N} \quad (2.63)$$

that would be reduced to 4.18 N when the pressure drop in the sinter bed was of 400 mm of mercury column and to 1.15 N when it was of 110 mm.

Exercise 2.2 The pseudoternary composition of a hematite sinter (CaO-Fe$_2$O$_3$-SiO$_2$ diagram) is: 70% Fe$_2$O$_3$, 20% CaO, and 10% SiO$_2$. Define the phases and temperatures at which they appear during the heating process in a reduction furnace.

The pseudoternary composition is located in Fig. 2.16, and it is possible to check:

(a) The primary recrystallization field, compatible with the chemical composition of the enunciation, corresponds to the hematite (Fe$_2$O$_3$). Thus, the hematite is the last compound that dissolves in the melt or the first in crystallizing during the cooling. The liquidus temperature, T_l, is approximately 1334 °C (above this temperature there is only liquid).

The temperature at which the formation of liquid phase starts (solidus temperature, T_s) is 1216 °C, which corresponds to the ternary eutectic point (Fig. 2.17):

$$Ca_2SiO_4\,(s) + CaFe_4O_7\,(s) + Fe_2O_3\,(s) \rightleftharpoons \text{liquid} \quad (2.64)$$

and the quantity of liquid that appears at 1216 °C is around 30% (Fig. 2.17).

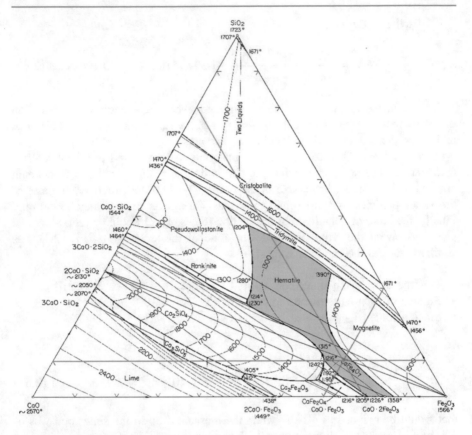

Fig. 2.16 Localization of the sinter composition in the CaO-Fe$_2$O$_3$-SiO$_2$ ternary diagram. The hematite crystallization field indicates that the hematite is the last solid in this sinter, and the first that appears during the cooling

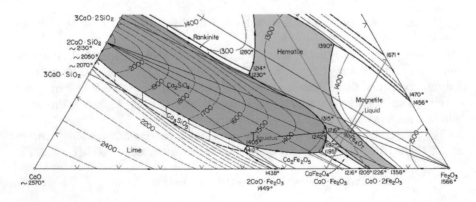

Fig. 2.17 Liquidus temperature of the sinter located in the CaO-Fe$_2$O$_3$.-SiO$_2$ ternary diagram

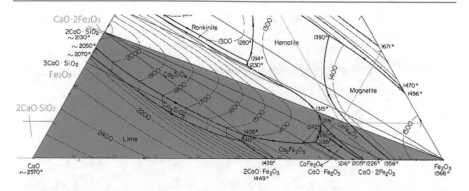

Fig. 2.18 Above 1216 °C, the calcium diferrite disappears, and there is only liquid, hematite, and dicalcium silicate

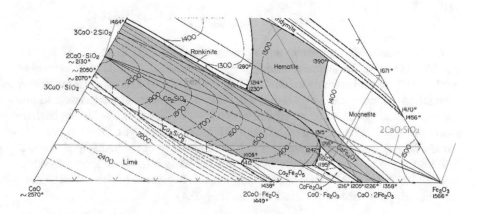

Fig. 2.19 Proportion of hematite and dicalcium silicate in the eutectic

Above 1216 °C, from the three solid phases that form the eutectic, the calcium diferrite (CaO·2Fe₂O₃) disappears (is the phase in lowest proportion in the eutectic, Fig. 2.18) and the quantity of solids in equilibrium with the eutectic is around 70% (Fig. 2.17).

Once concluded the eutectic reaction (1216 °C), the calcium diferrite disappears (this phase is in the lowest proportion in the eutectic), and there are liquid (around 30%) and solid (around 70%), being the proportion of the Fe_2O_3 and the $2CaO·SiO_2$ 71% and 29%, respectively (Fig. 2.19).

At approximately 1310 °C, the dicalcium silicate disappears, and there is only liquid and hematite in the sinter (Fig. 2.20).

The proportions of hematite and liquid at approximately 1310 °C are 17% and 83%, respectively. Finally, the softening interval or solid–liquid interval (difference between the liquidus and solidus temperatures) is 118 °C.

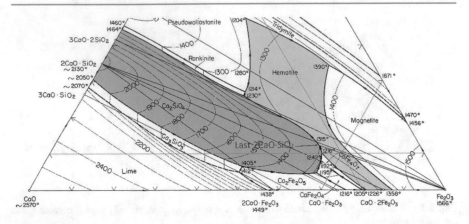

Fig. 2.20 Dissolution of the last dicalcium silicate in the liquid takes place at approximately 1310 °C

Commentaries: If compared with the magnetite sinter (78% Fe_2O_3, 13% CaO, and 9% SiO_2, Fig. 2.8 and neighboring pages and figures), the softening interval of the hematite sinter is almost reduced by half (from 226 to 118 °C). An interesting correlation between the solid–liquid interval of the sintered materials and the basicity index, I_B, can be found (Ballester et al. 2000, page 271), being the basicity index the relation between the CaO and the SiO_2 percentages in the sinter. The value of the I_B in the magnetite sinter is 1.44, and the I_B in the hematite sinter is 2. The bigger the basicity index, the shorter the softening interval and, consequently, smaller the difference of temperature between the beginning of the cohesive zone and the dripping zone of the blast furnace (Fig. 6.4).

Exercise 2.3 To produce a self-fluxing sinter, dolostone, D, with 75% $CaCO_3$ and 25% $MgCO_3$, is used to feed the blender of the plant at a rate of 12,000 kg h^{-1}. If the CaO content in the flow of return fines, RF, is 2.5% CaO and the CaO percentage in the sintered product, P, is 12.5%, estimate the capacity of production of the plant if the entrance and exit currents are (the thermal decomposition of the dolostone takes place at the temperature reached in the process):

- Entrance: dolostone (D), ferric charge of the ore (FC), and return fines (RF).
- Exit: gases (G) and sinter (P).

The corresponding equations for the global mass balance and the partial mass balance (referred to the CaO) are:

- Global balance:

$$12{,}000 + (FC + RF) = G + P \tag{2.65}$$

- Balance to the CaO at the entrance:

$$12{,}000 \cdot 0.75 \left(\frac{kg_{CaCO_3}}{h} \right) \cdot \left(\frac{56\ kg_{CaO}}{100\ kg_{CaCO_3}} \right) + (FC + RF) \frac{kg}{h} \cdot 0.025 \frac{kg_{CaO}}{kg} \tag{2.66}$$

- Balance to the CaO at the exit:

$$P \left(\frac{kg}{h} \right) \cdot 0.125 \ \frac{kg_{CaO}}{kg} \tag{2.67}$$

From Eqs. (2.65)–(2.67), the three unknowns are: G (quantity of gas generated), P (production of the plant), and the addition (FC + RF) (ferric charge and recirculated return fines, which are mixed and homogenized in the blender of the installation). Consequently, if the quantity of gas generated, G, is known, then we have a system of two equations and two unknowns: P and (FC + RF), which allows estimating the quantity of gas generated as a consequence of the thermal decomposition of the dolostone:

$$CaCO_3(s) \rightleftharpoons CO_2(g) + CaO(s) \tag{2.68}$$

$$MgCO_3(s) \rightleftharpoons CO_2(g) + MgO(s) \tag{2.69}$$

If we take as base for the calculation 1 kg of dolostone, 750 g of $CaCO_3$ generates the following quantity of CO_2:

$$750\ g\ CaCO_3 \cdot \frac{1\ mol\ CaCO_3}{100\ g\ CaCO_3} \cdot \frac{1\ mol\ CO_2}{1\ mol\ CaCO_3} \cdot \frac{44\ g\ CO_2}{1\ mol\ CO_2} = 330\ g\ CO_2 \tag{2.70}$$

and 250 g of $MgCO_3$ generates the following quantity of CO_2:

$$250\ g\ MgCO_3 \cdot \frac{1\ mol\ MgCO_3}{84.3\ g\ MgCO_3} \cdot \frac{1\ mol\ CO_2}{1\ mol\ MgCO_3} \cdot \frac{44\ g\ CO_2}{1\ mol\ CO_2} = 131\ g\ CO_2 \tag{2.71}$$

In total, per each kilogram of dolostone, 461 g of CO_2 is produced. For that reason, the production of carbon dioxide every hour is:

$$\frac{461\,g\;CO_2}{kg\;dolostone} \cdot \frac{12{,}000\,kg\;dolostone}{hour} \cdot \frac{1\;kg\;g\;CO_2}{1000\;g\;CO_2} = 5532\;kg\;CO_2 \cdot h^{-1} \quad (2.72)$$

From the mass global balance in Eq. (2.65), it is possible to calculate the value of the plant production, P, as a function of the addition of the group formed by the flows of the ferric charge, FC, and the return fines, RF:

$$12{,}000 + (FC+RF) = G+P \rightarrow P = 12{,}000 - 5532 + (FC+RF) \rightarrow P$$
$$= [6468 + (FC+RF)]\left(kg\,h^{-1}\right) \quad (2.73)$$

On the other hand, from the partial balance of the CaO (Eq. (2.74)), the value of the addition (FC + RF) can be calculated, considering Eq. (2.73).

$$12{,}000 \cdot 0.75\left(\frac{kg_{CaCO_3}}{h}\right) \cdot \left(\frac{56\;kg_{CaO}}{100\;kg_{CaCO_3}}\right) + (FC+RF)\frac{kg}{h} \cdot 0.025\frac{kg_{CaO}}{kg}$$
$$= P\left(\frac{kg}{h}\right) \cdot 0.125\,\frac{kg_{CaO}}{kg} \quad (2.74)$$

$$5040 + (FC+RF) \cdot 0.025 = [6468 + (FC+RF)] \cdot 0.125 \quad (2.75)$$

$$5040 - 808.5 = -(FC+RF) \cdot 0.025 + (FC+RF) \cdot 0.125 \rightarrow FC+RF$$
$$= 42{,}315\;kg\,h^{-1} \quad (2.76)$$

Consequently, the plant production, P, each hour is (using Eq. (2.73)):

$$P = [6468 + 42{,}315]\left(kg\;h^{-1}\right) = 48{,}783\;kg\,h^{-1} \quad (2.77)$$

Commentaries: The annual production of the plant will be of:

$$P = 48{,}783\frac{kg}{h} \cdot \frac{24\,h}{day} \cdot \frac{365\,days}{year} \cdot \frac{1\,t}{1000\,kg} = 427{,}339\,t\;sinter\,year^{-1} \quad (2.78)$$

Considering the size of the sinter plant (0.4 Mt $year^{-1}$), it is small for an integrated steel plant.

Exercise 2.4 A sintering machine produces, in a certain time, 1177 kg of sinter with the following composition: 64% Fe_2O_3, 6% FeO, 10% SiO_2, and 20% CaO. However, due to quality requirements (granulometry), 177 kg of sinter (with the same composition) is recycled to the feeding machine and to the homogenization drum. The raw materials used to produce this quantity of sinter include 806.862 kg of iron ore (containing only Fe_2O_3 and SiO_2), 60 kg of water, 8000 kg of air, 51.25 kg of CaO, 265 kg of $CaCO_3$, and 90 kg of coke (we assume that the coke is 100% carbon).

Calculate the quantity of gases per ton of sinter produced and the composition of the iron ore that is processed in the sinter plant.

Considering that the sinter produced in the process can be represented by a pseudoternary diagram with 70% Fe_2O_3, 10% SiO_2, and 20% CaO, calculate:

- the constitution and proportion of the phases compatible at room temperature with the composition.
- the temperature at which the formation of the first liquid phases starts in the heating.
- and, considering equilibrium conditions, calculate also the proportion and quantity of solid and liquid phases when the flame front reaches 1300 °C.

The quantity of gases (fumes) that are produced as a consequence of the environmental air suction is the result of a global mass balance of the flows at the entrance and exit (Fig. 2.21). The addition of the flows of entrance is 9273.11 kg (Point A, Fig. 2.21), the addition of the flows of exit is 1000 kg + gas (Point B, Fig. 2.21), and flow of return fines accounts for 177 kg. Consequently, the flow of exit for the gases is:

$$
\begin{aligned}
806.862\,\text{kg (iron ore)} &+ 60\,\text{kg (water)} + 8000\,\text{kg (air)} \\
&+ 51.25\,\text{kg (CaO)} + 265\,\text{kg (CaCO}_3) + 90\,\text{kg (coke)} \\
&= G_{exit} + 1000\,\text{kg (sinter)} \rightarrow G_{exit} = 8096.11\,\text{kg}
\end{aligned}
\tag{2.79}
$$

that referred to the total quantity of sinter produced in the sinter machine (1177 kg sinter, including the return fines) is:

$$
G_{exit} = \frac{8096.11\,\text{kg}}{1177\,\text{kg sinter}} \cdot \frac{1000\,\text{kg}}{t} = 6879\,\text{kg gas/t sinter}
\tag{2.80}
$$

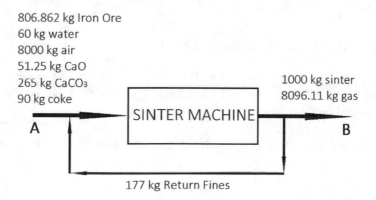

806.862 kg Iron Ore
60 kg water
8000 kg air
51.25 kg CaO
265 kg CaCO₃
90 kg coke

1000 kg sinter
8096.11 kg gas

SINTER MACHINE

A

B

177 kg Return Fines

Fig. 2.21 Flow diagram of the sintering machine

To obtain the composition of the ore that is fed and mixed with the other raw materials in the homogenization drum (Fig. 2.1), we make a balance to the quantity of iron that enters and leaves the system. Thus, the quantity of iron in the exit flow is:

- As Fe_2O_3:

$$1177 \text{ kg sinter} \cdot \frac{64 \text{ kg Fe}_2O_3}{100 \text{ kg sinter}} \cdot \frac{1 \text{ mol Fe}_2O_3}{159.7 \text{ kg Fe}_2O_3} \cdot \frac{2 \text{ mol Fe}}{1 \text{ mol Fe}_2O_3} \cdot \frac{55.85 \text{ kg Fe}}{1 \text{ mol Fe}}$$
$$= 526.87 \text{ kg Fe}$$

$$(2.81)$$

- As FeO:

$$1177 \text{ kg sinter} \cdot \frac{6 \text{ kg FeO}}{100 \text{ kg sinter}} \cdot \frac{1 \text{ mol FeO}}{71.85 \text{ kg FeO}} \cdot \frac{1 \text{ mol Fe}}{1 \text{ mol FeO}} \cdot \frac{55.85 \text{ kg Fe}}{1 \text{ mol Fe}}$$
$$= 54.89 \text{ kg Fe}$$

$$(2.82)$$

Altogether, 581.76 kg Fe is accounted, where 55.85, 71.85, and 159.7 are the atomic/molecular weights (g/mol) of the Fe, FeO, and Fe_2O_3, respectively.

As flows of entrance to the system, we have the iron ore (806.862 kg) and the recycled product or return fines (177 kg, which must not be included in the burden of the blast furnace because this sinter fraction does not accomplish the granulometric requirements, as it has a size <5 mm). For the recycled product (the procedure is the same that was used in the case of the sintered product), the quantity of iron in the form of Fe_2O_3 is:

$$177 \text{ kg sinter} \cdot \frac{64 \text{ kg Fe}_2O_3}{100 \text{ kg sinter}} \cdot \frac{1 \text{ mol Fe}_2O_3}{159.7 \text{ kg Fe}_2O_3} \cdot \frac{2 \text{ mol Fe}}{1 \text{ mol Fe}_2O_3} \cdot \frac{55.85 \text{ kg Fe}}{1 \text{ mol Fe}}$$
$$= 79.23 \text{ kg Fe}$$

$$(2.83)$$

and the quantity of iron in the form of FeO is:

$$177 \text{ kg sinter} \cdot \frac{6 \text{ kg FeO}}{100 \text{ kg sinter}} \cdot \frac{1 \text{ mol FeO}}{71.85 \text{ kg FeO}} \cdot \frac{1 \text{ mol Fe}}{1 \text{ mol FeO}} \cdot \frac{55.85 \text{ kg Fe}}{1 \text{ mol Fe}} \qquad (2.84)$$
$$= 8.26 \text{ kg Fe}$$

In total, the quantity of iron recycled to the sinter grate is 87.49 kg of Fe.

Finally, to calculate the percentage of Fe_2O_3 in the ore, we assume that all the iron is as hematite, and the following relation is used:

- Point A:

$$\frac{\%Fe_2O_3}{100 \text{ kg Iron Ore}} \cdot 806.862 \text{ kg Iron Ore} \cdot \frac{1 \text{ mol } Fe_2O_3}{159.7 \text{ kg } Fe_2O_3} \cdot \frac{2 \text{ mol Fe}}{1 \text{ mol } Fe_2O_3} \cdot \frac{55.85 \text{ kg Fe}}{1 \text{ mol Fe}} \tag{2.85}$$

- Point B:

$$581.76 \text{ kg Fe} \tag{2.86}$$

- Return Fines:

$$87.49 \text{ kg Fe} \tag{2.87}$$

and the global relation is:

$$\frac{\%Fe_2O_3}{100 \text{ kg Iron Ore}} \cdot 806.862 \text{ kg Iron Ore} \cdot \frac{1 \text{ mol } Fe_2O_3}{159.7 \text{ kg } Fe_2O_3} \\ \cdot \frac{2 \text{ mol Fe}}{1 \text{ mol } Fe_2O_3} \cdot \frac{55.85 \text{ kg Fe}}{1 \text{ mol Fe}} = 581.76 \text{ kg Fe} - 87.49 \text{ kg Fe} \tag{2.88}$$

and the percentage of Fe_2O_3 in the ore is:

$$\%Fe_2O_3 = 87.6\% \tag{2.89}$$

And the composition of the ore is: 87.6% Fe_2O_3 and 12.4% SiO_2.

The phases compatible, at room temperature, with the pseudoternary composition of the sinter indicated in the enunciation (70% Fe_2O_3, 10% SiO_2, 20% CaO) are:

- $2CaO \cdot SiO_2$-CaO-Fe_2O_3 (Fig. 2.22).
- $3CaO \cdot 2SiO_2$-CaO-Fe_2O_3 (Fig. 2.23).

According to the first alternative, the percentage of phases is: 28%, $2CaO \cdot SiO_2$; 2%, CaO; and, 70%, Fe_2O_3 (Fig. 2.22). In the case of the second alternative, the result is: 22%, $3CaO \cdot 2SiO_2$; 8%, CaO; and, 70%, Fe_2O_3 (Fig. 2.23). The softening

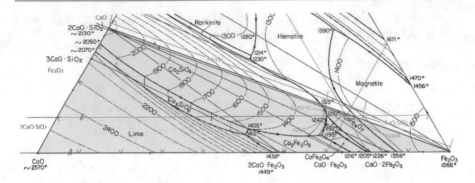

Fig. 2.22 Alternative 1: 2CaO·SiO₂-CaO-Fe₂O₃

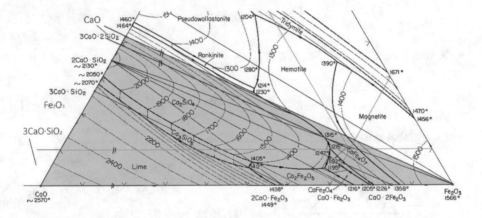

Fig. 2.23 Alternative 2: 3CaO·SiO₂-CaO-Fe₂O₃

of the material, which corresponds to the beginning of the liquid formation in the heating, will be consequence of the following eutectic ternary reaction:

$$2CaO\cdot SiO_2(s) + Fe_2O_3(s) + CaO\cdot Fe_2O_3(s) \overset{1216\,^{\circ}C}{\leftrightarrow} \text{liquid} \qquad (2.90)$$

Finally, the equilibrium is reached at 1300 °C (green line, Fig. 2.22) with 65% of liquid and 35% of solid with the following composition: 18.5% 2CaO·SiO₂ and 81.5% Fe₂O₃ (orange line, Fig. 2.22).

Commentaries: If the data provided by Babich et al. (2008) (page 59) is analyzed, the volume of gases, referred to the ton of sinter, for an iron ore sintering plant is 1400–2000 N m³ t⁻¹. If we did the calculation to express the 8096.11 kg of gas (6879 kg gas/t sinter, we assume that the quantity of CO₂ is very low as compared with the quantity of air) in cubic meters measured in standard conditions, N m³ of air at 273 K and one atmosphere (we use the ideal gas law, $P \cdot V = n \cdot R \cdot T$):

$$1 \text{ atm V} = \frac{6879 \text{ kg air/t sinter}}{28.84 \frac{\text{kg}}{\text{kmol}} \cdot \frac{1\,\text{kmol}}{1000\,\text{mol}}} \cdot 0.082 \frac{\text{atm} \cdot 1}{\text{mol} \cdot \text{K}} \cdot 273 \text{ K} \to \text{V}$$

$$= 5{,}339{,}573 \frac{1}{\text{t sinter}} \cdot \frac{1\,\text{m}^3}{10^3\,1} = 5339.6 \,\text{Nm}^3/\text{t sinter}$$

$$(2.91)$$

it is possible to check that the quantity of air suctioned by the sinter plant indicated in the enunciation will be 2.7–3.8 times above the normal. Equally, the lack of understanding on the operating values of the plant indicated in the enunciation is observed in the carbon-coke consumption used: 40-45 kg of coke per ton habitually used, while 90 kg of carbon was proposed in the enunciation of the exercise.

Exercise 2.5 An iron ore sintering plant has a production rate of 50 t h^{-1} measured the classification and return of the inadequate fraction of the sintered product. The efficiency of the installation is of 90%. Apart from the return fines, RF (recirculated sinter as a consequence of not accomplishing the size specifications), limestone, C, with 100% CaCO$_3$ and iron ore, M, with 2.0% CaO are fed to the blender of the plant at a rate of 40 t h^{-1}.

Calculate the rate at which the limestone should be fed to the blender to obtain a 12% CaO sinter.

The mass balance for the calcium oxide in the flows of entrance and exit to the blender drum has the following terms:

- Flows of entrance: iron ore, M, limestone, C, and return fines, RF.
- Flows of exit: sinter, S, and return fines, RF.

The addition of the produced sinter and the return fines in the flows of exit is 50 t h^{-1}. The quantities of sinter and return fines are:

$$S = \left(50\,\text{t h}^{-1}\right) \cdot 0.90 = 45\,\text{t h}^{-1} \qquad (2.92)$$

$$R = 50\,\text{t h}^{-1} - 45\,\text{t h}^{-1} = 5\,\text{t h}^{-1} \qquad (2.93)$$

Considering that the percentage of CaO in the CaCO$_3$ is 56 wt% (the molecular weights of the CaO and CaCO$_3$ are 56 and 100 g mol^{-1}, respectively). The balance to the CaO is the following one:

$$40\,\text{t h}^{-1} \cdot \frac{2\,\text{t CaO}}{100\,\text{t Iron Ore}} + C \cdot \frac{56\,\text{t CaO}}{100\,\text{t CaCO}_3} + 5\,\text{t h}^{-1} \cdot \frac{12\,\text{t CaO}}{100\,\text{t RF}}$$
$$= 5\,\text{t h}^{-1} \cdot \frac{12\,\text{t CaO}}{100\,\text{t RF}} + 45\,\text{t h}^{-1} \cdot \frac{12\,\text{t CaO}}{100\,\text{t sinter}}$$

$$(2.94)$$

thus the quantity of limestone is calculated as follows:

$$C = \cfrac{5\,h^{-1} \cdot \dfrac{12}{100} + 45\,h^{-1} \cdot \dfrac{12}{100} - 40\,h^{-1} \cdot \dfrac{2}{100} - 5\,h^{-1} \cdot \dfrac{12}{100}}{\dfrac{56}{100\,t\,CaCO_3}}$$

$$= \frac{0.6 + 5.4 - 0.8 - 0.6}{0.56}\,t\,CaCO_3\,h^{-1} = 8.2\,t\,CaCO_3\,h^{-1}$$

(2.95)

Exercise 2.6 Whether the reduction rate is uniform along all the sinter particle (model of ideal porous solid/particle), considering a particle of 30 mm in diameter in contact with CO at a pressure of 0.9 atm, and knowing that at 1050 °C the total reduction of the particle takes place after 4 h, calculate:

- The surface area of the sinter particle in cm^2/g.
- Compare the characteristics calculated in the previous point with those of a spherical pellet with the same geometrical and chemical properties.
- Indicate what could be the surface characteristics of other sinter particle that, having the same chemical behavior against the reductant gas, had a size of 15 mm.

Data: volumetric/bulk/apparent density of the sinter, 3.0 g/cm^3; real density of the sinter, 4.7 g/cm^3; molecular weight of the CO, 28 g mol^{-1}; partial pressure of CO in equilibrium with the sinter at 1050 °C: 0.40 atm; reaction rate constant of the chemical reduction process, by the CO, 0.028 cm s^{-1}. Reference temperature for the calculation of the reductant gas: 15 °C (288 K).

Equation (2.50) indicates that the value of $S_{e/M}$ (specific surface area expressed in surface area per unit of mass), when the reduction is complete, $X = 1$ and $k_q = 0.028\,cm\,s^{-1}$ is:

$$S_{e/M} = \frac{1}{k_q \cdot (C_{CO}^0 - C_{CO}^=) \cdot t}$$

(2.96)

where the values that the variables C_{CO}^0, $C_{CO}^=$ and ΔC take are:

$$C_{CO}^0 = \frac{0.9\,atm}{0.082\frac{atm\,l}{mol\,K} \cdot (15 + 273)K} = 3.81 \times 10^{-2}\,moles\ of\ CO\ per\ liter\ of\ gas$$

or 3.81×10^{-5} moles of CO per cm^3

(2.97)

$$C_{CO}^= = \frac{0.4\,atm}{0.082\frac{atm\,l}{mol\,K} \cdot (15 + 273)K} = 1.69 \times 10^{-2}\,moles\ of\ CO\ per\ liter\ of\ gas$$

or 1.69×10^{-5} moles of CO per cm^3

(2.98)

$$\Delta C = C_{CO}^0 - C_{CO}^{=} = 2.12 \times 10^{-5} \text{ moles of CO per cm}^3 \qquad (2.99)$$

This way, the value of $S_{e/M}$ in $cm^2 \, g^{-1}$ is:

$$S_{e/M} = \frac{1}{0.028\frac{cm}{s} \cdot 2.12 \times 10^{-5}\frac{mol}{cm^3} \cdot 28\frac{g}{mol} \cdot 4\,h \cdot \frac{3600\,s}{1\,h}} = 4.18\,cm^2\,g^{-1}$$

$$(2.100)$$

The surface characteristics of a sinter, $S_{e/M}$, which would have the same behavior in front of the reductant gas, CO, at the same temperature, 1050 °C, will be the same independently that the particle size was 30.0 mm or 15.0 mm or 45 mm.

This value of the specific surface area of the porous sinter contrasts with the prototype of dense particle (iron ore pellet) of the same size, that is to say, 30 mm in diameter. In the case of pellet of spherical surface, the area-to-volume ratio is:

$$\text{Area/Volume} = \frac{4 \cdot \pi \cdot \left(\frac{D_p}{2}\right)^2}{\frac{4}{3} \cdot \pi \cdot \left(\frac{D_p}{2}\right)^3} = \frac{4 \cdot \pi \cdot \left(\frac{3\,cm}{2}\right)^2}{\frac{4}{3} \cdot \pi \cdot \left(\frac{3\,cm}{2}\right)^3} = 2\,cm^2\,cm^{-3} \qquad (2.101)$$

And the $S_{e/M}$ is (considering the real density of the pellet, 4.7 g cm^{-3}):

$$S_{e/M} = 2\frac{cm^2}{cm^3} \cdot \frac{cm^3}{4.7\,g} = 0.43\,cm^2\,g^{-1} \text{ (100\% dense)} \qquad (2.102)$$

Better using the global/volumetric density: 3.0 g cm^{-3}. In this way, the surface characteristics of the pellet, $S_{e/M}$, would be comprised between the minimum value 0.43 cm^2 g^{-1} (calculated considering the real density of the pellet, 4.7 g cm^{-3}) and the maximum value of 0.67 cm^2 g^{-1} (calculated considering the global/volumetric density of the pellet, 2.3 g cm^{-3}).

Comment: The surface characteristics of a porous solid are expressed by means of the specific surface area, expressed in units of volume, $S_{e/V}$, or in units of area per unit of mass, $S_{e/M}$. If the specific surface area is expressed in units of volume, the reaction rate constant of the chemical reaction process indicated by Eq. (2.50) will have the units cm^4 g^{-1} s^{-1}. If the specific surface area is expressed in units of mass, the units of k_q will be cm s^{-1}.

Exercise 2.7 Calculate the concentration of reductant gas, CO, for a group of sinter porous particles of spherical geometry (30.0 mm in diameter) in the center, at a distance of 6.00 mm from the center and at 14.0 mm from the center. Working conditions: 1050 °C; total pressure of the gas 2.0 atm; partial pressure of the CO

0.9 atm; partial pressure of the CO in equilibrium 0.4 atm; specific surface area of the sinter 323 cm^2 cm^{-3}; reaction rate constant of the chemical reaction process 0.04 cm s^{-1}; diffusion coefficient of the CO inside of the porous particle 0.15 cm^2 s^{-1}.

The profile of concentration/penetration in the porous spherical particles can be calculated using Eq. (2.44). However, the enunciation of the exercise also proposes the calculation of the partial pressures of the reductant gas. Using the general gas equation of the ideal gas, it is possible to modify Eq. (2.44) in the following way:

$$P_{CO} = \left(P_{CO}^0 - P_{CO}^=\right) \cdot \frac{R_0 \cdot \sinh(R/h)}{R \cdot \sinh(R_0/h)} + P_{CO}^= \qquad (2.103)$$

where P_{CO}, P_{CO}^0 and $P_{CO}^=$ are, respectively (making equivalence between pressure and concentration as in Fig. 2.14): the partial pressure of the reductant gas inside of the porous solid, the partial pressure reached in particles with size R_0, and the partial pressure reached in the thermodynamic equilibrium between phases. The parameter h appears in the kinetics models of porous solids and quantifies the relative importance of the diffusional transfer in relation with the rate of chemical reaction inside of the solid (Habashi 1999). In this case, the values of h^2, in cm^2 (Eq. (2.34)), and of h, in cm, are:

$$h^2 = \frac{D_i}{k_q \cdot S_e} = \frac{0.15\,\mathrm{cm^2\,s^{-1}}}{0.04\,\mathrm{cm\,s^{-1}}\,323\,\mathrm{cm^2\,cm^{-3}}} = 0.0116\,\mathrm{cm^2} \qquad (2.104)$$

$$h = 0.1078\,\mathrm{cm} \qquad (2.105)$$

We show in Table 2.3 the relation between the pressure of the reductant gas, P_{CO}, in atm, and the R, in cm.

Consequently, apart from the pressure of the carbon monoxide in different zones of the particle, the results at 14 and 6 mm from the center are also shown. In the center of the sinter porous particles, the value of the gas pressure is that corresponding to the thermodynamic equilibrium. This profile indicates that, from the center of the particle to a radial position of 0.8 cm (size/radius of the unreacted nucleus), there is not reaction. Only in an external layer of the sphere of approximately 0.7 cm, reduction processes can take place.

Table 2.3 Relation $P_{CO} - R$

P_{CO} (atm)	R (cm)
0.612	1.4
0.439	1.2
0.407	1.0
0.401	0.8
0.400	0.6
0.400	0.4
0.400	0.2

Even though the values of the profiles of CO partial pressures cannot be verified *in situ*, the experimental evidence indicates that the reaction model takes place according to this mechanism, and it is possible to find an "unreacted sinter nucleus" of 0.8 cm that supports the utilization in the calculations of this reaction mechanism.

Commentaries: If the material has the same surface characteristics, $S_{e/V}$ of 323 cm^2 cm^{-3}, but with iron ores with less capacity of transformation against the CO or more difficultly reducible (k_q of 0.015 cm s^{-1}), almost all the particle will have a CO partial pressure above that of equilibrium of 0.4 atm, as it is possible to check in Table 2.4.

However, for distances shorter than 10 mm to the center (size/radius of the unreacted nucleus, 10 mm), the values of the driving force of the reaction, $P_{CO}^0 - P_{CO}^=$, become very small and also the corresponding transformation reduction rate.

Exercise 2.8 Estimate using the multivariate analysis the Tumbler, TI (% > 6.3 mm) and RDI (% < 3.15 mm) indices considering the experimental data collected in Table 2.5 (Formoso et al. 2000a, pages 244–253; Formoso et al. 2000b, pages 254–265).

As a first step of the process, the correlation and the functional dependencies between the Tumbler, TI (% > 6.3 mm) and RDI (% < 3.15 mm) indices and the twenty variables indicated in Table 2.5 for the eight sintered products are analyzed. This dependence can be quantified using the correlation coefficient R^2 (linear approach to the data) that takes values within 0 (dispersed data) and 1 (values over the regression line).

Table 2.6 collects the values of the correlation coefficients, R^2, for the 20 variables with respect to the Tumbler and RDI indices.

As the available data is very limited (only eight sintered products, S18, S20, S22, S24, S26, S28, S30, and S32), the correlation between the 20 variables and the TI and RDI indices is not as high as would be in the case of having more data: for instance results for 200–300 sintered products. Only in this example, we consider $R^2 > 0.3$ as criterion to indicate the existence of correlation for the TI and $R^2 > 0.14$ in the case of the RDI (to validate the existence at an industrial scale of

Table 2.4 Relation $P_{CO} - R$

P_{CO} (atm)	R (cm)
0.704	1.4
0.514	1.2
0.444	1.0
0.414	0.8
0.408	0.6
0.404	0.4
0.402	0.2

Table 2.5 Sinter properties

Ref. of the sinter mixture	S18	S20	S22	S24	S26	S28	S30	S32
Iron ore (kg) A	112.3	112.3	112.3	112.3	138.1	138.1	138.1	138.1
Dunite (kg) B	2.2	2	2	1.9	2.8	2.7	2.3	2.3
Coke (kg) C	5.0	4.8	4.4	4.5	5.8	5.9	5.5	5.5
Return fines (kg) D	27.5	27.5	38.5	38.5	33.8	33.8	47.3	47.3
Limestone (kg) E	10.1	13.1	8.9	12.9	11.9	16.0	10.8	15.6
Water (kg) F	8.5	8.7	9	9.5	11.0	11.0	11.5	11.9
Permeability (JPU[a]) G	2.01	2.04	2.02	2.02	2.02	1.57	1.55	1.53
Coke (kg t^{-1} sinter) H	44.2	42.0	40.2	39.4	42.8	42.8	41.0	39.2
Productivity (t m^{-3} day^{-1}) I	39.6	39.4	37.2	37.2	40.5	41.8	34.9	36.4
FeO (wt%)	5.7	5.6	3.7	3.6	4.7	5.0	3.5	3.9
SiO$_2$ (wt%)	5.4	5.3	5.4	4.9	5.2	5.7	5.0	5.8
Al$_2$O$_3$ (wt%)	1.22	1.17	1.19	1.16	1.20	1.10	1.19	1.10
CaO (wt%)	9.35	10.50	8.80	10.35	8.60	10.13	9.05	9.20
MgO (wt%)	1.70	1.68	1.60	1.61	1.62	1.66	1.56	1.45
MnO (wt%)	0.74	0.72	0.80	0.71	0.75	0.73	0.79	0.69
S (wt%)	0.011	0.014	0.013	0.018	0.012	0.014	0.012	0.015
P (wt%)	0.042	0.039	0.040	0.040	0.040	0.036	0.041	0.038
Na$_2$O (wt%)	0.039	0.037	0.036	0.031	0.032	0.036	0.040	0.039
K$_2$O (wt%)	0.071	0.071	0.072	0.068	0.066	0.072	0.075	0.072
Basicity (CaO/SiO$_2$) J	1.73	1.98	1.64	2.13	1.67	1.82	1.81	1.59
Tumbler (% > 6.3 mm)	66.7	70.0	60.0	64.0	67.3	72.7	58.0	66.7
RDI (% < 3.15 mm)	48.9	45.3	41.0	45.7	40.9	42.0	47.6	45.3

[a]JPU, Japanese Permeability Unit, N m^3 m^{-2} min^{-1} (Cores et al. 2013, page 160)

Table 2.6 Correlation coefficients

Variable	TI (R^2)	RDI (R^2)
A	0.012	0.0515
B	0.1677	**0.1756**
C	0.15	0.0259
D	**0.359**	0.0019
E	**0.4832**	0.0143
F	0.0011	0.0279
G	0	0.015
H	0.2169	0.0064
I	**0.7064**	**0.1631**
FeO (wt%)	**0.5497**	0.0153
SiO_2 (wt%)	0.2839	0.0588
Al_2O_3 (wt%)	0.2185	0.0517
CaO (wt%)	0.2644	0.0563
MgO (wt%)	0.1478	0.0007
MnO (wt%)	**0.4457**	0.0397
S (wt%)	0.0144	0.0038
P (wt%)	**0.4019**	**0.2602**
Na_2O (wt%)	0.0268	**0.2878**
K_2O (wt%)	0.112	**0.141**
J	0.0126	0.0786

correlation, it is assumed that $R^2 \gg 0.15$). In this exercise, the variables that have relation with the Tumbler and the RDI indices appear highlighted in bold font.

The multivariate correlation to forecast the Tumbler index, TI, without using experimental tests and, as indicated in Table 2.6, can be made for the index with the percentage of return fines (variable D), the percentage of limestone (variable E) and the productivity (variable I). Additionally, the variables related to the composition of the sintered product as the FeO, the MnO, and the P can be included in the correlation. That is to say, the following equation can be defined:

$$TI(\% > 6.3 \text{ mm}) = \beta_0 + \beta_1 \cdot (D) + \beta_2 \cdot (E) + \beta_3 \cdot (I) + \beta_4 \cdot (\%FeO) + \beta_5 \cdot (\%MnO) + \beta_6 \cdot (\%P)$$

(2.106)

β coefficients in Eq. (2.106) indicate the variation of the TI produced by the unitary change of each one of the variables (D, E, I, %FeO, %MnO and %P). Thus, the following system of equations can be obtained:

$$Y = X \cdot \beta + U$$

(2.107)

that can be solved using matrix analysis:

$$\beta = (X' \cdot X)^{-1} \cdot X' \cdot Y \tag{2.108}$$

where β is the column vector that comprises the seven coefficients (β_0 and the six coefficients associated with the variables D, E, I, %FeO, %MnO, and %P, which have influence on the value of the TI), Y is the column vector associated with the eight experimental values of the TI (66.7, 70.0, 60.0, 64.0, 67.3, 72.7, 58.0, 66.7) that can be found in Table 2.5, U is the column vector containing the seven residues (difference between the experimental values of the TI and the result that the model provides), X is the matrix that can be found below these lines, an eight columns matrix (X' is the transposed matrix of X):

$$\begin{pmatrix} 1 & 27.5 & 10.1 & 39.6 & 5.7 & 0.74 & 0.042 \\ 1 & 27.5 & 13.1 & 39.4 & 5.6 & 0.72 & 0.039 \\ 1 & 38.5 & 8.9 & 37.2 & 3.7 & 0.80 & 0.040 \\ 1 & 38.5 & 12.9 & 37.2 & 3.6 & 0.71 & 0.040 \\ 1 & 33.8 & 11.9 & 40.5 & 4.7 & 0.75 & 0.040 \\ 1 & 33.8 & 16.0 & 41.8 & 5.0 & 0.73 & 0.036 \\ 1 & 47.3 & 10.8 & 34.9 & 3.5 & 0.79 & 0.041 \\ 1 & 47.3 & 15.6 & 36.4 & 3.9 & 0.69 & 0.038 \end{pmatrix} \tag{2.109}$$

Consequently, the values of the column vector are: $\beta_0 = 75.8344$, $\beta_1 = 0.0086$, $\beta_2 = 0.2190$, $\beta_3 = 0.8100$, $\beta_4 = 1.8860$, $\beta_5 = -37.4804$ and $\beta_6 = -630.6164$, and the Tumbler index can be estimated using the following equation:

$$\text{TI } (\% > 6.3\,\text{mm}) = 75.8344 + 0.0086 \cdot (D) + 0.2190 \cdot (E) + 0.8100 \cdot (I) + 1.8860 \cdot (\%\text{FeO})$$
$$- 37.4804 \cdot (\%\text{MnO}) - 630.6174 \cdot (\%\text{P})$$

$$\tag{2.110}$$

In the case of the RDI index, the studied dependent variables are five: the percentage of dunite in the charge (variable B), the productivity of the installation (variable I) and the percentages of phosphorus, sodium oxide, and potassium oxide in the sintered product:

$$\text{RDI } (\% < 3.15\,\text{mm}) = \beta_0 + \beta_1 \cdot (B) + \beta_2 \cdot (I) + \beta_3 \cdot (\%\text{P}) + \beta_4 \cdot (\%\text{Na}_2\text{O}) + \beta_5 \cdot (\%\text{K}_2\text{O})$$

$$\tag{2.111}$$

In this case, the column vector β comprises six values: β_0 and the β coefficients associated with the variables B, I, %P, %Na$_2$O, and %K$_2$O, which have influence on the value of the RDI, Y is the column matrix associated with the eight RDI experimental results for the sintered samples (48.9, 45.3, 41.0, 45.7, 40.9, 42.0, 47.6, 45.3), U is the column vector of the six residues (difference between the experimental values of the RDI and the result that the model provides) and X is the six columns and eight rows matrix:

$$
\begin{pmatrix}
1 & 2.2 & 39.6 & 0.042 & 0.039 & 0.071 \\
1 & 2.0 & 39.4 & 0.039 & 0.037 & 0.071 \\
1 & 2.0 & 37.2 & 0.040 & 0.036 & 0.072 \\
1 & 1.9 & 37.2 & 0.040 & 0.031 & 0.068 \\
1 & 2.8 & 40.5 & 0.040 & 0.032 & 0.066 \\
1 & 2.7 & 41.8 & 0.036 & 0.036 & 0.072 \\
1 & 2.3 & 34.9 & 0.041 & 0.040 & 0.075 \\
1 & 2.3 & 36.4 & 0.038 & 0.039 & 0.072
\end{pmatrix}
\tag{2.112}
$$

The values of the column vector β are: $\beta_0 = 33.85$, $\beta_1 = -2.7712$, $\beta_2 = -0.0362$, $\beta_3 = 471.6843$, $\beta_4 = 693.7159$ and $\beta_5 = -357.66265$, and the value of the RDI can be calculated using the following equation:

$$
\begin{aligned}
\text{RDI } (\% < 3.15\,\text{mm}) = {} & 75.85 - 2.7712 \cdot (B) - 0.0362 \cdot (I) + 471.6843 \\
& \cdot (\%P) + 693.7159 \cdot (\%Na_2O) - 357.6626 \cdot (\%K_2O)
\end{aligned}
\tag{2.113}
$$

Commentaries: Even with the limited data, it is possible to confirm that there are variables linked to the process that have influence on the TI and RDI indices in a negative or positive way (parameter β negative) and others that have positive or negative influence (parameter β positive), respectively. Consequently, both for the TI and for the RDI, the ideal situation would be that where the percentages >6.3 mm and <3.15 mm were high and low respectively.

Exercise 2.9 Going deeper in the results obtained in Exercise 2.8 for the estimation, using linear multivariate correlation, of the TI and RDI quality indices for the sintered product, calculate the uncertainty associated with the value of both parameters, TI ($\% > 6.3\,\text{mm}$) and RDI ($\% < 3.15\,\text{mm}$), in the eight samples that were analyzed.

To estimate the uncertainty of the TI and RDI indices, it is necessary to decompose the dependent variable Y (TI and RDI) into two components or variability sources: One component represents the variability or deviation associated with the regression model that was used, and the other component represents the non-explained by the model variability and, for that reason, that which is associated with random factors.

$$SCT_{n-1 \text{ degreesoffreedom}} = \sum_{i=1}^{n}(Y_i - \bar{Y})^2 = SCR_{eg} \; (k \text{ degrees of freedom})$$

$$+ SCE \; (n - k - 1 \text{ degrees of freedom}) \qquad (2.114)$$

$$= \sum_{i=1}^{n}(\hat{Y}_i - \bar{Y})^2 + \sum_{i=1}^{n}(Y_i - \hat{Y}_i)^2$$

where n is the number of samples that were studied (eight in this case) and k is the number of variables involved in the estimation of the values of the TI (six) and RDI (five). The calculations of the total source of variation (SCT, total addition of squares), of regression (SCR_{eg}, addition of squares of regression) and residual (SCE, addition of squares residual), are collected in Table 2.7.

Once the values of the TI and RDI were estimated, it is convenient to obtain a measure of the quality of the regression. A statistical parameter adequate to calculate the reliability of the estimations is the correlation coefficient, R^2, which is defined as:

$$R^2 = \frac{SCR_{eg}}{SCT} = 1 - \frac{SCE}{SCT} \qquad (2.115)$$

If all points are located over the regression line, the non-explained variance will be zero and the correlation coefficient will be one. In the rest of the situations, it is possible to make a classification as a function of the value of R^2, which is shown below:

- Very bad <0.3.
- Bad 0.3–0.4.
- Regular 0.4–0.5.
- Good 0.5–0.85
- Very good >0.85.

In the estimation of the Tumbler index, the calculation of the statistic parameters can be found in Table 2.8.

Table 2.7 Calculation of the source of variation

Source of variation	Addition of squares	Degrees of freedom
SCT	$\sum_{i=1}^{n}(Y_i - \bar{Y})^2$	$n - 1$
SCR_{eg}	$\sum_{i=1}^{n}(\hat{Y}_i - \bar{Y})^2$	k
SCE	$\sum_{i=1}^{n}(Y_i - \hat{Y}_i)^2$	$n - k - 1$

Table 2.8 Calculation of the statistical parameters (TI)

TI (% > 6.3 mm) measured	TI (% > 6.3 mm) model	SCT $(TI - TI_{mean})^2$	SCE $(TI - TI_{model})^2$	SCR_{eg} $(TI_{model} - TI_{mean})^2$
66.7	66.8876	1.05062	0.0352	1.4703
70.0	69.8354	18.7056	0.0271	17.3092
60.0	60.0158	32.2056	0.0002	32.0267
64.0	64.0764	2.8056	0.0058	2.5555
67.3	66.8224	2.6406	0.0550	1.9332
72.7	72.8542	49.3506	0.0238	51.5404
58.0	58.0116	58.9056	0.0001	58.7285
66.7	66.6720	1.0506	0.0008	0.9941
TI_{mean}	65.6773			
SCT		166.7150		
SCE			0.1481	

Table 2.9 Calculation of the statistical parameters (RDI)

RDI (% < 3.15 mm) measured	RDI (% < 3.15 mm) model	SCT $(RDI - RDI_{mean})^2$	SCE $(RDI - RDI_{model})^2$	SCR$_{eg}$ $(RDI_{model} - RDI_{mean})^2$
48.9	47.7915	18.6128	1.2289	10.2766
45.3	45.5505	0.5102	0.0627	0.9307
41.0	45.0504	12.8575	16.4057	0.2159
45.7	43.2896	1.2416	5.8101	1.6800
40.9	42.0851	13.5847	1.4044	6.2533
42.0	41.0573	6.6860	0.8887	12.4499
47.6	46.4759	9.0858	1.2637	3.5725
45.3	45.3858	0.5102	0.0074	0.6401
RDI$_{mean}$	44.5858			
SCT		63.0888		
SCE			27.0716	

Considering the data collected in Table 2.8, the calculation of the R^2 for the estimation of the TI shows:

$$R^2 = 1 - \frac{SCE}{SCT} = 1 - \frac{0.1481}{165.715} = 0.9991 \text{ (very good)} \tag{2.116}$$

Finally, Table 2.9 collects the calculations of the statistical parameters for the RDI index.

Considering the data collected in Table 2.9, the value of the R^2 for the estimation of the RDI shows:

$$R^2 = 1 - \frac{SCE}{SCT} = 1 - \frac{27.0716}{63.0888} = 0.571 \text{ (good)} \tag{2.117}$$

Commentaries: While in the estimation of the TI, operational and chemical variables are involved, the estimation of the RDI is subjected to statistical variables. Moreover, in the estimation of the RDI two chemical variables appear (%Na_2O and %K_2O) that should be affected in the same direction, but they have β coefficients of opposite sign. The low correlation coefficient of the RDI variables can give origin to second-order interactions, with two variables affecting the index at the same time:

$$\text{RDI } (\% < 3.15 \text{ mm}) - \beta_0 + \beta_1 \cdot (\%Na_2O)(\%K_2O) + \sum_{i=2}^{i=n} \beta_i \cdot x_i \tag{2.118}$$

References

Adamson AW (1982) Physical chemistry of surfaces, 4th edn. Wiley, New York, USA, pp 6–93

Babich A, Senk D, Gudenau HW, Mavrommatis TTh (2008) Ironmaking. RWTH Aachen University, Department of Ferrous Metallurgy, Aachen, Germany

Ballester A, Verdeja LF, Sancho J (2000) Metalurgia Extractiva. Volumen I. Fundamentos. Síntesis, Madrid, Spain

Cores A, Verdeja LF, Ferreira S, Ruiz-Bustinza I, Mochón J (2013): Iron ore sintering. Part 1. Theory and practice of the sintering process. DYNA 80(180):152–171

Cores A, Verdeja LF, Ferreira S, Ruiz-Bustinza I, Mochón J, Robla JI, González-Gasca C (2015) Iron ore sintering. Part 3: automatic and control systems. DYNA Colombia 82(190):227–236

Coudurier L, Hopkins DW, Wilkomirsky I (1985) Fundamentals of metallurgical processes, 2nd edn. Pergamon Press, Oxford, United Kingdom

Fernández-González D, Ruiz-Bustinza I, Mochón J, González-Gasca C, Verdeja LF (2017a) Iron ore sintering: raw materials and granulation. Miner Process Extr Metall Rev 38(1):36–46

Fernández-González D, Ruiz-Bustinza I, Mochón J, González-Gasca C, Verdeja LF (2017b) Iron ore sintering: process. Miner Process Extr Metall Rev 38(4):215–227

Fernández-González D, Ruiz-Bustinza I, Mochón J, González-Gasca C, Verdeja LF (2017c) Iron ore sintering: quality indices. Miner Process Extr Metall Rev 38(4):254–264

Fernández-González D, Ruiz-Bustinza I, Mochón J, González-Gasca C, Verdeja LF (2017d) Iron ore sintering: environment, automatic and control techniques. Miner Process Extr Metall Rev 38(4):238–249

Fernández-González D, Piñuela-Noval J, Verdeja LF (2018) Iron ore agglomeration technologies. In: Shatokha V(ed) Iron ores and iron oxide materials. Intechopen, Londres, pp 61–80 (chapter 4)

Formoso A, Moro A, Fernández-Pello G, Muñiz M, Jiménez J, Moro A, Cores A (2000a) Estudio de la granulación de la mezcla de minerales de hierro en proceso de sinterización. I parte. Granulacion. Revista de Metalurgia 36:244–253

Formoso A, Moro A, Fernández-Pello G, Muñiz M, Jiménez J, Moro A, Cores A (2000b) Estudio de la granulación de la mezcla de minerales de hierro en proceso de sinterización. II parte. Índice de granulacion. Revista de Metalurgia 36:254–265

Gaye H, Welfringer J (1984) Modelling of the thermodynamic properties of complex metallurgical slags. In: Fine UA, Gaskell DR (eds) Proceedings of the second international symposium on metallurgical slags and fluxes. Metallurgical Society of AIME, Warrendale

Habashi F (1999) Kinetics of metallurgical processes. Métallurgie Extractive Quebec, Canada

Levin EM, Robbins CR, McMurdie HF, Reser MK (1969) Phase diagrams for ceramists, 1969 Suppl. American Ceramic Society, Columbus, OH, USA

Sancho J, Verdeja LF, Ballester A (2000) Metalurgia Extractiva. Volumen II: Procesos de Obtención. Síntesis, Madrid, Spain, pp 22–41

Verdeja LF, Huerta MA, Blanco F, Viera J, del Campo JJ (1993) Nociones de plasticidad, resistencia mecánica en verde, migración, de aglomerantes y distribución de partículas durante el tratamiento de productos refractarios/cerámicos previos a la sinterización. Revista de Minas 8:109–119

Verdeja LF, Sancho J, Ballester A, González R (2014) Refractory and ceramic materials. Síntesis, Madrid, Spain, p 249

Pelletizing

3

3.1 Introduction

Pelletizing, in ironmaking, is a process used to agglomerate iron ore concentrates with a granulometry <150 μm and low concentration of impurities. Iron ore concentrates are mixed with water, bentonite, and hydrated lime (cold agglomeration) and are treated in a rotary disk to form round agglomerates (*green pellets*) with sizes comprised within 10 and 20 mm in diameter. These green pellets are hardened in a furnace at temperatures of around 1200 °C to obtain a product with the suitable physical, chemical, and mechanical properties to be used in the blast furnace. The sintering of material with high proportion of fines (particles with less than 150 μm) causes a dramatic reduction in the efficiency of the sintering machine, known as Dwight-Lloyd machine (Fig. 2.1, sintering machine), where the charge moves continuously, and thus the only method to obtain stable agglomerates with efficiency and metallurgical performance is using the pelletizing.

Chemically, the typical composition of a pellet is: 94% Fe_2O_3. 3.5% SiO_2, 0.5% CaO, 0.5% MgO, 1.0% Al_2O_3, 0.3% MnO, and 0.2% of P, S, and alkalis (Babich et al.2008, p. 75). Figure 3.1 shows a micrograph (10X) of a pellet.

From the physical–chemical point of view, the pellet might be considered a product with the following characteristics:

- Great uniformity as regarding the size (10–20 mm in diameter).
- High mechanical strength.
- Inert to the action of the water (it does not contain free lime as the sinter). This allows the transportation and storage of the pellets exposed to the open air.
- Good reducibility (predominance of either the hematite, Fe_2O_3, or the magnetite, Fe_3O_4, also known as magnetite taconite).
- High iron concentration.
- High tendency to the swelling.

J. I. Verdeja González et al., *Operations and Basic Processes in Ironmaking*, Topics in Mining, Metallurgy and Materials Engineering, https://doi.org/10.1007/978-3-030-54606-9_3

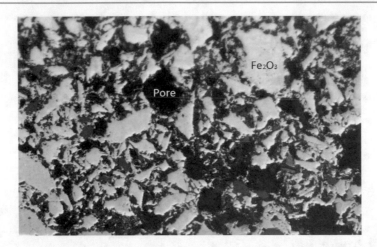

Fig. 3.1 Micrograph of a pellet: pores in dark gray, Fe_2O_3 in white and matrix in light gray (bentonite)

While in the sinter the formation of the liquid phases that agglomerate the iron particles takes advantage of the coke fines combustion in a conveyor belt furnace (Fig. 3.2), in the pellets the development of liquid phases is achieved with the supply of energy from an external source (fuel, natural gas, or pulverized coal) in conveyor belt, shaft, or rotary furnaces (Fig. 3.2).

The current processes of agglomeration of powdered materials also use briquetting and plastic extrusion techniques in the presence (in most of the cases) of carbonaceous substances (Fernández-González et al. 2018). The presence of carbon diminishes the swelling that they naturally have during the first steps of the iron ore reduction from the hematite–magnetite to the wustite, FeO.

The standard used to quantify the swelling of the pellet or any other agglomerated ferric product is the *Reduction Swelling Index*, RSI, or swelling index of the pellet after its partial reduction. The RSI test is performed at 900 °C under CO/N_2 (30/70 vol%) atmosphere, and it is quantitatively expressed as follows:

$$RSI = \frac{V_a - V_b}{V_a} \qquad (3.1)$$

where V_a and V_b are, respectively, the volume of the pellet after and before the reduction process.

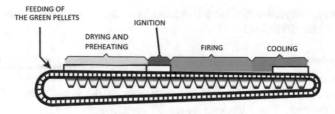

Schematic diagram of the system of moving grates
to produce pellets

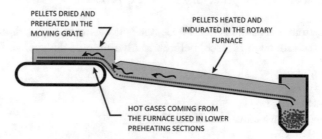

Schematic diagram of the system of moving grates
to produce pellets

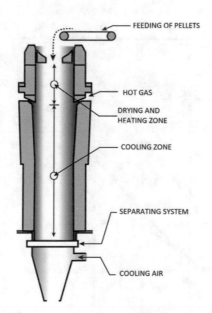

Schematic diagram of the system of moving grates
to produce pellets

Fig. 3.2 Detail of the thermal agglomeration of the green pellets

3.2 Thermodynamic Considerations About the Process and the Product

Thermodynamically, the succession of phases that are produced during the reduction of the pellets can be analyzed using an Ellingham diagram as that shown in Fig. 3.3, where the stability zones for each one of the compounds of the Fe-O system are defined. Two situations might be considered:

- Reduction at low temperature: $T < 387$ °C (660 K).
- Reduction at high temperature: $T > 387$ °C (660 K).

Starting from a hematite pellet, Fe_2O_3, the succession of iron compounds that are formed in the reduction furnace, before the obtaining of the sponge iron, will be:

(a) When $T < 387$ °C:

1. Zones with Fe_2O_3.
2. Layers with mixture of Fe_2O_3-Fe_3O_4.
3. Fe_3O_4 oxide.

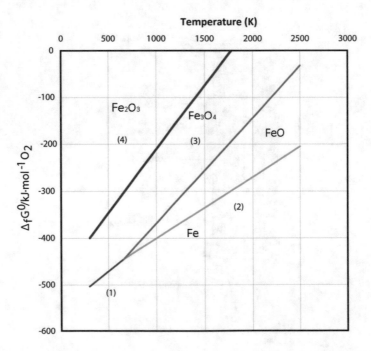

Fig. 3.3 Diagram $\Delta_f G^0 / T$ for the iron compounds of the Fe-O system. (1) $\Delta_f G^0$ of the Fe_3O_4; (2) $\Delta_f G^0$ of the FeO; (3) $\Delta_f G^0$ of the Fe_3O_4 from the FeO; and (4) $\Delta_f G^0$ of the Fe_2O_3 from the Fe_3O_4

   4. Zones with mixture of Fe_3O_4 and sponge iron.
   5. Sponge iron.

(b) When $T > 387$ °C:

   1. Zones with Fe_2O_3.
   2. Layer with mixture of Fe_2O_3-Fe_3O_4.
   3. Fe_3O_4 oxide.
   4. Zones with mixture of Fe_3O_4-FeO.
   5. FeO oxide.
   6. Mixture of FeO–Sponge iron ($T < 1150$ °C).
   7. Sponge iron ($T > 1150$ °C).

The same conclusions might be extracted if the Fe-O phase diagram shown in Fig. 3.4 was used, although in this thermodynamic calculation, the temperature, at which the invariant equilibrium between the phases Fe-FeO-Fe_3O_4-O_2 (gas) takes place, is 570 °C, that is to say, 183 °C above the estimated temperature calculated using the thermodynamic data in Appendix 1 (Fig. 3.3), where the following reactions and the corresponding standard free energies are collected:

$$Fe(s) + \frac{1}{2}O_2(g) \rightarrow FeO(s) \tag{3.2}$$

$$\Delta G^0(Eq.\ 3.2) = -265.0 + 0.065 \cdot T\,(kJ\ mol^{-1}FeO) \tag{3.3}$$

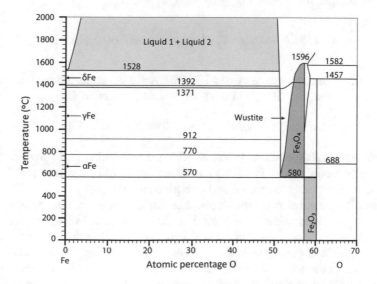

Fig. 3.4 Fe-O binary diagram (elaborated with data of Massalski et al. 1990)

$$3Fe(s) + O_2(g) \rightarrow Fe_3O_4(s) \tag{3.4}$$

$$\Delta G^0(Eq.\,3.4) = -1105.0 + 0.328 \cdot T \left(kJ\ mol^{-1}Fe_3O_4\right) \tag{3.5}$$

$$3FeO(s) + \frac{1}{2}O_2(g) \rightarrow Fe_3O_4(s) \tag{3.6}$$

$$\Delta G^0(Eq.\,3.6) = -296.0 + 0.112 \cdot T \left(kJ\ mol^{-1}Fe_3O_4\right) \tag{3.7}$$

$$\frac{2}{3}Fe_3O_4(s) + \frac{1}{6}O_2(g) \rightarrow Fe_2O_3(s) \tag{3.8}$$

$$\Delta G^0(Eq.\,3.8) = -80.2 + 0.045 \cdot T \left(kJ\ mol^{-1}Fe_2O_3\right) \tag{3.9}$$

These last reactions are not Ellingham's reactions, that is to say, they are not adjusted to the mol of oxygen, as the standard free energies of formation of the different chemical compounds (per mol of product) usually appear.

To find the temperature of the invariant point of Fig. 3.3, we must equalize the standard energies of the corresponding Ellingham's straight lines. For example, if we equalize the straight lines 2 and 1 (Eqs. 3.2 and 3.4) of Fig. 3.3:

$$(-1105.0 + 0.328 \cdot T) \cdot \left(\frac{1}{2}\right) = (-265.0 + 0.650 \cdot T) \cdot (2) \tag{3.10}$$

we obtain as result the temperature of 662 K. For the equilibriums described by Eq. (3.4), straight line 1, and Eq. (3.6), straight line 3:

$$(-1105.0 + 0.328 \cdot T) \cdot \left(\frac{1}{2}\right) = (-296.0 + 0.112 \cdot T) \tag{3.11}$$

which indicates a temperature of 658 K. Finally, for the equilibriums between the straight line 2, reaction (3.2), and the straight line 3, reaction (3.6):

$$(-265.0 + 0.065 \cdot T) = (-296.0 + 0.112 \cdot T) \tag{3.12}$$

In this case, the temperature is 660 K, which is interpreted as the temperature of the invariant point of the Fe-O system where an equilibrium between the four phases can take place (three solid and one gaseous, oxygen).

In Sect. 2.1, we described the degradation indices of a ferric material after the process of partial reduction (*RDI* index Fig. 2.5). The objective of this index was the quantification of the iron compounds degradation during the reduction process as a consequence of the appearance of different oxides or stable phases compatibles with the reduction potential of the gas. The density changes in the particles can generate values of expansion coefficient incompatibles with the critical stress of fracture for the same material. In the case of products with well-defined geometrical characteristics (spherical or rectangular), a quality index usually employed is, as it was mentioned, the *Reduction Swelling Index* (*RSI*).

The simulation of the heating process of a pellet in a furnace can be studied in the same way as in the case of a sintered product, using the SiO_2-CaO-Fe_2O_3 ternary diagram (Fig. 2.7 Levin et al. 1964) (Pero-Sanz et al. 2019). A possible pseudoternary composition of the sinter is: 94% Fe_2O_3, 5% SiO_2, and 1% CaO (material of acid characteristics) (Fig. 3.5). At room temperature, the thermodynamic phases compatibles with the global composition are the wollastonite, the SiO_2 (cristobalite–tridymite) and the hematite, Fe_2O_3 as it is possible to see in the corresponding triangle of compatibility (Fig. 3.5).

The beginning of the pellet softening, solidus temperature T_S, starts with the ternary eutectic reaction at 1204 °C (i.e., the point with the lowest temperature inside of the compatibility triangle) as it is shown in Fig. 3.6:

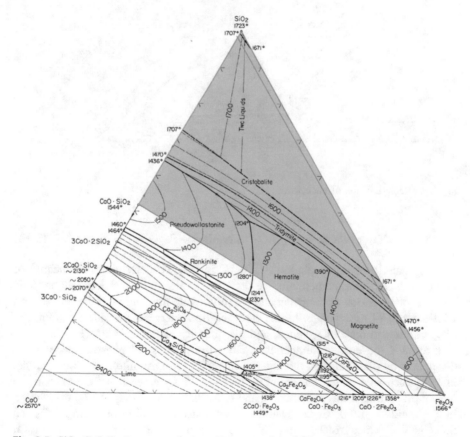

Fig. 3.5 SiO_2-CaO-Fe_2O_3 ternary diagram. Representation of the pellet composition and triangle of compatibility that defines the phases that are stable at room temperature: wollastonite, tridymite–cristobalite, and hematite

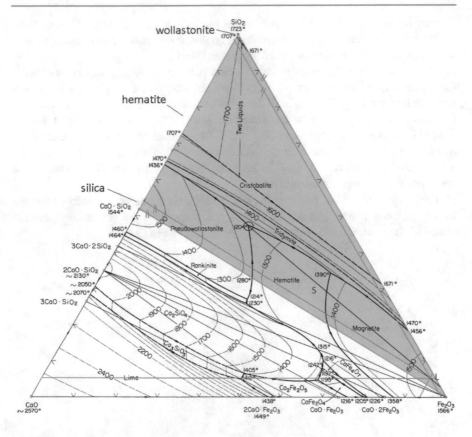

Fig. 3.6 Representation of the eutectic

$$CaO \cdot SiO_2(\text{wollastonite}) + SiO_2(\text{tridymite}) + Fe_2O_3(s) \rightleftharpoons \text{liquid} \qquad (3.13)$$

The quantity of liquid phase at 1204 °C is 5%. To determine the quantity of liquid at the eutectic temperature, we draw the straight line that joins the eutectic with the composition of the pellet and we prolongate this line until the intersection with the side of the triangle. In this case, the line intersects the side that joins the SiO_2-vertex and the Fe_2O_3-vertex, and the quantity of liquid is 5% (Fig. 3.6).

Exceeded the eutectic temperature, from the three solid phases, one of them disappears by dissolution in the liquid: $CaO \cdot SiO_2$ (wollastonite). We draw parallel lines to each side of the compatibility triangle, and we see that the phase in the lowest proportion is the wollastonite, and thus, this phase disappears by dissolution in the liquid. At 1300 °C, the quantity liquid phase in equilibrium with Fe_2O_3 and SiO_2 is 6%. Once reached the 1390 °C, the peritectic reaction of the Fe_2O_3 decomposition (Fig. 3.7) takes place:

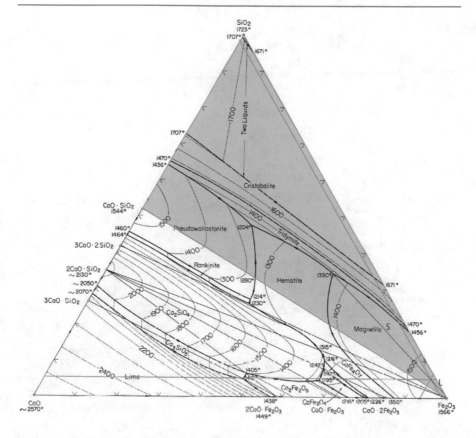

Fig. 3.7 At 1390 °C the transformation of the Fe_2O_3 into Fe_3O_4 takes place

$$Fe_2O_3(s) + liquid + SiO_2(trydimite) \rightarrow Fe_3O_4(s) \qquad (3.14)$$

and the corresponding proportion of liquid phase is around 10%.

At around 1420 °C, the dissolution of all the SiO_2 in the liquid phase takes place, and the only solid that is still not dissolved is the Fe_3O_4 (Fig. 3.8).

At 1560 °C, the liquidus temperature of the system, T_L, is reached (temperature that corresponds to the isotherm that passes through the composition of the pellet in the ternary diagram), and the softening interval of the pellet is 356 °C.

Even though it is not usual, basic pellets, with pseudoternary compositions of 90.5% Fe_2O_3, 4.5% SiO_2, and 5.0% CaO are being commercialized (Fig. 3.9). The phases thermodynamically compatibles at room temperature with the global composition are the rankinite, $3CaO \cdot 2SiO_2$, the wollastonite, $CaO \cdot SiO_2$, and the hematite, Fe_2O_3 (Fig. 3.9).

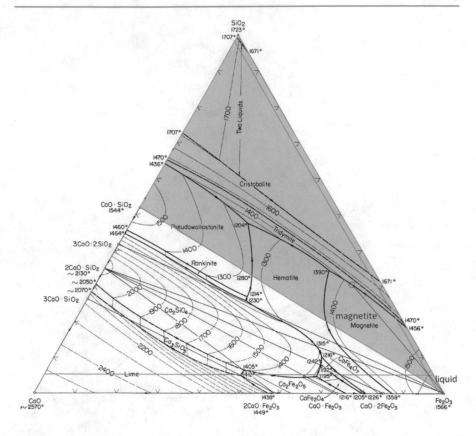

Fig. 3.8 At around 1420 °C the dissolution of the SiO₂ in the liquid takes place

The beginning of the pellet softening, solidus temperature T_S, starts with the ternary eutectic reaction at 1214 °C (Fig. 3.9), which is the point with the lowest temperature in the ternary diagram:

$$3CaO \cdot 2SiO_2(s) + CaO \cdot SiO_2(s) + Fe_2O_3(s) \xrightarrow{1214\,^{\circ}C} liquid \qquad (3.15)$$

The quantity of liquid phase is, at 1124 °C, 4.5%. As if we prolongate the line that joins the eutectic point with the pellet composition until the intersection with the side of the compatibility triangle, we see that the intersection is produced in the side that joins the vertex of the wollastonite and the vertex of the hematite. For that reason, once exceeded the eutectic temperature of the three solid phases, one disappears by dissolution in the liquid, 3CaO·2SiO₂ (rankinite). The composition of the solid phase is defined by Fe₂O₃ (93%) and wollastonite (7%), obtained by the prolongation of the line that joins the eutectic point with the pellet composition until the intersection with the side of the compatibility triangle (the side that joins the

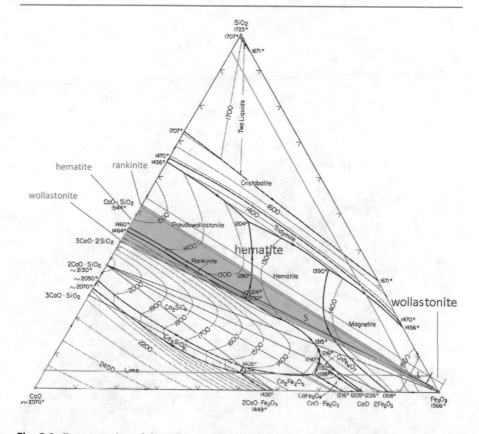

Fig. 3.9 Representation of the pellet composition in the SiO$_2$-CaO-Fe$_2$O$_3$ ternary diagram, and the triangle of compatibility, which defines the phases that are compatibles at room temperature. The eutectic point is also highlighted in the image, as well as it is indicated how it is possible to calculate the proportion of each phase (85% Fe$_2$O$_3$, 5% rankinite, and 10% wollastonite)

vertex of the wollastonite and the vertex of the hematite) (Fig. 3.9). Slightly below the 1280 °C, the disappearance, by dissolution in the melt, of the wollastonite happens, and the last solid to be dissolved is the hematite (Fig. 3.10). This point is determined if we prolongate the line that joins the Fe$_2$O$_3$-vertex with pellet composition until the intersection with the solid line that joins the eutectic of 1204 °C and the eutectic of 1214 °C.

The increase in the quantity of liquid during the heating is 14.3% (85.7% hematite), 26.3% (73.7% magnetite), and 74.0% (26% magnetite) at the corresponding temperatures of 1300, 1400, and 1500 °C. In this interval of temperature, close to 1375 °C, the peritectic reaction of thermal decomposition of the Fe$_2$O$_3$ takes place (Fig. 3.11 and Eq. 3.14).

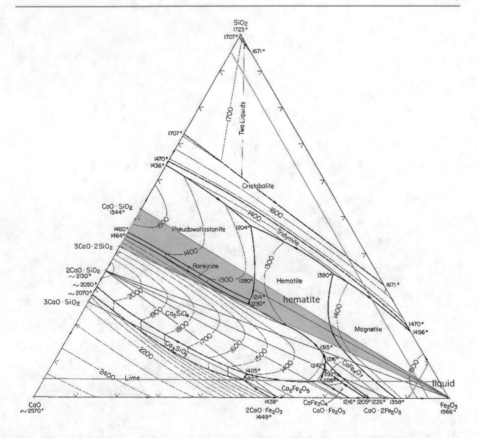

Fig. 3.10 At a temperature slightly below the 1280 °C, the wollastonite dissolves in the liquid

Finally, at temperatures of around 1520 °C, the liquidus temperature of the product, T_L, is reached (this temperature corresponds to the isotherm that passes through the pellet composition), and the softening interval of the basic pellet is 306 °C.

3.3 Solid–Gas Kinetics in Dense Solids. Unreacted Core Model

Kinetically, the behavior of a pellet will be similar to that of a dense particle that, after a time of reaction with the reductant gas, has a porous layer of reduced iron (T < 1150 °C) that surrounds the unreacted core of Fe_2O_3/Fe_3O_4.

The equations that enclose the position of the reaction front as a function of the time, in spherical particles, are the same as those used in the roasting of metallic sulfides or in the calcination of the limestone, $CaCO_3$ (Ballester et al. 2000,

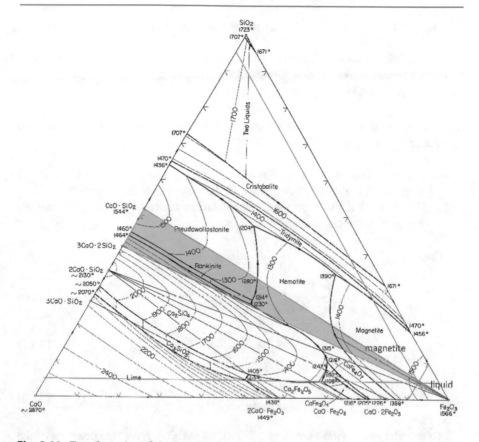

Fig. 3.11 Transformation of the hematite into magnetite

pp. 232–235). Consequently, the solid–gas reduction reactions in a dense pellet particle are kinetically similar to the reactions of oxidizing roasting or sulfating roasting of a metallic sulfide or to the decomposition reactions of calcium and magnesium carbonates. The conceptual unreacted core model considers the following operating hypothesis:

(a) Three compositions of reductant gas are considered:

- C_i^0, concentration of reductant gas in the gaseous phase, experimentally calculated,
- C_i, concentration in the solid–gas interface and,
- $C_i^=$, concentration derived from the thermodynamic equilibrium.

(b) The global reduction rate, v_g, might be considered as:

- Diffusional rate (diffusional control, $v_g = v_d$),
- Chemical reaction rate (chemical control, $v_g = v_q$),
- Mixed $(v_g = v_d + v_q)$.

(c) The time of reaction, t, according to the unreacted core model, has two addends:

- The time required for the diffusion of the reductant gas from the gas to the position of the solid–gas interface in the pellet, diffusional time, t_d,
- The time that the heterogeneous reduction chemical reaction (solid–gas) requires, chemical reaction time, t_q.

Thus,

$$t = \frac{\rho \cdot R_0}{k_q \cdot \Delta C} \cdot f + \frac{R_0^2 \cdot \rho}{6 \cdot D_i \cdot \Delta C} \cdot f^2 \cdot (3 - 2 \cdot f) \qquad (3.16)$$

$$t_q = \frac{\rho \cdot R_0}{k_q \cdot \Delta C} \cdot f \qquad (3.17)$$

$$t_d = \frac{R_0^2 \cdot \rho}{6 \cdot D_i \cdot \Delta C} \cdot f^2 \cdot (3 - 2 \cdot f) \qquad (3.18)$$

To calculate the relation between the front of reaction, f, and the progress of the reaction, X, in spherical particles (Eq. 2.48), we must take into account that:

$$X = \frac{V_0 - V_i}{V_0} = \frac{R_0^3 - R_i^3}{R_0^3} = 1 - \left(\frac{R_i}{R_0}\right)^3 \qquad (3.19)$$

On the other hand, the relation between the front of reaction and radius size of the particle is:

$$f = 1 - \frac{R_i}{R_0} \qquad (3.20)$$

The two previous equations lead to that the relation between the reaction front, f, and the fraction of solid that was reduced, X, will be:

$$f = 1 - (1 - X)^{1/3} \qquad (3.21)$$

When the controlling step in the reduction process is the diffusional transfer, the chemical reaction is very fast:

$$t = t_q + t_d \simeq t_d \qquad (3.22)$$

Thus, Eq. (3.16) is transformed into:

$$t = \frac{R_0^2 \cdot \rho}{6 \cdot D \cdot \Delta C} \cdot f^2 \cdot (3 - 2 \cdot f) \qquad (3.23)$$

Replacing in Eq. (3.23), the functional relation between the reaction front f and the progress of the reduction reaction X in spherical particles, Eq. (3.21), we obtain the following equation:

$$t = \frac{R_0^2 \cdot \rho}{6 \cdot D \cdot \Delta C} \cdot \left[3 \cdot \left(1 - (1-X)^{1/3}\right)^2 - 2 \cdot \left(1 - (1-X)^{1/3}\right)^3 \right] \qquad (3.24)$$

Developing each one of the brackets of the second term and regrouping them, we have:

$$t = \frac{R_0^2 \cdot \rho}{6 \cdot D \cdot \Delta C} \cdot \left[3 - 2 \cdot X - 3 \cdot (1-X)^{2/3} \right] = K \cdot \left[3 - 2 \cdot X - 3 \cdot (1-X)^{2/3} \right] \qquad (3.25)$$

Finally, if we multiply and divide by 3 the second term of the equation, the Ginstling–Brounshtein equation can be obtained (Rahaman 2003, p. 68):

$$k \cdot t = \left[1 - \frac{2}{3} \cdot X - (1-X)^{2/3} \right] \qquad (3.26)$$

where the value of k will be:

$$k = \frac{2 \cdot D \cdot \Delta C}{R_0^2 \cdot \rho} \qquad (3.27)$$

3.4 Solved Exercises

Exercise 3.1 Discuss if the reduction of iron oxides is possible using only the temperature, always and when the standard free energy of formation of iron oxides, as that of any other chemical compound, grows (it becomes more negative) when the temperature is greater (Appendix 1 and Fig. 3.3).

Using the data of Appendix 1, where the standard free energies of formation of different combinations iron–oxygen are indicated, it is possible to evaluate the obtaining of 6 mol of Fe_2O_3 from the Fe_3O_4, see Eqs. (3.8) and (3.9) ($4Fe_3O_4+O_2 = 6Fe_2O_3$). The temperature at which the standard free energy of formation is zero is calculated as follows:

$$\Delta G^0(T) = 0 \rightarrow -481.2 + 0.270 \cdot T(K) = 0 \tag{3.28}$$

$$T(K) = \frac{481.2}{0.270} = 1782K(1509\,^\circ C) \tag{3.29}$$

This indicates that, above 1509 °C, the oxide that is thermodynamically stable is the magnetite, Fe_3O_4, when the pressure of oxygen in the gas is one atmosphere. In the most frequent situation of treatment in air atmospheres, the oxygen partial pressure is 0.21 atm and the free energy must be corrected because one reagent, the oxygen, is not in standard conditions:

$$
\begin{aligned}
\Delta G = {}& \Delta G^0 + R \cdot T \cdot \ln K_{eq} \\
& \rightarrow (-481.2 + 0.270 \cdot T(K))(kJ/mol) \\
& + 8.314 \left(\frac{J}{mol\,K}\right) \frac{1kJ}{1000\,J} \cdot \ln\left(\frac{a_{Fe_2O_3}^6}{P_{O_2} \cdot a_{Fe_3O_4}^4}\right) \cdot T(K) = 0 \\
& \rightarrow (-481.2 + 0.270 \cdot T(K))(kJ/mol) \\
& + 8.314 \left(\frac{J}{mol\,K}\right) \frac{1\,kJ}{1000\,J} \cdot \ln\left(\frac{1}{0.21\,atm \cdot 1}\right) \cdot T(K) = 0
\end{aligned}
\tag{3.30}
$$

$$T(K) = \frac{481.2}{0.270 + 0.0129} = 1701K(1428^\circ C) \tag{3.31}$$

The presence of an atmosphere with low quantity of oxygen facilitates the thermal decomposition of the hematite, Fe_2O_3:

$$6Fe_2O_3(s) \rightarrow 4Fe_3O_4(s) + O_2(g) \tag{3.32}$$

In the case of the magnetite, to decompose the oxide without using a reductant reagent, the following reaction of formation must be accomplished (Eqs. 3.6 and 3.7):

$$6FeO(s) + O_2(g) \rightarrow 2Fe_3O_4(s) \tag{3.33}$$

By setting to zero, the standard free energy of formation of the magnetite from the FeO, when the oxygen pressure in the gas is equal to one atmosphere, is:

$$\Delta G^0(T) = 0 \rightarrow -592.0 + 0.224 \cdot T(K) = 0 \tag{3.34}$$

$$T(K) = \frac{592.0}{0.224} = 2643K(2370°C) \tag{3.35}$$

While the result in the air atmosphere will be:

$$\Delta G = \Delta G^0 + R \cdot T \cdot \ln K_{eq}$$
$$\rightarrow (-592.0 + 0.224 \cdot T(K))(kJ/mol)$$
$$+ 8.314 \left(\frac{J}{mol\ K}\right) \frac{1\ kJ}{1000\ J} \cdot \ln\left(\frac{1}{0.21\ atm \cdot 1}\right) \cdot T(K) = 0 \tag{3.36}$$

$$T(K) = \frac{592.0}{0.224 + 0.0129} = 2498\ K(2225\ °C) \tag{3.37}$$

Commentaries: The kinetics (reaction rate) associated with the thermal decomposition of the Fe_2O_3, in front of that of disappearance of the hematite by reduction (gas or coke), can be favored by the presence or formation of high-temperature zones, plasm arcs, flame fronts or spaces affected by an electric arc. One of the advantages of using high temperatures in the reduction of oxides is that the diminishing of reductant consumptions is favored.

Moreover, the thermal decomposition of the Fe_2O_3 into Fe_3O_4 is influenced by both the oxygen partial pressure and the presence of other components in the system. In the analysis of Sect. 3.2 about the heating of an acid and a basic pellet, we indicated that the temperatures of the peritectic transformation–decomposition are modified with the basicity of the system: An increase of the basicity leads to a slight decrease of the transformation temperature.

Exercise 3.2 Calculate the volume and linear thermal expansion coefficients for the following iron oxides: Fe_2O_3, Fe_3O_4 and FeO within the temperatures of 575 °C and 625 °C.

Demonstrate that the stresses that are produced in a particle of spherical geometry (pellet) of 20 mm in diameter, consequence of the existence of an unreacted core nucleus of hematite, Fe_2O_3, and of an adhered surface layer of 500 μm in thickness, can be sufficient to cause superficial cracks–breaks and the corresponding degradation of the material at low temperature (600 °C).

Data: The elasticity modulus of the pellet is 50 GPa. The difference of temperature between the reductant atmosphere and the pellet is 50 °C. The variation of the density of the iron oxides with the temperature according to the TAPP database 1994 is:

$$\rho(Fe_2O_3, kg\ m^{-3}) = 5332 - 0.187 \cdot T(K)(298 - 985\ K) \tag{3.38}$$

$$\rho(Fe_3O_4, kg\ m^{-3}) = 5275 - 0.231 \cdot T(K)(298 - 900\ K) \tag{3.39}$$

$$\rho(FeO, kg\ m^{-3}) = 6077 - 0.378 \cdot T(K)(843 - 1697\ K) \tag{3.40}$$

Table 3.1 Specific volume for the iron oxides at 575 and 625 °C

	V_e (m^3 kg^{-1}) $\times$ 10^6 (575 °C)	V_e (m^3 kg^{-1}) $\times$ 10^6 (625 °C)
Fe_2O_3	193.30	193.65
Fe_3O_4	196.88	197.33
FeO	173.72	174.29

The value of the volume thermal expansion coefficient for the three proposed oxides at 600 °C is calculated by means of the values of the specific volume (m^3 kg^{-1}) for each one of the oxides at the temperatures of 575 °C (848 K) and 625 °C (898 K) (Table 3.1).

The value of the volume thermal expansion coefficient is calculated as follows:

$$\alpha_v\left(K^{-1}\right) = \frac{V_e(625\ °C) - V_e(575\ °C)}{V_e(575\ °C)} \tag{3.41}$$

In the case of isotropic materials, it is possible to demonstrate (see Verdeja et al. 2014, pp. 101) that the relation between the volume thermal expansion coefficient and the linear thermal expansion coefficient is:

$$\alpha_v = 3 \cdot \alpha_e \tag{3.42}$$

Consequently, the values of the volume and linear thermal expansion coefficients for the proposed oxides at the temperature of 600 °C are collected in Table 3.2.

Finally, to calculate the stresses accumulated in the interface that separates the unreacted nucleus of hematite, Fe_2O_3, and the layer of reduced material (FeO and Fe_3O_4) adhered to the unreacted nucleus, we must use the equation developed in Verdeja et al. 2014 (page 224), where it is verified that:

$$E_{pellet} \cdot (section\ of\ Fe_2O_3) \gg E_{pellet} \cdot (section\ reduced) \tag{3.42}$$

the stresses of compression (negatives) in the FeO and of tension (positives) in the unreacted Fe_2O_3 are:

$$\sigma_{compression}(FeO) = -E_{pellet} \cdot \Delta T \cdot (\alpha_{l-FeO} - \alpha_{l-Fe_2O_3}) \tag{3.43}$$

$$\sigma_{tension}(Fe_2O_3) = E_{pellet} \cdot \Delta T \cdot (\alpha_{l-FeO} - \alpha_{l-Fe_2O_3}) \cdot \left[\frac{Reduced\ section}{Fe_2O_3\ section}\right] \tag{3.44}$$

Table 3.2 Volume and linear thermal expansion coefficient

	Fe_2O_3	Fe_3O_4	FeO
$\alpha_v(K^{-1})$	$36.21 \cdot 10^{-6}$	$45.58 \cdot 10^{-6}$	$65.88 \cdot 10^{-6}$
$\alpha_l\ (K^{-1})$	$12.07 \cdot 10^{-6}$	$15.19 \cdot 10^{-6}$	$21.96 \cdot 10^{-6}$

where the "reduced section" is identified with the area of the circular crown (FeO) adhered to the unreacted nucleus of Fe_2O_3. Assuming that the circular section of the pellet has a radius of 10.0 mm and that the circular thickness of the FeO layer is of 500 μm (0.5 mm):

$$Section - Fe_2O_3\left(m^2\right)(core) = \pi \cdot \left((10{,}000 - 500)\ \mu m \cdot \frac{1\ mm}{1000\ \mu m} \cdot \frac{1\ m}{1000\ mm}\right)^2$$
$$= 2.835 \cdot 10^{-4}\ m^2$$

$$(3.45)$$

$$Section - Reduced\left(m^2\right)(FeO) = \pi \cdot \left(10{,}000\mu m \cdot \frac{1\ mm}{1000\ \mu m} \cdot \frac{1\ m}{1000\ mm}\right)^2 - 2.835 \cdot 10^{-4}\ m^2$$
$$= 3.066 \cdot 10^{-5}\ m^2$$

$$(3.46)$$

The results are obtained in two situations:

- Unreacted nucleus of Fe_2O_3 surrounded by a layer of FeO.
- Unreacted nucleus of Fe_2O_3 surrounded by a layer of Fe_3O_4*.

*In the calculation of the tension and compression stresses for the unreacted nucleus of Fe_2O_3 surrounded by a layer of Fe_3O_4, we used the same equations that in the case of the Fe_2O_3-FeO, but, in this case, we replaced the value of the linear expansion coefficient of the FeO by the value of that of the Fe_3O_4.

Situation A:

Compression stresses in the zone of the unreacted nucleus:

$$\sigma_{compression}(Fe_2O_3) = -50{,}000\ MPa \cdot 50\ K$$
$$\cdot \left(21.96 \cdot 10^{-6}\ K^{-1} - 12.07 \cdot 10^{-6}\ K^{-1}\right) \qquad (3.47)$$
$$= -24.72\ MPa$$

Tension stresses in the FeO layer:

$$\sigma_{tension}(FeO) = 50{,}000\ MPa \cdot 50\ K \cdot \left(21.96 \cdot 10^{-6}\ K^{-1} - 12.07 \cdot 10^{-6}\ K^{-1}\right) \cdot \left[\frac{3.066 \cdot 10^{-5}\ m^2}{2.835 \cdot 10^{-4}\ m^2}\right]$$
$$= 2.67\ MPa$$

$$(3.48)$$

Situation B:

Compression stresses in the zone of the unreacted nucleus:

$$\sigma_{compression}(Fe_2O_3) = -50,000 \text{ MPa} \cdot 50 \text{ K} \cdot \left(15.19 \cdot 10^{-6} \text{ K}^{-1} - 12.07 \cdot 10^{-6} \text{ K}^{-1}\right)$$
$$= -7.8 \text{ MPa}$$

$$(3.49)$$

Tension stresses in the FeO layer:

$$\sigma_{tension}(FeO) = 50,000 \text{ MPa} \cdot 50 \text{ K} \cdot \left(15.19 \cdot 10^{-6} \text{ K}^{-1} - 12.07 \cdot 10^{-6} \text{ K}^{-1}\right) \cdot \left[\frac{3.066 \cdot 10^{-5} \text{ m}^2}{2.835 \cdot 10^{-4} \text{ m}^2}\right]$$
$$= 0.84 \text{ MPa}$$

$$(3.50)$$

Commentaries: The tension and compression stresses developed in both sides of the interface unreacted nucleus-reduced layer (Fe_3O_4 or FeO) can cause the cracking–breaking–degradation of the pellet from inside to outside. The resistance to compression in the pellets is around 12 MPa while the corresponding resistance in tension is 15 times lower ($\simeq 1$ MPa).

Compression or tension stresses are developed inside of a pellet particle (Fe_2O_3) affected by a partial reduction into Fe_3O_4 or into FeO. Under conditions of maximum adhesion, the variation in length (Δl) experienced either by the Fe_2O_3, Fe_3O_4 or FeO is equivalent:

$$\Delta l(Fe_2O_3) = \Delta l(Fe_3O_4) = \Delta l(FeO) \tag{3.51}$$

Under these equilibrium conditions, the equations associated with the compression ($-$) or tension ($+$) forces previously mentioned are derived. The variations in length that the different iron oxides can experience must be compensated with the elastic component positive, derived from the tension stresses, or negative, related with the compression forces.

Exercise 3.3 When the oxygen potential in the atmosphere of the furnace at 1273 K (1000 °C) is controlled by the CO-CO_2 equilibrium in the gas at a total pressure of 1.0 atm and the relation of partial pressures of carbon monoxide to carbon dioxide in the gas is 2.15, determine what of the following iron oxides is the most stable: stoichiometric FeO, Fe_2O_3, or Fe_3O_4.

Even though there are different methods of solving this exercise, the fundamental aspects of the thermodynamic equilibrium are the adequate form. Considering the data of Appendix 1, we obtain the standard free energy of the reaction of the CO_2 from the CO, considering the linear combination of standard free energies of formation of both compounds:

$$CO(g) + \frac{1}{2}O_2(g) \rightarrow CO_2(g) \tag{3.52}$$

$$\Lambda_r G^0(T) = -277.0 + 0.085 \cdot T(K)kJ\ mol^{-1}CO_2 \tag{3.53}$$

At 1273 K:

$$\Delta_r G^0(1273\ K) = -168,795J\ mol^{-1} = -8.314\frac{J}{mol\ K} \cdot 1273K \cdot \ln K_{eq} \tag{3.54}$$

$$K_{eq} = 8.433 \cdot 10^6 \tag{3.55}$$

$$P(O_2) = \left(\frac{1}{K_{eq}}\right)^2 \cdot \left(\frac{P(CO_2)}{P(CO)}\right)^2 = \left(\frac{1}{8.433 \cdot 10^6}\right)^2 \cdot \left(\frac{1}{2.15}\right)^2 = 3.04 \cdot 10^{-15} atm \tag{3.56}$$

Using the data of Appendix 1, and by linear combination of the reactions, the values for the formation of iron oxides adjusted, either per mol of oxygen or mol of iron, can be obtained:

$$Fe(s) + \frac{1}{2}O_2(g) \rightarrow FeO(s) \tag{3.57}$$

$$\Delta_f G^0(T) = -265.0 + 0.065 \cdot T(K)kJ\ mol^{-1}Fe \tag{3.58}$$

$$Fe(s) + \frac{2}{3}O_2(g) \rightarrow \frac{1}{3}Fe_3O_4(s) \tag{3.59}$$

$$\Delta_f G^0(T) = -368.3 + 0.1093 \cdot T(K)kJ\ mol^{-1}Fe \tag{3.60}$$

$$Fe(s) + \frac{3}{4}O_2(g) \rightarrow \frac{1}{2}Fe_2O_3(s) \tag{3.61}$$

$$\Delta_f G^0(T) = -403.75 + 0.125 \cdot T(K)kJ\ mol^{-1}Fe \tag{3.62}$$

The criterion to know if a reaction or process is thermodynamically favorable is the diminishing of its standard free energy, or that the standard free energy was negative. In this case, we calculate the free energy at 1273 K for each one of the reactions of formation when the oxygen partial pressure (reductant atmosphere) is 3.04×10^{-15} atm:

- For the wustite, FeO:

$$\Delta_f G(T, FeO) = -182{,}255\,\frac{J}{molFe} + 8.314\,\frac{J}{mol\,K} \cdot 1273K \cdot \ln\left[\frac{1}{(3.04 \cdot 10^{-15}\,atm)^{\frac{1}{2}}}\right]$$

$$= -182{,}255\,\frac{J}{molFe} + 176{,}891\,\frac{J}{molFe}$$

$$= -5364\,\frac{J}{molFe}$$

$$(3.63)$$

- For the magnetite, Fe_3O_4:

$$\Delta_f G(T, Fe_3O_4) = -229{,}161\,\frac{J}{molFe} + 8.314\,\frac{J}{mol\,K} \cdot 1273K \cdot \ln\left[\frac{1}{(3.04 \cdot 10^{-15}atm)^{\frac{2}{3}}}\right]$$

$$= -229{,}161\,\frac{J}{molFe} + 235{,}854\,\frac{J}{molFe}$$

$$= 6693\,\frac{J}{molFe}$$

$$(3.64)$$

- For the hematite, Fe_2O_3:

$$\Delta_f G(T, Fe_2O_3) = -244{,}625\,\frac{J}{molFe} + 8.314\,\frac{J}{mol\,K} \cdot 1273K \cdot \ln\left[\frac{1}{(3.04 \cdot 10^{-15}atm)^{\frac{3}{4}}}\right]$$

$$= -244{,}625\,\frac{J}{molFe} + 265{,}336\,\frac{J}{molFe}$$

$$= 20{,}711\,\frac{J}{molFe}$$

$$(3.65)$$

From the three values of free energy, two are positives and one, that of the stoichiometric wustite, negative. Precisely, the FeO, under these conditions of temperature and oxygen partial pressure, is the stable oxide. The other two oxides, the hematite and the magnetite, as they have positive free energies of formation, are not thermodynamically stables under the same conditions of pressure and temperature.

Exercise 3.4 Calculate for the hematite pellets reduction from hydrogen at 1100 °C under the conditions listed below:

1. The hydrogen partial pressure in equilibrium with Fe and Fe_2O_3 considering that:

   (a) the pellet is formed by pure Fe_2O_3 ($a = 1$; Raoult),
   (b) the total pressure of the system is 1.5 atm and,
   (c) the partial pressure of the nitrogen in the gas is 0.15 atm.

2. Specify and quantify if a heat supply is required to maintain the reduction process at 1100 °C.
3. If we assume a pellet completely homogeneous and formed by spherical particles of 1.5 cm in diameter, calculate the time required to obtain a 100% metallization if the partial pressure of hydrogen in the gas is 1.30 atm at 1100 °C.
4. Determine, if possible, the controlling step in the reduction process.
5. If the pellet is formed by 86.0% Fe_2O_3, 2.0% CaO and 12.0% SiO_2, calculate the phases that can appear at 1100 °C before the significative manifestation of the reduction process.

Use the following data:

- Density of the pellet: 4500 kg m^{-3}.
- $D(H_2) = 0.76 \cdot (T/273)^{1.8}$ where $D(H_2)$ is the diffusion coefficient of the molecular hydrogen in the gas in cm^2 s^{-1} and T is the temperature of the gas in K for total pressures of the gas close to one atmosphere (Poirier and Geiger 1994, p. 456).
- $k_q = 10^4 \cdot \exp(-\Delta E/R \cdot T)$ expressed in cm s^{-1}, where ΔE is the activation energy, equal to 73,220 J mol^{-1}, k_q is the reaction rate constant for the interface controlled chemical reaction process, and T is the temperature in K.
- Consider for the measurement of the gas concentration that the standard conditions are 273 K and 1 atm of pressure.

The standard free energy of formation of one mol of hematite (from Fe and $O_2(g)$), considering that the path to reach the Fe_2O_3 goes through FeO and Fe_3O_4 (Fig. 3.3), which allows analyzing the formation of the hematite with the data of Appendix 1. From the same Appendix 1, the standard free energy of formation of one mol of water is obtained. The standard free energy for the reduction of the hematite with hydrogen is:

$$Fe_2O_3(s) + 3H_2(g) \rightarrow 2Fe(s) + 3H_2O(g) \tag{3.66}$$

$$\Delta_f G^0(T) = 60.5 - 0.079 \cdot T(K) kJ \, mol^{-1} Fe_2O_3 \tag{3.67}$$

The equilibrium constant at 1373 K is:

$$\Delta_f G^0(T) = R \cdot T \cdot \ln K_{eq} \rightarrow -47,967 \frac{J}{mol} = -8.314 \frac{J}{mol\,K} \cdot 1373K \cdot \ln K_{eq}$$

(3.68)

$$K_{eq} = 66.81 = \left(\frac{P(H_2O)}{P(H_2)}\right)^3$$

(3.69)

$$\left(\frac{P(H_2O)}{P(H_2)}\right)^{=} = 4.06$$

(3.70)

The addition of the gases partial pressures will be equal to the total pressure, P_T:

$$P_T = P(N_2) + P(H_2) + P(H_2O) + P(O_2)$$

(3.71)

In the reductant systems, controlled by the partial pressures of carbon monoxide–carbon dioxide or controlled by the hydrogen–water vapor, the partial pressure of the oxygen is very small and can be disregarded in the previous equation when the total pressure is calculated. This way:

$$1.5atm = 0.15atm + P(H_2)^{=} + 4.06 \cdot P(H_2)^{=} + 0 \rightarrow P(H_2)^{=} = \frac{1.35atm}{5.06}$$
$$= 0.27atm$$

(3.72)

The value of the hydrogen partial pressure obtained corresponds to the minimum equilibrium partial pressure, $P(H_2)^{=}$, that, if it is not reached but exceeded, the reduction of the hematite will not be possible. In the data of the exercise, it is indicated that the hydrogen pressure in the reduction environment reaches 1.3 atm.

In the second section of the exercise, we asked for the endothermic or exothermic nature of the reduction with hydrogen. Habitually, the reduction processes require the supply of thermal energy, although there are some reduction processes (as the reduction of hematite with CO) that are endothermic. While the enthalpies or heat associated with the reduction processes are explained in Chap. 5, in this exercise we used the standard enthalpy of reaction, $\Delta_r H^0$ (T), for the formation of different oxidized combinations of the elements of the periodic table. In the case of the reduction of the hematite using hydrogen, the variation of the standard enthalpy with the temperature (Appendix 2) is:

$$\Delta_r H^0(T) = 80.3 - 0.0274 \cdot T(K)kJ\,mol^{-1}Fe_2O_3$$

(3.73)

At 1373 K, the positive value (endothermic) of the standard enthalpy of the reduction is:

$$\Delta_r H^0(1373K) = 42,680J\,mol^{-1}\,Fe_2O_3 \qquad (3.74)$$

$$\Delta_r H^0(1373K) = 42.68\frac{kJ}{molFe_2O_3} \cdot \frac{1molFe_2O_3}{2molFe} \cdot \frac{1molFe}{55.85gFe} \cdot \frac{10^6gFe}{1tFe} \qquad (3.75)$$
$$= 382,095kJ\,t^{-1}Fe$$

$$\Delta_r H^0(1373K) = 382,095\frac{kJ}{tFe} \cdot \frac{1kWh}{3600kJ} = 106kWh\,t^{-1}Fe \qquad (3.76)$$

Consequently, to keep constant the temperature of reduction at 1100 °C, it is necessary a thermal energy supply, and thus avoid changes in the reduction rate.

The reduction rate of a pellet particle using gas can be analyzed using the unreacted core model (Eqs. 3.16–3.18), where the reaction time to reach a total metallization degree, f = 1, will be the addition of the time consumed in the hydrogen diffusion through the stationary layer of gas that surrounds the pellet, t_d, and the time consumed in the chemical reaction of the gas molecule with the oxide in the interface–interphase controlled with the unreacted core, t_q. The driving force for the reaction, the difference of hydrogen concentration, ΔC, and the reaction rate constant of the diffusional and chemical mechanisms are:

$$\Delta C = C_{H_2}^0 - C_{H_2}^= \qquad (3.77)$$

$$C_{H_2}^0 = \frac{1.30atm \cdot 2\frac{g}{mol}}{0.082\frac{atm\cdot dm^3}{mol\cdot K} \cdot \frac{10^3 cm^3}{1dm^3} \cdot 273K} = 1.161 \cdot 10^{-4}g\,cm^{-3} \qquad (3.78)$$

$$C_{H_2}^= = \frac{0.27atm \cdot 2\frac{g}{mol}}{0.082\frac{atm\cdot dm^3}{mol\cdot K} \cdot \frac{10^3 cm^3}{1dm^3} \cdot 273K} = 2.412 \cdot 10^{-5}g\,cm^{-3} \qquad (3.79)$$

$$\Delta C = C_{H_2}^0 - C_{H_2}^= = 1.161 \cdot 10^{-4}g\,cm^{-3} - 2.412 \cdot 10^{-5}g\,cm^{-3}$$
$$= 9.198 \cdot 10^{-5}g\,cm^{-3} \qquad (3.80)$$

$$D(H_2, 1373K) = 0.76 \cdot \left(\frac{1373K}{273}\right)^{1.8} = 13.92cm^2\,s^{-1} \qquad (3.81)$$

$$k_q = 10^4 \cdot exp\left(-\frac{\Delta E}{R \cdot T}\right) = 10^4 \cdot exp\left(-\frac{73,220J\,mol^{-1}}{8.314\frac{J}{mol\cdot K} \cdot 1373K}\right) = 16.38cm\,s^{-1}$$

$$(3.82)$$

Consequently, at 1373 K (1100 °C), the total reduction time will be equal to the addition of the diffusional time and the corresponding to the chemical reaction:

$$t = \frac{4.5(\mathrm{g\,cm^{-3}}) \cdot (1.5/2)(\mathrm{cm})}{16.38(\mathrm{cm\,s^{-1}}) \cdot 9.198 \cdot 10^{-5}(\mathrm{g\,cm^{-3}})} + \frac{4.5(\mathrm{g\,cm^{-3}}) \cdot (1.5/2)^2(\mathrm{cm})^2}{6 \cdot 13.92(\mathrm{cm^2\,s^{-1}}) \cdot 9.198 \cdot 10^{-5}(\mathrm{g\,cm^{-3}})}$$

$$= 2240\mathrm{s} + 329\mathrm{s} = 2569\mathrm{s} \cdot \frac{1\mathrm{h}}{3600\mathrm{s}}$$

$$= 0.71\mathrm{h}(43\mathrm{min})$$

$$(3.83)$$

Under the conditions of temperature and pressure indicated in the enunciation, the controlling mechanism of the reduction is by far the chemical reaction. If instead of being the H_2 the reductant gas, it was de CO, as the diffusion coefficients for the CO is 5–6 time smaller than that of the H_2, the controlling mechanism of the reduction might be the mixed regime.

Finally, to estimate the phases that form the pellet that are reduced with gas at 1373 K, the composition of the pellet indicated in the enunciation must be located in the CaO-Fe_2O_3-SiO_2 ternary diagram (Fig. 2.7) (86% Fe_2O_3, 2.0% CaO and 12% SiO_2) and we must check that the compatibility triangle (phases-compounds compatible with the global composition) would be that formed by the hematite, wollastonite, and the silicon oxide. The percentage of the phases at 1100 °C is 86% Fe_2O_3, 10% SiO_2, and 4% CaO·SiO_2.

Commentaries: The rate of reduction processes is strongly influenced by the temperature. For example, at 500 °C, keeping the same composition of reductant gas, the slow step or controlling step will be also the chemical reaction (147.91 h) while the hydrogen diffusion will be faster (25.26 min). In this case, the values of the variables involved in the unreacted core model are:

$$\Delta_f G^0(T) = R \cdot T \cdot \ln K_{eq} \rightarrow -567\frac{J}{mol} = -8.314\frac{J}{mol \cdot K} \cdot 773K \cdot \ln K_{eq} \quad (3.84)$$

$$K_{eq} = 1.09 = \left(\frac{P(H_2O)}{P(H_2)}\right)^3 \quad (3.85)$$

$$\left(\frac{P(H_2O)}{P(H_2)}\right)^{=} = 1.03 \quad (3.86)$$

$$1.5\mathrm{atm} = 0.15\mathrm{atm} + P(H_2)^{=} + 1.03 \cdot P(H_2)^{=} + 0 \rightarrow P(H_2)^{=} = \frac{1.35\mathrm{atm}}{2.03}$$

$$= 0.67\mathrm{atm}$$

$$(3.87)$$

$$C_{H_2}^0 = \frac{1.30\text{atm} \cdot 2\frac{\text{g}}{\text{mol}}}{0.082\frac{\text{atm}\cdot\text{dm}^3}{\text{mol}\cdot\text{K}} \cdot \frac{10^3\text{cm}^3}{1\text{dm}^3} \cdot 273\text{K}} = 1.161 \cdot 10^{-4}\text{g cm}^{-3} \qquad (3.88)$$

$$C_{H_2}^= = \frac{0.67\text{atm} \cdot 2\frac{\text{g}}{\text{mol}}}{0.082\frac{\text{atm}\cdot\text{dm}^3}{\text{mol}\cdot\text{K}} \cdot \frac{10^3\text{cm}^3}{1\text{dm}^3} \cdot 273\text{K}} = 5.986 \cdot 10^{-5}\text{g cm}^{-3} \qquad (3.89)$$

$$\Delta C = C_{H_2}^0 - C_{H_2}^= = 1.161 \cdot 10^{-4}\text{g cm}^{-3} - 5.986 \cdot 10^{-5}\text{g cm}^{-3}$$
$$= 5.624 \cdot 10^{-5}\text{g cm}^{-3} \qquad (3.90)$$

$$D(H_2, 773K) = 0.76 \cdot \left(\frac{773K}{273}\right)^{1.8} = 4.95\text{cm}^2\,\text{s}^{-1} \qquad (3.91)$$

$$k_q = 10^4 \cdot \exp\left(-\frac{\Delta E}{R \cdot T}\right) = 10^4 \cdot \exp\left(-\frac{73,220\text{J mol}^{-1}}{8.314\frac{\text{J}}{\text{mol K}} \cdot 773K}\right) = 0.1127\text{cm s}^{-1}$$
$$(3.92)$$

Consequently, at 773 K (500 °C), the total reduction time will be equal to the addition of the diffusional time and the corresponding to the interface controlled chemical reaction:

$$t = \frac{4.5(\text{g cm}^{-3}) \cdot (1.5/2)(\text{cm})}{0.1127(\text{cm s}^{-1}) \cdot 5.624 \cdot 10^{-5}(\text{g cm}^{-3})} + \frac{4.5(\text{g cm}^{-3}) \cdot (1.5/2)^2(\text{cm})^2}{6 \cdot 4.95(\text{cm}^2\,\text{s}^{-1}) \cdot 5.624 \cdot 10^{-5}(\text{g cm}^{-3})}$$
$$= 532,482\text{s} + 1515\text{s} = 533,997\text{s} \cdot \frac{1\text{h}}{3600\text{s}}$$
$$= 148.33\text{h}(8899.95\text{min}) \qquad (3.93)$$

If the kinetics of the reduction are analyzed, the H_2 has a bigger diffusion coefficient than the CO (at 1373 K the value is 2.56 cm^2 s^{-1} against 13.92 cm^2 s^{-1} of the H_2). That is to say, from both the thermodynamics and the kinetics point of view, the hydrogen is the preferred reductant at high temperatures (Sects. 2.2 and 5.1).

References

Babich A, Senk D, Gudenau HW, Mavrommatis TT, Ironmaking RWTH, (2008) Aachen University, Department of Ferrous Metallurgy, Aachen
Ballester A, Verdeja LF, Sancho J (2000) Metalurgia extractiva: fundamentos, vol 1. Síntesis, Madrid

Fernández-González D, Piñuela-Noval J, Verdeja LF (2018) Iron ore agglomeration technologies. In: Shatokha V (ed) Iron ores and iron oxide materials. Intechopen, Londres, pp 61–80 (Chapter 4)

Levin E, Robbins CR, McMurdie HF (1964) Phase diagrams for ceramists. The American Ceramic Society, Columbus, Ohio

Massalski TB, Okamoto H, Subramanian PR, Kacprzac L (1990) Binary alloy phase diagrams. ASM International, Materials Park, Ohio

Pero-Sanz JA, Fernández-González D, Verdeja LF (2019) Structural materials: properties and selection. Springer, Cham

Poirier DR, Geiger GH (1994) Transport phenomena in materials processing. TMS, Warrendale

Rahaman HN (2003) Ceramic processing and sintering. In: Dekker M, (ed) CRC press, New York

Verdeja LF, Ballester A, Sancho JP, González R (2014) Refractory and ceramic materials. In: Síntesis. Madrid, Spain

Ironmaking Coke

4

4.1 Introduction

Coke used in the ironmaking is a porous solid that is obtained by distilling of non-caking coal with 22–30% of volatile matter, in closed ovens isolated from the air (pyrolysis) at temperatures in the range 1000–1300 °C. From the chemical point of view, the coke has the following chemical composition: 86–91% fixed carbon, 3.0% or less of moisture, up to 11.5% of ashes, and <1.0% of sulphur. Finally, its heat of combustion is around 29,700 kJ kg^{-1}. Several definitions of "coke" have been proposed. Here, we indicate two from relevant authors:

- Butts 1943 (p. 53) defined the coke as follows: if coal is heated out of contact with air, it cannot oxidize or burn. It will, however, partly decompose, and volatile matter contained in it will be expelled in gaseous form. The volatile matter will consist of various hydrocarbon gases (such as CH_4, C_2H_4, C_6H_6), hydrogen, tar, ammonia, light oils and moisture. The residue will consist (coke) of fixed carbon and ash, together with some volatile matter not completely expelled. The amount of volatile matter expelled depends on the composition of the coal and the temperature to which it is heated. When the process is carried out in by-product coke oven, the residue is coke, and the volatile matter is led through pipes, condenser, etc., to separate its various constituents and recover them as by-products.
- Brady and Clauser 1986 (p. 210) defined coke as follows: the porous, gray, infusible residue left after the volatile matter is driven out of bituminous coal. The coal is heated to a temperature of 1200–1400 °C, without allowing air to burn it (pyrolysis). The residue, which is mainly fixed carbon and ash, is a cellular mass of greater strength than the original coal. Its nature and structure make it a valuable fuel for blast furnace, burning rapidly and supporting a heavy charge of metal without packing. Soft, or bituminous, coals are designed as coking and non-coking according to their capacity for being converted into coke. Coal low in carbon and high in ash will produce a coke that is friable, or the ash

J. I. Verdeja González et al., *Operations and Basic Processes in Ironmaking*, Topics in Mining, Metallurgy and Materials Engineering, https://doi.org/10.1007/978-3-030-54606-9_4

may have low-melting constituents that leave glassy slag in the coke.... The fixed carbon of good coke should be at least 86%, and sulphur not more than 1%. The porosity may vary from 40 to 60%, and the apparent specific gravity should not less than 0.8. Foundry coke should have an ignition point about 538 ° C (1000 °F), with sulphur below 0.7% and the pieces should be strong enough to carry the burden of ore and limestone.

On the other hand, it is typical for the coke ashes the following composition: 45.0% SiO_2, 34.0% Al_2O_3, 9.5% Fe_2O_3, 5.0% CaO, 2.0% MgO, 0.8% S, 0.2% P_2O_5, 3.0% the addition of K_2O and Na_2O, and 0.5% of other oxides (Babich 2008, p. 120). Moreover, the physical characteristics of the coke used in the ironmaking must be sufficient to guarantee the following points:

- Good resistance to compression at high temperature.
- At room temperature, the resistance to compression must be greater than 100 kg cm^{-2} ($\approx$10 MPa).
- Porosity must be available to favor the kinetics of the Boudouard gasification reaction.
- High resistance to the abrasion and wear to support the friction with the refractory lining, with the ferric charge and with the rest of the coke available in the furnace. In this way, the materials entering the iron blast furnace consists of the iron ore, the flux (limestone), the fuel (nearly always coke) and the air blown through the tuyeres. The ore and flux without the fuel is sometimes termed the "burden".

In Table 4.1, the most important physical-chemical characteristics of the coke used in the ironmaking are detailed. The value of the mechanical resistance of the coke at room temperature allows understanding the effect of the degradation during its transportation. The *IRSID* index consists in studying the mechanical degradation that the coke suffers after four minutes of treatment in a cylinder of one meter in diameter and one meter in height that rotates at 125 rpm. At the end of the test, it is possible to determine the percentage <10 mm (%I10 index) or the percentage with a size >20 mm (%I20) or >40 mm (%I40). This index together with the *TUMBLER* (very similar to the *IRSID*) can be used to understand and quantify the cold degradation of the coke (Fig. 2.4).

Regarding the reactivity of the coke (Fig. 2.5), a typical test consists in putting a calibrated coke (20 mm of grain size) in a tubular furnace at 1100 °C under a flow of 5 L of CO_2 per minute for 60 min to gasify it. The mass loss that the coke experiences during the test, in percentage, is the chemical reactivity, *CRI (Coke Reactivity Index)*. However, other quality index, which adapts better to the changes and transformations that the coke experiences in a blast furnace, is the test that quantifies the percentage of coke with a grain size >10 mm after subjecting it to a test of reactivity and mechanical degradation (usually less demanding, with less revolutions per minute than in the cold degradation test, at room temperature): *CSR-Coke Strength after Reaction.*

Table 4.1 Physical-chemical properties required to the coke used in the ironmaking

Property	Value (%)
Moisture	<2.5
Ashes	<10
Granulometry	
>70 mm	<14
70–20 mm	>84
<20 mm	<2
IRSID test	
>20 mm	>79
<10 mm	<18.5
Reactivity	
CSR	>52
CRI	<33

4.2 Coking Process

The manufacture of the coke used in the blast furnace is performed in the *coke battery*, succession of individual coking chambers (or coking ovens), also disposed as a *Multi-Chamber-System* (*MCS*), which are externally heated using the gas generated (recovered) in the pyrolysis of the carbonaceous material.

Figure 4.1 shows a scheme with the main parts of the coke battery, which is a good example of the *MCSR* (*Multi-Chamber-System-Recovery*):

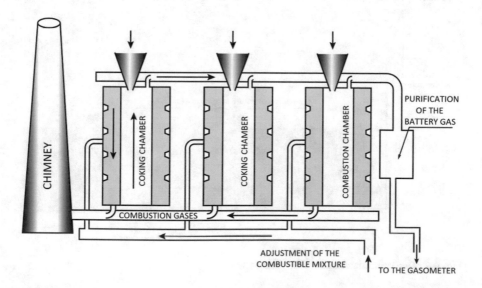

Fig. 4.1 General scheme of a coke battery

- Coking chambers.
- Combustion chamber.
- Cleaning system of the coke battery gas.

The typical dimensions of the coking chambers are: width, 450–600 mm; height, 4–8 m; and, length, 12–18 m. With these dimensions, the useful volume of the furnace is 40–70 m³ and the productivity is within 6200 and 17,000 tons per year or 25–36 kg m⁻³ h⁻¹. Nowadays, many batteries of the *MCS* type are built with bigger length and height: width, 590 mm; height, 8.3 m; and, length, 20 m. The useful volume of the chamber with these dimensions is 93 m³.

The control of the process of coking coals is indirectly performed by means of the measurement of the temperature of the gas that leaves the coking chambers. Figure 4.2 represents the curve temperature-time for one oven of the coke battery, where it is indicated that, after charging the oven, the temperature of the gases at the exit of the coking chamber starts to grow until reaching an equilibrium value; the value of this parameter will depend on the system and the location of the temperature acquisition equipment. After around 15 h, the temperature at the exit of the chamber starts to drop and, in this moment, it is possible to say that the coking of the coal has ended. However, and foreseeing that the coke will be used in the blast furnace, the coke production plants over-coke the material for 1.5–2 h. The objective of overcoking is the obtaining of a product kinetically few reactive at the temperature of the chemical reserve zone of the blast furnace (around 950 °C).

The cyclical operation of each oven of the battery leads to the thermal shock problems in those zones, as doors and frames, sensitives to the abrupt change of temperature. The material used in the construction of the ovens is the silica acid refractory (tridymite), which has an excellent behavior on the condition that the process must not be below 600 °C (Pero-Sanz et al. 2019).

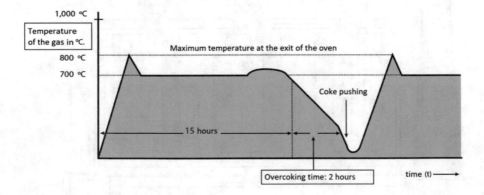

Fig. 4.2 Curves heating/cooling in a coking oven

The reparation of the refractory of these chambers must be performed, if it is possible, at temperatures above 600 °C, with high refractoriness ceramic materials and without using water during the process. The reparations at room temperature require special attention during the cooling and the heating of the oven to prevent mechanical stresses due to the diffusionless transformation (military) of the silica allotropes (Verdeja et al. 2014, pp. 201–204).

The mass balance in each one of the battery ovens indicates that, from 1000 kg of bituminous coal, 750 kg of coke, 300 Nm^3 of gas with heat of combustion lower than 18,300 kJ Nm^3, 20 kg of pitch, 30 kg of coal tar, 3 kg of ammonia gas and, 6 L of benzol are produced. The volumetric composition of the coke-oven gas MCS, free of other paraffin, alkene and aromatic hydrocarbon, might be 57% H_2, 27% CH_4, 8% CO, 4% CO_2, 3.5% N_2 and 0.5% O_2.

Facing to the traditional MCS process, the alternatives that are currently proposed in the ironmaking are:

(a) *NRS, Non Recovery Systems.* In this design, the recovery, treatment, purification, reparation and storage of the gas in gasometers for its utilization in the ironmaking process are not considered.
(b) *SCS, Single Chamber System.*
(c) *SCOPE21* process.

From the processes without recovery system, the simplest consists in taking the reductant gases produced in the partial combustion of the coal, which is being coked, and sending them to the vapor or electric power plants. One of the versions recently developed for the systems of coke production without recovery is based on taking the heat from the coking gases, which can have concentrations of 60% H_2 and 30% CO (*HNRS, Heat Non Recovery Systems*). Part of these gases can be burnt or oxidized in the upper zone of the chambers, supporting, this way, the energy requirements of the coking process. Other part of these gases can be used as iron ore reductant reagent. The size of each one of the furnaces of the HNRS process is usually the following one: width, 4.6 m; height, 1.2 m; and length, 14.0 m. The productivity of this type of designs to produce blast furnace coke is 4500 tons per year with a coking time of 48 h (against the 15–25 h of the MCS process). If we compare the MCS process with the HNRS process, the second one has clear environmental advantages due to the lack of emissions through the doors of the oven, and also, the HNRS process operates in a negative pressure environment and the ambient air enters in the coking chambers. Moreover, it is necessary to manipulate gases neither toxic components, benzo(a)pyrenes, dioxins nor hydrocarbons derived from the condensation of benzene rings. On the other hand, the main disadvantages of the NRS and HNRS processes against the MCS process are the low productivity, the requirement of a large extension of land for the installation and the important investments required for the installation.

From the other two alternatives to the MCS traditional process, one of them, the SCS (*Single Chamber Systems*) is a development of the European ironmaking industry while the other one is an initiative of the Japanese ironmaking industry

(*SCOPE21*). Both of them are still in the stages of development and tests at semi-industrial scale. A common objective of the *SCS* and *SCOPE21* processes is alleviating the conceptual deficiencies of the *MCS* in the following points:

- Increasing the rate of heat transfer to the coking chambers.
- Obtaining a suitable coke quality using different coals from those available in the market.
- Minimizing the gas emissions increasing the size of the chambers and diminishing the number of coking ovens.
- Avoiding the contact with the environment during the process of coke discharge, by pushing of the coked material and its subsequent quenching and cooling.

From 1993, the *Prosper Coke Plant* operates in Germany, a *SCS* reactor (also known as *Jumbo Coking Reactor*) with the following dimensions: width, 850 mm; height, 9.6 m; and, length, 10 m.

4.3   Coke Reactivity: Thermodynamic and Kinetic Aspects

In the C-O system, the equilibrium of the CO_2 and CO gaseous compounds with carbon-graphite is represented by the reaction (Boudouard equilibrium):

$$C(s) + CO_2(g) \rightarrow 2CO(g) \tag{4.1}$$

The $\Delta_r G^0$ of the Eq. (4.1), according to the data of the Appendix 1 is:

$$\Delta_r G^0(4.1) = 159000 - 169 \cdot T \left(\frac{J}{mol\ C} \right) \tag{4.2}$$

As the activity of the carbon is one, the variation of the volumetric composition of the CO in equilibrium with the graphite, at the total pressure of one atmosphere, is:

$$C_{CO}/\%vol. = 100 \cdot \left[1 - \frac{P_{CO_2}}{\exp\left(-\frac{159,000}{R \cdot T} + \frac{169}{R}\right)} \right] \tag{4.3}$$

The ironmaking coke is one of the fundamental raw materials in the charge of the blast furnace. The roles of the blast furnace coke, irreplaceable nowadays, are (Fernández-González et al. 2018a):

- Mechanical: coke supports the charge at high temperature under reductant conditions.
- Energetic: partial combustion of the coke in the zone of tuyeres.

- Reductant: coke generates the reductant gas required for the solid-gas reduction process in the shaft of the furnace.

To accomplish with the above-mentioned roles, the coke used in the ironmaking (ironmaking or blast furnace coke) must have suitable physical and chemical characteristics to be charged in the blast furnace. These characteristics are particular of this type of coke. Apart from in the blast furnace, coke is also used as energy supply in the cupola furnace to manufacture grey cast irons (Pero-Sanz et al. 2018) and also as reductant reagent in the manufacture of manganese, silicon and chromium ferroalloys (Fernández-González et al. 2018b, 2019a, b). Even when the elemental chemical analysis of any coke can be comparable, independently of the utility given to the coke, the reactivity or the physical and mechanical properties open, for each one of the coke's possible applications, a wide range of potential standards. Concretely, the blast furnace coke must accomplish a series of chemical, physical and mechanical specifications with the purpose of reaching a high level of efficiency in the three roles that the coke has in the blast furnace (mechanical, energetic and reductant). To achieve it, a certain chemical inertia must be guaranteed in the coke at high temperature (900 °C) because, on the contrary, if the gasification is very high, the coke will not efficiently accomplish its objectives.

Assuming that the controlling mechanism (slow step) of the coke gasification (Eq. 4.1) is the absorption-chemical reaction of the carbon dioxide in the coke surface, a kinetics equation of gasification can be proposed, where a new variable is involved: the concentration of active centers in the surface of the coke $[C_T]$.

Empirically, it is possible to demonstrate that the gasification of the coke at temperatures lower than 1100 °C follows the next equation:

$$-\frac{dm_C}{dt} = \frac{a \cdot P_{CO_2}}{b \cdot P_{CO_2} + c \cdot P_{CO} + 1} \qquad (4.4)$$

where a, b and c are constants and dm_C/dt is the coke gasification rate. If the reaction mechanism progresses according to the following steps:

$$coke_f(s) + CO_2(g) \xrightarrow{k_1} coke_O(s) + CO(g) \qquad (4.5)$$

$$coke_O(s) + CO(g) \xrightarrow{k_2} coke_f(s) + CO_2(g) \qquad (4.6)$$

$$coke_O(s) \xrightarrow{k_3} coke_f(s) + CO(g) \qquad (4.7)$$

where: $coke_O(s)$ is the solid carbon (coke) with a certain concentration of active points on its surface, over which, oxygen atoms are already absorbed (concentration of occupied centers); and, $coke_f(s)$ is the solid carbon (coke) with a certain concentration of free active points on its surface, over which, an oxygen atom coming from the dissociation of a CO_2 molecule can adhere (concentration of free centers). Thus:

$$[C_T] = [C_f] + [C_O] \tag{4.8}$$

where $[C_T]$ is the total concentration of active centers available on the surface of the coke. The rate equations in each one of the reactions (4.5), (4.6) and (4.7) are:

$$V_{(reaction 4.5)} = V_1 = k_1 \cdot [C_f] \cdot P_{CO_2} \tag{4.9}$$

$$V_{(reaction 4.6)} = V_2 = k_2 \cdot [C_O] \cdot P_{CO} \tag{4.10}$$

$$V_{(reaction 4.7)} = V_3 = k_3 \cdot [C_O] \tag{4.11}$$

Once reached the equilibrium, at a certain temperature, the following equality will be verified:

$$\frac{d[C_O]}{dt} = 0 = k_1 \cdot [C_f] \cdot P_{CO_2} - k_2 \cdot [C_O] \cdot P_{CO} - k_3 \cdot [C_O] \tag{4.12}$$

The coke gasification rate corresponds to the third step of the mechanism (Eq. 4.11). The value of $[C_O]$ can be calculated using Eq. (4.12) and knowing that $[C_T]$ is also related with $[C_f]$ by Eq. (4.8), thus:

$$[C_O] = \frac{k_1 \cdot [C_T] \cdot P_{CO_2}}{k_1 \cdot P_{CO_2} + k_2 \cdot P_{CO} + k_3} \tag{4.13}$$

resulting in:

$$-\frac{dm_C}{dt} = \frac{k_3 \cdot k_1 \cdot [C_T] \cdot P_{CO_2}}{k_1 \cdot P_{CO_2} + k_2 \cdot P_{CO} + k_3} \tag{4.14}$$

As the reaction (4.7) is the controlling step of the process (the slowest):

$$k_1 \cdot P_{CO_2} \gg k_3 \tag{4.15}$$

$$k_2 \cdot P_{CO} \gg k_3 \tag{4.16}$$

If the numerator and the denominator of Eq. (4.14) are divided by k_3, an equation equivalent to that of the (4.4) is obtained, where:

$$a = k_1 \cdot [C_T], \quad b = k_1/k_3, \quad c = k_2/k_3 \tag{4.17}$$

The coke gasification rate is directly proportional to the total concentration of active centers on its surface. An active center is a zone of maximum free energy where the carbon dioxide molecule preferentially establishes. The less ordered the carbon atoms in the coke, the bigger the superficial concentration of active centers. The overcoking of the coals in the batteries has as objective achieving a more ordered coke both at atomic and microscopic scales. The structure of an overcoked coal tries to reach, if the time is enough or the temperature is high, full graphitic symmetry.

Whereas at low temperatures, and small particle sizes (around 1 cm), the slow step of the coke gasification is the chemical reaction, at high temperatures ($T > 1100\ °C$), and big particle sizes, the coke gasification (or the gasification of the graphite electrodes in an electric furnace) is controlled by the diffusion inside of the porous particles. On the other hand, all the equations for solid-gas reaction in porous particles (Sect. 2.3) can be also used in the coke gasification process.

4.4 Solved Exercises

Exercise 4.1 In the C-O system, one of the most important reactions is the Boudouard equilibrium (Eq. 4.1). Analyze the influence of both the temperature and the total pressure of the system in the composition of the gas in equilibrium with the thermodynamically most stable phase of the carbon at room temperature, the graphite.

To analyze the Boudouard equilibrium (coke gasification using an oxidizing gas, carbon dioxide), it is necessary to consider that the carbon in the reaction (4.1) is as carbon-graphite. All the carbon structures in the nature or those artificially obtained develop, with the time and the temperature, into the thermodynamically stable allotropic form known as graphite.

In the C-O system, there are two components (carbon and oxygen) and two phases (graphite and gas), thus, the number of degrees of freedom is also two, for instance, temperature and pressure. If the value of the total pressure of the system is known, it is only necessary to set the temperature and the molar fraction of the CO in the gas to determine the equilibrium.

In the study of the reaction, it is possible to make an analysis relating the volume fraction or the CO partial pressure in the gas and the temperature. If the addition of partial pressures in the gas is:

$$P_T = P_{CO} + P_{CO_2} + P_{O_2} \tag{4.18}$$

$$P_T \simeq P_{CO} + P_{CO_2} \tag{4.19}$$

because $P_{O_2} \rightarrow 0$ and the relation of CO to CO_2 partial pressures in the gas is:

$$\frac{P_{CO}^2}{P_{CO_2}} = \exp\left(-\frac{159,000}{8.314 \cdot T} + \frac{169}{8.314}\right) \tag{4.20}$$

the following quadratic equation is obtained:

$$A \cdot C_{CO}^2 + B \cdot C_{CO} - C = 0 \tag{4.21}$$

where C_{CO} is the molar fraction of CO in the gas or the volumetric composition of the CO per unit, the value of A is the total pressure, P_T, while the constants B and C are:

$$B = C = \exp\left(-\frac{159,000}{8.314 \cdot T} + \frac{169}{8.314}\right) \tag{4.22}$$

We collect in Table 4.2 the composition of the CO in the gas as a function of the temperature for different values of the total pressure (15, 1.0 and 0.01 atm): the graphical plot would fit to a sigmoid curve with two different zones:

- Zone of the gas stability (CO).
- Zone of the coke-graphite carbon stability (C/CO$_2$).

If a gas with high CO content, for instance the gas generated in the tuyeres of the blast furnace, is rapidly cooled, very small particles (nanoparticles) are formed, and they precipitate as "cinder".

Commentaries: the Boudouard reaction is basic in all metallurgical processes associated to the carbothermal reduction. Besides, the obtaining of non-equilibrium carbon structures, as fullerenes (3D), nanotubes (2D) and graphene (1D), over

Table 4.2 Composition of the CO in the gas as a function of the temperature for different values of the total pressure (15, 1.0 and 0.01 atm)

T (K)	273	373	473	573	673	773	873	973
For 15 atm								
C_{CO}	0.000	0.000	0.000	0.000	0.005	0.028	0.111	0.302
T (K)	1073	1173	1273	1373	1473	1573	1673	1773
C_{CO}	0.582	0.820	0.935	0.976	0.990	0.996	0.998	0.999
For 1.0 atm								
T (K)	273	373	473	573	673	773	873	973
C_{CO}	0.000	0.000	0.000	0.001	0.017	0.104	0.362	0.729
T (K)	1073	1173	1273	1373	1473	1573	1673	1773
C_{CO}	0.929	0.983	0.995	0.998	0.999	1.000	1.000	1.000
For 0.01 atm								
T (K)	273	373	473	573	673	773	873	973
C_{CO}	0.000	0.000	0.000	0.015	0.160	0.651	0.956	0.995
T (K)	1073	1173	1273	1373	1473	1573	1673	1773
C_{CO}	0.999	1.000	1.000	1.000	1.000	1.000	1.000	1.000

different substrates that allow the clustering and growth of these structures, has the thermodynamic foundation of this reaction.

Exercise 4.2 The coke reactivity is analyzed using the *CRI* (*Coke Reactivity Index*), which was explained in the Sect. 4.1 (Babich et al. 2008, p. 108; ASTM 2010, pp. 1–4). Assuming that the kinetics model of porous particles can be applied to the blast furnace coke samples with different granulometric characteristics (Table 4.3), calculate the corresponding CRI values.

The other physical and chemical properties are the same in both samples.

Data: coke particles have a shape factor, f, of 1.390 and an apparent density of 0.999 g cm^{-3} (Verdeja et al. 2014, pp. 91–93). A particle of 20 mm suffers a mass loss of 30% after 2 h at 1100 °C in a CO_2 atmosphere at a pressure of 1 atm. Finally, at 1100 °C, the partial pressure of the CO_2 in equilibrium is 0.783 atm and the CO_2 diffusion coefficient is 2.34 cm^2 s^{-1} (Poirier et al. 1994, p. 456).

From Eqs. (2.34) and (2.50), the reaction rate constant of the coke gasification process, k_q^*, can be expressed as follows:

$$k_q^* = \frac{D_i \cdot A}{\rho_a \cdot V_0 \cdot h} \tag{4.23}$$

being the variation of the transformed/reduced solid with the time:

$$X = k_q^* \cdot \left(C_i^0 - C_i^=\right) \cdot t \tag{4.24}$$

The units of this constant in the CGS system are cm^3 g^{-1} s^{-1}. When X is 0.30, the variable t is 7200 s and the driving force of the reaction (difference of concentrations) is 4.265×10^{-4} g cm^{-3}, then k_q^* is 9.769×10^{-2} cm^3 g^{-1} s^{-1}. Using this value, it is possible to calculate other kinetic parameters for the porous particles, the value of h and k_q in cm s^{-1}:

$$h = \frac{D_i \cdot A}{\rho_a \cdot V_0 \cdot k_q^*} = \frac{D_i \cdot S_{e/V}}{\rho_a \cdot k_q^*} = \frac{2.34(\text{cm}^2/\text{s}) \cdot 4.17(\text{cm}^2/\text{cm}^3)}{0.999(\text{g/cm}^3) \cdot 9.769 \times 10^{-2}(\text{cm}^3/\text{g s})}$$
$$= 100.19 \text{ cm} \tag{4.25}$$

Table 4.3 Particle-size distribution for the two cokes

Size (mm)	A (%)	B (%)
25	10	9
20	80	90
15	10	
1		1

where the specific surface area of the coke per unit of volume, $S_{e/V}$, is:

$$S_{e/V} = \frac{6 \cdot f}{D_p} = \frac{6 \cdot 1.390}{2 \text{ cm}} = 4.17 \text{ cm}^2 \text{ cm}^{-3} \qquad (4.26)$$

where f is the shape factor and D_p is the particle size. Finally, the value of k_q in cm s^{-1} is:

$$k_q = \frac{D_i}{h^2 \cdot S_{e/V}} = \frac{2.34(\text{cm}^2/\text{s})}{(100.19 \text{ cm})^2 \cdot 4.17(\text{cm}^2/\text{cm}^3)} = 5.602 \cdot 10^{-5} \text{ cm s}^{-1} \quad (4.27)$$

The parameter the most widely used to characterize the solid-gas kinetics of porous particles is h. Dimensionally, h, has unities of length but at the same time it is a variable that groups other parameters involved in the kinetics of porous particles:

- Characteristics of the reductant gas through the D_i.
- Chemical properties of the solid with k_q.
- Specific surface of the solid that reacts, represented by $S_{e/V}$.

For the same temperature, reductant gas and chemical characteristics of the solid, the variable that should be controlled will be the specific surface of the solid. In this second part of the Exercise 4.2, we want to relate how the value of h is influenced by $S_{e/V}$ when k_q and D_i are constant.

In the case of the two samples of coke considered in the exercise, the physical (granulometric) and chemical characteristics are similar and the main difference between both cokes is the value of $S_{e/V}$, which also determines the value of h in each sample. It is also possible to define a specific surface area for each particle size, referred to the unit of volume:

$$S_{e/V} = \frac{6 \cdot f}{D_p} \qquad (4.28)$$

For the particle-size distribution of the samples, and calculating the weighted arithmetic mean:

- Sample A:

$$S_{e/V} = 4.17(\text{cm}^2/\text{cm}^3) \cdot 0.80 + 5.56(\text{cm}^2/\text{cm}^3) \cdot 0.10 + 3.34(\text{cm}^2/\text{cm}^3) \cdot 0.10$$
$$= 4.23(\text{cm}^2/\text{cm}^3)$$

$$(4.29)$$

– Sample B:

$$S_{e/V} = 4.17(\text{cm}^2/\text{cm}^3) \cdot 0.90 + 83.4(\text{cm}^2/\text{cm}^3) \cdot 0.01 + 3.34(\text{cm}^2/\text{cm}^3) \cdot 0.09$$
$$= 4.89(\text{cm}^2/\text{cm}^3)$$

$$(4.30)$$

Thus, the values of the parameter h are:

- Sample A:

$$h = \sqrt{\frac{2.34(\text{cm}^2/\text{s})}{4.23(\text{cm}^2/\text{cm}^3) \cdot 5.602 \times 10^{-5}(\text{cm/s})}} = 99.37 \text{ cm} \qquad (4.31)$$

- Sample B:

$$h = \sqrt{\frac{2.34(\text{cm}^2/\text{s})}{4.89(\text{cm}^2/\text{cm}^3) \cdot 5.602 \times 10^{-5}(\text{cm/s})}} = 92.42 \text{ cm} \qquad (4.32)$$

And, the values of the CRI are:

- Sample A:

$$\text{CRI} = \frac{2.34(\text{cm}^2/\text{s}) \cdot 4.23(\text{cm}^2/\text{cm}^3) \cdot 4.265 \cdot 10^{-4}(\text{g/cm}^3) \cdot 3600(s) \cdot 100}{0.999(\text{g/cm}^3) \cdot 99.37(\text{cm})}$$
$$= 15.3\%$$

$$(4.33)$$

- Sample B:

$$\text{CRI} = \frac{2.34(\text{cm}^2/\text{s}) \cdot 4.89(\text{cm}^2/\text{cm}^3) \cdot 4.265 \cdot 10^{-4}(\text{g/cm}^3) \cdot 3600(s) \cdot 100}{0.999(\text{g/cm}^3) \cdot 92.42(\text{cm})}$$
$$= 19.03\%$$

$$(4.34)$$

Commentaries: CRI results show the importance of working with a well-calibrated size of sample (very fitted to the 20 mm that the standard indicates). If it is not the case, the values of both the CRI and the CSR (*Coke Strength after Reaction*) could be seriously affected. Apart from that previously indicated, the new characterization techniques allow improving the kinetics models of porous particles, and improved equations to evaluate both the coke quality and the h, k_q^* and k_q parameters are defined.

Exercise 4.3. The coal coking is performed using an MCS process in the coke batteries (Sect. 4.2). The dimensions of one of the ovens of the coke battery has a length of 13.25 m, a height of 4.50 m, an average width of 0.35 m and a chamber taper of 60 mm.

From the heat transfer point of view, the process can be simulated as an infinite source of heat that transfers its energy to the coal located in the chamber. Calculate the evolution of the coking flame front as a function of the time (Verdeja et al. 2014, pp. 102–104). Data:

(a) The temperature in the walls of the chamber at any time is 1250 °C.
(b) After 15 h, the temperature reached in the center of the chamber is 1000 °C when the thermal diffusivity of the charge is 4.55×10^{-5} m^2 s^{-1}.

The following analytical equation must be used in the calculation of the error function:

$$\text{erf}(z) = A + B \cdot z + C \cdot z^2 + D \cdot z^4 + E \cdot z^5 + F \cdot z^6 + G \cdot z^7$$
$$= -0.00191 + 1.17 \cdot z - 0.206 \cdot z^2 - 0.338 \cdot z^4 + 0.303 \cdot z^5 \qquad (4.35)$$
$$- 0.0967 \cdot z^6 + 0.0108 \cdot z^7$$

Table 4.4 shows the values of the error function.

During the coking process of a mixture of coals, it is possible to consider the following structures as a function of the temperature (Babich et al. 2008, pp. 99–100):

(a) Below 100 °C: the zone where the moisture of the coals is dried can be identified.
(b) Within 100 and 350 °C: the release of nitrogen, methane and carbon monoxide takes place and the appearance of hydrogen by pyrolysis begins.
(c) Between 350 and 480 °C: the formation of a plastic state, whose characteristics and properties will define the final ones of the coke, sets.
(d) Within 480 and 600 °C: the formation of semicoke, with the corresponding appearance of cracks associated to the contraction, is promoted.
(e) Final period, at 600–1100 °C, where the final structure of the coke is developed.

The model of infinite heat sources that supply the heat required for the advance of the coking front from the chamber wall (origin of the coordinate system) to the center of the chamber ($x = 0.175$ m $= 17.5$ cm) is:

Table 4.4 Values of the error function

z	erf(z)	erf (z) Eq. 4.35	z	erf(z)	erf(z) Eq. 4.35	z	erf(z)	erf (z) Eq.4.35
0	0.00000	−0.00191	0.55	0.56332	0.56332	1.20	0.91031	0.90849
0.05	0.05637	0.05607	0.60	0.60386	0.60386	1.30	0.93401	0.93162
0.10	0.11246	0.11300	0.65	0.64203	0.64203	1.40	0.95229	0.94922
0.15	0.16800	0.16878	0.70	0.67780	0.67780	1.50	0.96611	0.96243
0.20	0.22270	0.22331	0.75	0.71116	0.71116	1.60	0.97635	0.97235
0.25	0.27633	0.27640	0.80	0.74210	0.74210	1.70	0.98379	0.97997
0.30	0.32863	0.32781	0.85	0.77067	0.77067	1.80	0.98909	0.98607
0.35	0.37938	0.37729	0.90	0.79691	0.79691	1.90	0.99279	0.99120
0.40	0.42839	0.42449	0.95	0.82089	0.82089	2.00	0.99532	0.99569
0.45	0.47548	0.46905	1.00	0.84270	0.84270	2.50	0.99959	1.00571
0.50	0.52050	0.51054	1.10	0.88021	0.88021	3.00	0.99997	1.03039

$$T(x,t) = A + B \cdot \mathrm{erf}(z) \qquad (4.36)$$

$$z = \frac{x}{\sqrt{4 \cdot \alpha_e \cdot t}} \qquad (4.37)$$

$$\alpha_e = \frac{\lambda_e}{c_p \cdot \rho_g} \qquad (4.38)$$

where A and B are constants that depend on the boundary conditions defined in the simulation, α_e is the thermal diffusivity, t is the time, c_p is the specific heat of the charge, λ_e is the effective thermal conductivity and ρ_g is the global density of the charge (Verdeja et al. 2014, p. 91). At 1250 °C, the value of the constant A is:

$$T(0,t) = A + B \cdot \mathrm{erf}(0) = 1250\,°C \rightarrow A = 1250\,°C \qquad (4.39)$$

On the other hand, the value of the constant B is related with the temperature reached in the center of the chamber after 15 h, when $\alpha_e = 4.55 \times 10^{-5}$ m^2 s^{-1}:

$$B = \frac{1100 - 1250}{\mathrm{erf}(0.05582)} = -\frac{150}{0.6275} = -2390.5 \qquad (4.40)$$

The profile of the 1100 °C isotherm progression with the time from the walls to the center of the chamber is showed in Table 4.5.

The overcoking of the material (Fig. 4.2), from 1100 °C, might be performed in a period of five hours, during which, the temperature in the center of the chamber (0.175 m) rises from 1100 to 1200 °C. In this case, the diffusivity of the charge will arise due to the contribution of the radiation heat transfer mechanism (Table 4.6).

Table 4.5 Position of the isotherm of 1100 °C as a function of the time

Time (h)	2	4	6	8	10	12	14	15
Position (cm)	6.4	9.0	11.1	12.8	14.3	15.7	16.9	17.5

Table 4.6 Diffusivity of the charge as a function of the temperature and the time

t (h)	T (°C)	α_e (m^2 s^{-1})
15	1100	4.55×10^{-5}
16	1154	4.00×10^{-4}
17	1172	9.00×10^{-4}
18	1187	3.00×10^{-3}
19	1195	9.00×10^{-3}
20	1200	3.50×10^{-2}

Commentary: it is possible to simulate the progress of the plastic zone in the chamber if the time that the center (0.175 m) takes in reaching 480 °C is known, 6 h. The profile concentration-penetration of the plastic front as a function of the time is showed in Table 4.7.

Exercise 4.4. Two coke qualities, either inside of the C-O or the C-O-N systems, have the following gasification rates (m min^{-1}) at 850 °C and at 1 atm of total pressure:

- **Quality A:**

$$\frac{dX}{dt} = \frac{0.85 \cdot P_{CO_2}}{1 + [7500 \cdot (P_T - P_{N_2}) - 1300 \cdot P_{CO_2}]} \tag{4.41}$$

- **Quality B:**

$$\frac{dX}{dt} = \frac{0.90 \cdot P_{CO_2}}{1 + [5300 \cdot (P_T - P_{N_2}) - 1700 \cdot P_{CO_2}]} \tag{4.42}$$

Table 4.7 Position of the plastic front as a function of the time

Time (h)	0.2	0.5	1.0	2.0	3.0	4.0	5.0	6.0
Position (cm)	3.2	5.1	7.1	10.1	12.4	14.3	16.0	17.5

where X is the fraction of gasified carbon in the coke and P_{CO_2} is the carbon dioxide pressure in the gas in atm, calculate:

1. The reaction rate constants of the coke gasification process for the above-mentioned qualities assuming that the rate constant of the specific step of the carbon atoms gasification at 850 °C k_3 is 1.00×10^{-5} m^2 min^{-1}:
   i. Adsorption constant of the CO_2 gas on the surface of the coke,
   ii. Desorption constant of the CO_2 gas on the surface of the coke and,
   iii. Total concentration of active centers per square meter of surface.
2. At 850 °C, when the total pressure is 2.0 atm (C-O-N system), the CO_2 partial pressure is 0.60 atm and that of the N_2 is 0.60 atm, determine what of the qualities will be more adequate for its utilization in the blast furnace. Calculate, in these conditions, what would be the concentration of free sites of CO_2 in the surface of both cokes.
3. Calculate the time that would be required for the gasification of a coke particle of 40 mm in diameter for the qualities A and B at 850 °C under the following conditions (C-O-N system): total pressure 3.0 atm, P(CO) = 0.40 atm and P(CO_2) = 0.80 atm.

From the obtained results, indicate if there are or not any advantages for the blast furnace if we use high pressurized gases inside of the shaft.

Considering Eqs. (4.4), (4.8) and (4.14), different situations can be defined in the coke gasification:

- *Situation I*: C-O system, total pressure of 1 atm and:

$$P_{CO} + P_{CO_2} = 1 \tag{4.43}$$

where:

$$\frac{dX}{dt} (\text{min}^{-1}) = \frac{k_1 \cdot C_T \cdot P_{CO_2}}{1 + \frac{k_2}{k_3} + \left(\frac{k_1 - k_2}{k_3}\right) \cdot P_{CO_2}} = \frac{A \cdot C_T \cdot P_{CO_2}}{1 + B + C \cdot P_{CO_2}} \tag{4.44}$$

- *Situation II*: C-O system, total pressure, P_T, different from 1 atm and:

$$P_{CO} + P_{CO_2} = P_T \tag{4.45}$$

where:

$$\frac{dX}{dt} (\text{min}^{-1}) = \frac{k_1 \cdot C_T \cdot P_{CO_2}}{1 + \left(\frac{k_2}{k_3}\right) \cdot P_T + \left(\frac{k_1 - k_2}{k_3}\right) \cdot P_{CO_2}} \tag{4.46}$$

- *Situation III*: C-O-N (air) system, total pressure, P_T, different from 1 atm and:

$$P_{CO} + P_{CO_2} + P_{N_2} = P_T \tag{4.47}$$

where:

$$\frac{dX}{dt} \left(min^{-1} \right) = \frac{k_1 \cdot C_T \cdot P_{CO_2}}{1 + \left(\dfrac{k_2}{k_3} \right) \cdot (P_T - P_{N_2}) + \left(\dfrac{k_1 - k_2}{k_3} \right) \cdot P_{CO_2}} \tag{4.48}$$

However, in any of the previous situations, the concentration of active centers, C_O, on the coke surface, susceptible of entering in reaction with the CO_2 gas molecules, is:

$$C_O = \frac{k_1 \cdot C_T \cdot P_{CO_2}}{k_1 \cdot P_{CO_2} + k_2 \cdot P_{CO} + k_3} \tag{4.49}$$

In the case of the coke quality A, the calculation of the reaction rate constants related with the coke gasification is:

$$\frac{k_2}{k_3} = 7500 \tag{4.50}$$

$$\frac{k_1 - k_2}{k_3} = -1300 \tag{4.51}$$

$$k_1 \cdot C_T = 0.85 \tag{4.52}$$

If we know the value of k_3 (1.0×10^{-5} m^2 min^{-1}), the values of the other variables are:

$$k_2 = 0.075 \ m^2 \ min^{-1} \ atm^{-1} \tag{4.53}$$

$$k_1 = 0.062 \ m^2 \ min^{-1} \ atm^{-1} \tag{4.54}$$

and:

$$C_T = 13.7 \tag{4.55}$$

number of active centers per square meter.

If the same calculations are repeated for the coke quality B, the values are:

$$k_2 = 0.053 \ m^2 \ min^{-1} \ atm^{-1} \tag{4.56}$$

$$k_1 = 0.036 \ \mathrm{m}^2 \ \mathrm{min}^{-1} \ \mathrm{atm}^{-1} \tag{4.57}$$

$$C_T = 25.0 \tag{4.58}$$

The calculation of these reaction rate constants is correct as far as the 850 °C temperature was not modified.

- For a total pressure of 2 atm, the gasification rate of the coke A is:

$$\frac{dX}{dt} \left(\mathrm{min}^{-1}\right) = \frac{0.85 \cdot 0.60}{1 + [7500 \cdot (2.0 - 0.60) - 1300 \cdot 0.6]} = 5.24 \times 10^{-5} \ \mathrm{min}^{-1} \tag{4.59}$$

And in the case of the coke quality B is:

$$\frac{dX}{dt} \left(\mathrm{min}^{-1}\right) = \frac{0.90 \cdot 0.60}{1 + [5300 \cdot (2.0 - 0.60) - 1700 \cdot 0.6]} = 8.43 \cdot 10^{-5} \ \mathrm{min}^{-1} \tag{4.60}$$

If we know that the total concentration, C_T, of active points on the surface of the coke where the CO_2 molecules can interact, the concentration or number of positions that does not absorb CO_2 is calculated using Eq. (4.13). In the case of the coke A, the concentration of occupied positions, C_0, is:

$$C_O$$
$$= \frac{0.062 \ \mathrm{m}^2 \ \mathrm{min}^{-1} \ \mathrm{atm}^{-1} \cdot 13.7 \ \text{positions} \ \mathrm{m}^{-2} \cdot 0.60 \ \mathrm{atm}}{0.062 \ \mathrm{m}^2 \ \mathrm{min}^{-1} \ \mathrm{atm}^{-1} \cdot 0.60 \ \mathrm{atm} + 0.075 \ \mathrm{m}^2 \ \mathrm{min}^{-1} \ \mathrm{atm}^{-1} \cdot 0.80 \ \mathrm{atm} + 1.0 \times 10^{-5} \ \mathrm{m}^2 \ \mathrm{min}^{-1}}$$
$$= 5.24 \ \text{positions} \ \mathrm{m}^{-2} \tag{4.61}$$

while in the case of the coke B:

$$C_O$$
$$= \frac{0.036 \ \mathrm{m}^2 \ \mathrm{min}^{-1} \ \mathrm{atm}^{-1} \cdot 25.0 \ \text{positions} \ \mathrm{m}^{-2} \cdot 0.60 \ \mathrm{atm}}{0.036 \ \mathrm{m}^2 \ \mathrm{min}^{-1} \ \mathrm{atm}^{-1} \cdot 0.60 \ \mathrm{atm} + 0.053 \ \mathrm{m}^2 \ \mathrm{min}^{-1} \ \mathrm{atm}^{-1} \cdot 0.80 \ \mathrm{atm} + 1.0 \times 10^{-5} \ \mathrm{m}^2 \ \mathrm{min}^{-1}}$$
$$= 8.43 \ \text{positions} \ \mathrm{m}^{-2} \tag{4.62}$$

Thus, the quantity of free positions on the surface of the coke A is:

$$C_T = C_f + C_O \rightarrow 13.7 \ \text{positions} \ \mathrm{m}^{-2} = C_f + 5.24 \ \text{positions} \ \mathrm{m}^{-2} \rightarrow C_f$$
$$= 8.46 \ \text{positions} \ \mathrm{m}^{-2} \tag{4.63}$$

And in the coke B:

$$C_T = C_f + C_O \rightarrow 25.0 \text{ positions m}^{-2} = C_f + 8.43 \text{ positions m}^{-2} \rightarrow C_f$$
$$= 16.57 \text{ positions m}^{-2} \tag{4.64}$$

The best blast furnace coke will be that with the slower gasification rate, or smaller values of both C_O and C_T, that is to say, the coke A.

Finally, to calculate the time required for the gasification in the C-O-N system of a coke particle of 40 mm in diameter, we apply Eqs. (4.47), (4.59) and (4.60) when $P_{CO} = 0.4$ atm, $P_{CO_2} = 0.8$ atm and $P_{CO} + P_{CO_2} = 1.2$ atm. The integration of the differential equations allows obtaining the time required for the complete gasification of a coke particle when X is unity:

- Coke A:

$$\frac{dX}{dt} \left(\text{min}^{-1}\right) = \frac{0.85 \cdot 0.8}{1 + (7500 \cdot 1.2 - 1300 \cdot 0.8)} = 8.54 \times 10^{-5} \text{ in}^{-1} \tag{4.65}$$

gasification in 195 h.

- Coke B:

$$\frac{dX}{dt} \left(\text{min}^{-1}\right) = \frac{0.9 \cdot 0.8}{1 + (5300 \cdot 1.2 - 1700 \cdot 0.8)} = 1.44 \times 10^{-4} \text{ min}^{-1} \tag{4.66}$$

gasification in 116 h.

According to the proposed kinetics model, the increase of the total pressure of the gases in the shaft of the furnace has influence in the coke gasification rate: the rate in the shaft decreases if the total pressure (C-O system) or if the difference between the total pressure and that of nitrogen (C-O-N system) in the furnace increases.

Commentaries: the proposed kinetics model for the coke gasification, controlled by the gasification of the CO_2 molecules absorbed by the coke surface, does not consider the size of the particles. The calculations made in the point 3 of the exercise estimate the time that would be required for the gasification of a coke particle of 40 mm, although the calculations for coke particles of different size will be the same. The time of gasification will not depend, thus, on the size of the ironmaking coke particles.

References

ASTM Standard Test Method D5341-99 (2010) Measuring coke reactivity index (CRI) and coke strength after reaction (CSR)

Babich A, Senk D, Gudenau HW, Mavrommatis TTH (2008) Ironmaking. RWTH Aachen University, Department of Ferrous Metallurgy, Aachen

Brady GS, Clauser HR (1986) Materials handbook: an encyclopedia for managers, technical professionals, purchasing and production managers. 12th edn. Technicians, supervisors and foremen, McGraw Hill, London

Butts A (1943) Metallurgical problems. 2nd edn. McGraw Hill, London

Fernández-González D, Prazuch J, Ruiz-Bustinza I, González-Gasca C, Piñuela-Noval J, Verdeja LF (2018a) Iron metallurgy via concentrated solar energy. Metals 8(11):873

Fernández-González D, Sancho-Gorostiaga J, Piñuela-Noval J, Verdeja LF (2018b) Anodic lodes and scrapings as a source of electrolytic manganese. Metals 8(3):162

Fernández-González D, Prazuch J, Ruiz-Bustinza I, González-Gasca C, Piñuela-Noval J, Verdeja LF (2019a) Transformations in the Si-O-Ca system: silicon-calcium via solar energy. Sol Energy 181(3):414–423

Fernández-González D, Prazuch J, Ruiz-Bustinza I, González-Gasca C, Piñuela-Noval J, Verdeja LF (2019b) Transformations in the Mn-O-Ca system using concentrated solar energy. Sol Energy 184(5):148–152

Pero-Sanz JA, Fernández-González D, Verdeja LF (2018) Physical metallurgy of cast irons. Springer, Cham

Pero-Sanz JA, Fernández-González D, Verdeja LF (2019) Structural materials: properties and selection. Springer, Cham

Poirier DR, Geiger GH (1994) Transport phenomena in materials processing. TMS, Warrendale, Pennsylvania

Verdeja LF, Ballester A, Sancho JP, González R (2014) Refractory and ceramic materials. síntesis, Madrid, Spain

Production of Iron by Reduction with Gas

5

5.1 Introduction

The scrap or recycled steel is a basic raw material for the electric steel works. The main problem in certain regions of the world is the shortage of this resource because of both the constant increase in the construction of electric steel plants and the simultaneous closure of blast furnaces, although in recent periods, furnaces of bigger diameter of hearth are being rebuilt together with the construction of new blast furnaces of smaller size (100 500 m^3 of capacity) and productions in the range of 100,000–350,000 ton/year located in China, India, and Brazil, which can help in solving the shortage of this raw material.

The installations of direct reduction are a technological alternative for the production of high-quality synthetic scrap free of residual impurities (mainly Cu, Sn, Ni, and to a lesser extent, V, Ti, Cr, and Mo). All the processes of direct reduction of iron ore (*DRI, Direct Reduction Iron*) are based on the following points:

- Use of natural gas or gasified coal as reductant agents. That is to say, the simulation of the process of reduction with gas can be performed using gases: methane (CH_4), hydrogen (H_2), and carbon monoxide (CO).
- Working at T < 1150 °C mean carbon percentages in the iron sponge lower than 1.5% (this temperature of 1150 °C is related with the temperature of the eutectic point in the Fe-C diagram, see Fe-C diagram (p. 3) in Pero-Sanz et al. 2018).

Table 5.1 presents a summary of the different processes used to reduce iron ore as a function of the ferric charge, of the reductant agent and of the available reactive phases "solid–gas" or "solid–liquid (metallic melt)–gas" (Babich et al. 2008, pp. 371–402; Naito et al. 2015, pp. 31–33). Despite Table 5.1 assigns a certain type of ferric charge being fed to the processes Midrex and HYL-III, the types of raw materials or residues different from the iron that they can nowadays process are very broad.

© The Editor(s) (if applicable) and The Author(s), under exclusive license to
Springer Nature Switzerland AG 2020
J. I. Verdeja González et al., *Operations and Basic Processes in Ironmaking*,
Topics in Mining, Metallurgy and Materials Engineering,
https://doi.org/10.1007/978-3-030-54606-9_5

Table 5.1 Alternatives for the reduction of the iron ore

Product (ferric charge)	Smelting reduction		Direct reduction (DRI)		
	Pig iron		Sponge iron DRI/HBI		Iron carbide (Fe₃C)
	Coke + others	Coal/O₂	Natural gas	Coal	Natural gas
Coarse fraction	Blast furnace	–	Midrex	SL/RN	–
Pellets	Blast furnace	Corex	HYL-III	DRC	–
Sinter	Blast furnace	AISI	Arex	Fastmet	–
Green pellets	–	–	–	Inmetco	–
Fines of iron ore		Finex	–	–	Iron carbide
		CCF	Fior	–	–
		DIOS	Finmet	Circofer	–
		Hismelt	Circored	Comet	–
		–	Ausmelt	Spirex	–
		–	Romelt	–	–

The mechanisms of reduction of the ferric charge are two:

- Direct reduction with carbon or coke:

$$Fe_2O_3(s) + 3C\ (s) \rightarrow 2Fe\ (l) + 3CO\ (g) \tag{5.1}$$

- Indirect reduction with gas (H_2, CO, CH_4):

$$Fe_2O_3(s) + 3H_2(g) \rightarrow 2Fe\ (s) + 3H_2O\ (g) \tag{5.2}$$

$$2Fe_2O_3(s) + 3C\ (g) \rightarrow 4Fe\ (s) + 3CO_2(g) \tag{5.3}$$

$$4Fe_2O_3(s) + 3CH_4(g) \rightarrow 8Fe\ (s) + 3CO_2(g) + 6H_2O(g) \tag{5.4}$$

Nevertheless, despite Eq. (5.1) is correct for calculations of free energy or enthalpy of the hematite direct reduction, from the kinetics point of view, this reaction cannot be taken into practice with a high degree of metallization. The metallization of the solid–solid and solid–liquid reactions is low, the temperature of the solid–liquid reactions is identified with the liquidus temperatures of the SiO_2-CaO-Fe_2O_3 ternary diagram. For that reason, the Boudouard mechanism is involved in the kinetic mechanism of direct reduction with coke, and the reaction takes place in two steps:

(1) Reduction of the hematite with CO:

$$Fe_2O_3(s) + 3CO(g) \rightarrow 2Fe\ (l) + 3CO_2(g) \tag{5.5}$$

(2) Gasification of the carbon (Boudouard equilibrium):

$$3CO_2(g) + 3C\ (s) \rightarrow 6CO(g) \tag{5.6}$$

The addition of the reactions (5.5) and (5.6) is equal to that of the process that thermodynamically expresses the mechanism of reduction with coke (Eq. 5.1).

Under the suitable reduction potentials and working temperatures of $\simeq$1100 °C, the thermodynamics and the kinetics of the processes of sponge iron obtaining are developed according to the nature of the reactive phases, dense solids (Chap. 3, pelletization) or porous solids (Chap. 2, sintering, and Chap. 4, coking). To achieve that a reaction could take place, apart from a negative value of free energy, it is necessary an excess of reactive (reductant gas) that could ensure the total conversion of the charge. This leads to lay out the equations of reduction with excess of reductant gas, and the subsequent possibility of modifying the extensive thermodynamic properties, concretely the enthalpy of reaction.

If we want to approach the real energetic values to those obtained with the reduction equations, it is necessary to set out and calculate the chemical equations considering the excess of reductant gas that is available in the process. In Appendix 2, it is indicated the variation of standard enthalpy with the temperature for different ironmaking compounds.

The mechanism of direct reduction of the hematite at 1700 K (1427 °C), using an excess of reductant gas, considering a CO/CO_2 ratio in the products of reaction equal to one, is represented by means of the following chemical reaction:

$$Fe_2O_3(s) + 2C(s) \rightarrow CO(g) + CO_2(g) + 2Fe(s) \tag{5.7}$$

According to the data of Appendix 2, the variation of the standard enthalpy for the reaction (5.7) is:

$$\Delta_r H^0_{(reaction5.7)}\left(kJ\ mol^{-1}\right) = 334.3 - 0.0306 \cdot T/K \tag{5.8}$$

$$\Delta_r H^0(1700\ K) = 282.28 kJ\ mol^{-1}\ of\ Fe_2O_3 \tag{5.9}$$

which referred to one ton of iron is:

$$\Delta_r H^0_{(reaction\ 5.7)}(1700\ K) = 2.52 \times 10^6 kJ\ t^{-1}\ of\ Fe \tag{5.10}$$

When the excess of reductant agent implies that the CO/CO_2 ratio in the gases is equal to 4, we have:

$$2Fe_2O_3(s) + 5C(s) \rightarrow 4CO(g) + CO_2(g) + 4Fe(s) \tag{5.11}$$

Therefore, the value of the standard enthalpy is:

$$\Delta_r H^0_{(reaction\ 5.11)}\left(kJ\ mol^{-1}\right) = 835.6 - 0.0745\ T/K \tag{5.12}$$

$$\Delta_r H^0_{(reaction\ 5.11)}(1700\ K) = 708.95\ kJ\ mol^{-1}\ of\ Fe_2O_3$$
$$= 3.165 \times 10^6 kJ\ t^{-1}\ of\ Fe \tag{5.13}$$

For the mechanism of indirect reduction or with gas, which can be exemplified with the reduction of the Fe_2O_3 at 1200 K, and if the excess of reductant gas causes a CO/CO_2 ratio equal to one:

$$Fe_2O_3(s) + CO(g) \rightarrow 3CO(g) + 3CO_2(g) + 2Fe(s) \tag{5.14}$$

we have:

$$\Delta_r H^0_{(reaction\ 5.14)}\left(kJ\ mol^{-1}\right) = -35.7 - 0.004\ T/K \tag{5.15}$$

$$\Delta_r H^0_{(reaction\ 5.14)}(1200\ K) = -40.5\ kJ\ mol^{-1} of\ Fe_2O_3$$
$$= -361 \cdot 10^6 kJ\ t^{-1} of\ Fe \tag{5.16}$$

In the reduction with gas, when the excess of reductant gas leads to a CO/CO_2 ratio equal to 4:

$$Fe_2O_3(s) + 15CO(g) \rightarrow 12CO(g) + 3CO_2(g) + 2Fe(s) \tag{5.17}$$

when processed a quantity of iron equal to that indicated in Eq. (5.14), the heat released to the system or to the furnace (exothermic reaction) per ton of iron is the same.

As it was previously indicated, the reduction of hematite with coal is an endothermic process, while with gas is slightly exothermic. However, the reduction with carbon monoxide is an exception. The enthalpies of reduction are endothermic both with the hydrogen and with the methane (reduction with gases). All those reduction processes where either hydrogen or natural gas is involved are endothermic. With the purpose of maintaining the temperature of the solid–gas reaction, and also to avoid a low productivity in the installation due to the diminishing of the process rate, systems to supply heat to the furnace or to the reduction reaction must be incorporated in the DRI processes.

Finally, Figs. 5.1 and 5.2 show the phase diagrams of the Fe-O-C system, compatibles with the volumetric fraction of carbon monoxide C_{CO}(%vol.) and

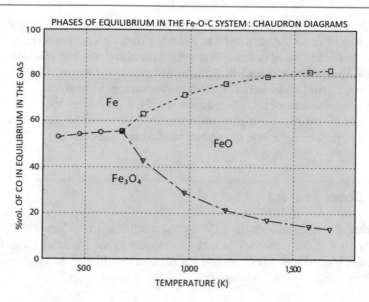

Fig. 5.1 Fe-O-C system. Domains with predominance of Fe-FeO-Fe$_3$O$_4$ in equilibrium with the CO at different temperatures

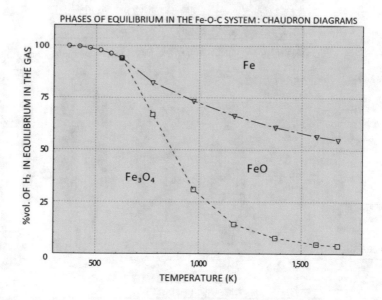

Fig. 5.2 Fe-O-C system. Domains with predominance of Fe-FeO-Fe$_3$O$_4$ in equilibrium with the H$_2$ at different temperatures

C_{H_2} (%vol.) in the gas as a function of the temperature, respectively. Diagrams that are also known as Chaudron diagrams. From the data collected in Appendix 2 and represented in Fig. 3.3, it is possible to obtain the corresponding equations for the Fe_3O_4-Fe, Fe_3O_4-FeO and FeO-Fe equilibriums in the presence of CO and H_2. The Fe_2O_3-Fe_3O_4 equilibrium is not represented because, principally, the curve is not distinguished from the x-axis. Thermodynamically, it is possible to confirm that the hydrogen is the reductant at high temperature for the iron oxides, and kinetically, it has a better performance than the carbon monoxide (Exercise 3.4).

5.2 Midrex Process

This process was developed in 1950 by the Midland Ross Corporation (USA). Later, in 1974, Korf Industries bought the division to the Midland Ross. Finally, Kobe Steel (Japan) took the control of Midrex in 1984. Figure 5.3 summarizes the main characteristics of this technology:

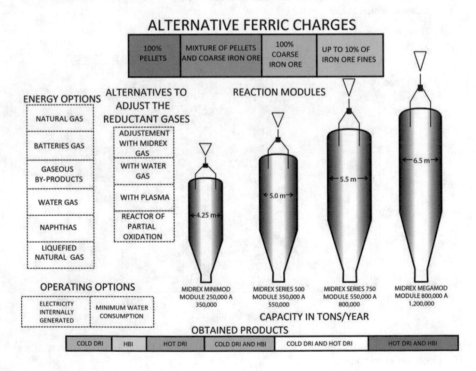

Fig. 5.3 General characteristics of the midrex process

- Different alternatives for the processing of ferric charges with varied characteristics.
- Different reductant and energy options.
- It is possible the production of different products with the same equipment: DRI (Direct Reduced Iron) and HBI (Hot Briquetted Iron).
- It has different modules of reduction: from 250,000 t/year to 1.2×10^6 t/year.

The Midrex is the most important commercial alternative to produce iron by reduction with gas. The main advantages of the Midrex process are:

- Good technological and operational reliability of the process.
- Acceptable operating costs.
- Low values of contamination.
- Utilization of broad number of energy resources.

All previously indicated contrast with some of its main drawbacks:

- The price of the natural gas in some regions.
- The high-quality that is required for the ferric charge (coarse fraction of the iron ore or pellets).

The order of magnitude of the investment required to install a Midrex unity of 1.2×10^6 t/year is around 190 million of dollars, which is considered a competitive quantity if compared with the ~ 1000 million of dollars of investment required to produce 2.5 millions of tons of pig iron (sinter plant, coking ovens, and blast furnace).

The potential market of the process includes those geographical areas that have both high-quality iron ores and natural gas resources at a competitive price (Latin America, Middle East, India, Australia, and USA).

5.3 HYL-III Process

Hojalata y Lámina (HYLSA) from Monterrey, Mexico, developed a discontinuous process for the reduction of the iron ore with gas (HYL-I) whose first commercial plant to produce sponge iron (DRI) with a capacity of 100,000 t/year was built in 1957.

The most relevant development of the process took place in 1980 with the replacement of the discontinuous reactors by a pressurized shaft furnace (Fig. 5.4), which led to the current name of the process, HYL-III.

The HYL-III process is, after the Midrex process, the most important alternative to produce DRI or HBI. One of the most noticeable differences with respect to the Midrex process is that the production of reductant gas and the reduction of the ferric charge are two independent operations. One of the possibilities that the HYL-III

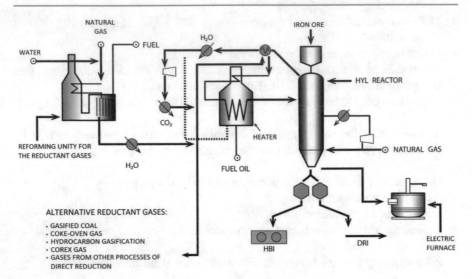

Fig. 5.4 HYL-III process: layout

process offers is that it can operate under the option of zero energy consumption (producing only the required gas) to satisfy the energy requirements of the endothermic reactions of reduction with gas.

Both the characteristics and the potential markets of the HYL-III process are similar to that of the Midrex process, although the investment required for the construction of a 1.2×10^6 t/year plant is slightly higher, ~ 220 millions of dollars.

5.4 Finmet Process

Sivensa (Venezuela) and Voest Alpine (Austria) developed a process to produce HBI using iron ore fines with a particle size <125 μm (Fig. 5.5). The construction of the first two Finmet plants ended in 1998, both with a capacity of two millions of tons of HBI per year: one in Australia and the other in Venezuela.

The Finmet process has as precedent 20 years of work in the development of the Fior project. It is possible to say that the Finmet is the result of transferring to a commercial scale, with some improvements, the laboratory work and the experience of the pilot plant done under the Fior project. The main advantages are:

- Possibility of treating iron ore fines.
- High capacity of the installations (>2 millions of tons per year).

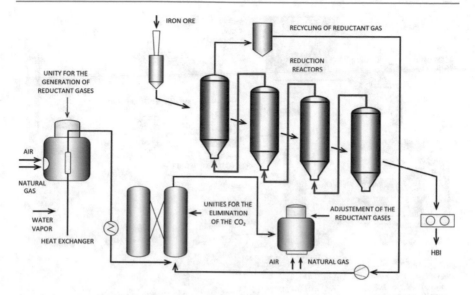

Fig. 5.5 Finmet process: flow diagram

The main disadvantages of the process are:

- High investment (~ 300 dollars per ton-year produced, which are comparable with that of the blast furnace installation).
- Evaluation of the final operating results that could be reached in the two installations existing nowadays.

5.5   Fastmet Process

Midrex started in 1980 a research program in its Technological Research Centre of Charlotte (USA) dedicated to the direct reduction of iron ore fines using non-coking coals. Later, Kobe Steel built a pilot plant that was able to process 2.5 tons per hour. The installation was patented in 1995, and pilot plant tests were performed to check the operating parameters and the quality of the materials. The most noticeable characteristics of the process are:

- Very fast reduction reactions.
- Iron ore fines are used (previously pelletized).
- Non-coking coals are employed.

The first industrial installation, 0.9 Mt per year, will be soon built in the USA by the Kobe Steel and the Mitsui (Fig. 5.6).

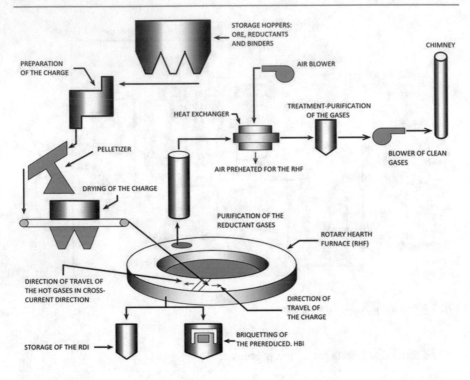

Fig. 5.6 Fastmet process: flow diagram

The Fastmet process has an installation cost of around 150 dollars per ton-year produced, the lowest from all the DRI production processes. However, the process must be improved in the following points:

- The pollutant emissions, which will mean an increase in the installation and production investment.
- Possible limited quality of the produced DRI (as a result of the sulfur and ash coming from the coal).
- Excessive costs of the organic binder.

5.6   Iron Carbide Process

The inclusion of techniques of oxygen injection in the refining standards of the electric furnace makes necessary the utilization, apart from the scrap or the DRI as ferric charge supply, of other raw materials carrying carbon: the pig iron or the Fe_3C (iron carbide, also known as cementite).

Developed in the seventies by the Iron Carbide Development Corporation and the Hazen Research Institute from Colorado (USA), the iron carbide process obtains an intermetallic compound of cementite in a fluidized bed using iron ore fines and natural gas enriched with hydrogen.

The fluidized bed works at 600 °C and 4 atmospheres. The reductant gas constantly flows through the 2000 tuyeres of that fluidized bed that the Iron Carbide Holdings and the Nucor Steel have in operation since July 1994, in Trinidad–Tobago. The factory with a capacity to produce 300,000 t/year was able to produce only 10,000 t in 1995 and 100,000 t in 1996. The problems resulting from the materials design for the heat exchangers of the plant (they must resist the abrasive action at high temperature of the Fe_3C) caused shutdowns of the process that were reflected in the low annual productions. However, the Iron Carbide and the Nucor built (1998) a new plant in Corpus Christi (Texas, USA) with a capacity of 600,000 t/year. The investment was around 150 dollars per ton produced every year.

The weakness of the process arises from the values that the investment and operating cost might reach (high specific energy consumption), while the positive aspect is the low cost of the iron ore fines.

The total conversion of the hematite, Fe_2O_3, into cementite, Fe_3C, takes place in a fluidized bed after a holding time of 16–20 h. The chemical reaction is:

$$3Fe_2O_3(s) + 5H_2(g) + 2CH_4(g) \rightarrow 2Fe_3C(s) + 9H_2O(g) \qquad (5.18)$$

From the kinetic point of view, the reaction is slow (long times of permanence of the material in the fluidized bed) and the controlling step might be the diffusion of the carbon inside of the particles of iron oxide whose size must not exceed the millimeter.

As in most of the reduction processes, the enthalpy is positive (endothermic reaction). The first estimations point out that to obtain a ton of iron carbide, the energy required would be of 13 GJ t^{-1}, and studying the possible catalytic effects that certain heavy metals, which were harmful neither for the process nor for the subsequent operations nor for the final quality of the steel, might be convenient to improve the knowledge and the kinetics of the process of the iron carbide formation at low temperature.

5.7 Solved Exercises

Exercise 5.1 Hydrogen with 1% CO_2 is used, 40,000 kg h^{-1}, to reduce the hematite according to the following reaction:

$$Fe_2O_3(s) + 3H_2(g) \rightarrow 3H_2O(g) + 2Fe(s) \qquad (5.19)$$

In the circuit, the hydrogen is recycled and the release, P, of carbon dioxide, CO_2, is performed to avoid that its concentration was higher than 2.5 wt% in the gas. If the relation of recirculated gas flow to the reduction reactor, R, with respect to the feeding of reductant gas, F, is 4/1, calculate the quantity and composition of the gases that is necessary to purge or remove from the circuit.

Data: Although it is assumed that the treated mineral is 100% hematite, the metallization yield by the gas is of 95%.

The current of entrance of 100% hematite iron ore, and the stoichiometry of the reduction reaction provide the composition of the solid charge that leaves the reduction with gas-DRI reactor, iron sponge and non-transformed hematite:

- Sponge iron obtained:

$$\frac{40,000\,\text{kg}\,\text{Fe}_2\text{O}_3}{h} \cdot \frac{2 \cdot 55.85\,\text{kg}\,\text{Fe}}{(2 \cdot 55.85 + 3 \cdot 16)\,\text{kg}\,\text{Fe}_2\text{O}_3} \cdot 0.95 = 26,579\,\text{kg}\,\text{Fe}\,\text{h}^{-1}$$

(5.20)

- Non-transformed hematite ore:

$$\frac{40,000\,\text{kg}\,\text{Fe}_2\text{O}_3}{h} \cdot 0.05 = 2000\,\text{kg}\,\text{Fe}_2\text{O}_3\,\text{h}^{-1} \qquad (5.21)$$

The requirements of hydrogen for the hematite reduction are calculated as follows:

$$\frac{40,000\,\text{kg}\,\text{Fe}_2\text{O}_3}{h} \cdot \frac{3 \cdot 2\,\text{kg}\,\text{H}_2}{(2 \cdot 55.85 + 3 \cdot 16)\,\text{kg}\,\text{Fe}_2\text{O}_3} \cdot 0.95 = 1428\,\text{kg}\,\text{H}_2\,\text{h}^{-1} \quad (5.22)$$

The feeding of gas supplied to the system, F, has a composition in mass of 99% H_2 and 1% CO_2. For that reason, the quantity of gas, F, is:

$$1428\,\text{kg}\,\text{H}_2\,\text{h}^{-1} \cdot \frac{100\,\text{kg of reductant gas}\,F}{99\,\text{kg}\,\text{H}_2} = 1442\,\text{kg of reductant gas}\,F\,\text{h}^{-1}$$

(5.23)

To calculate the current of gas recirculated, R, to the reduction reactor, the ratio indicated in the enunciation is used, where R/F is equal to four: the mass flow of R is of 5768 kg h^{-1} of gas.

With the above-calculated values, the calculation of the gas current that leaves the reduction reactor can be performed because the values of the three currents that enter–leave the DRI reactor are known. The mass flow of the current of gas, G-DRI, enriched with water vapor that leaves the furnace is:

$$G - DRI = \frac{40000 \, \text{kg Fe}_2\text{O}_3}{h} + \frac{5768 \, \text{kg gas R}}{h} - \frac{2000 \, \text{kg Fe}_2\text{O}_3}{h} - \frac{26579 \, \text{kg Fe}}{h}$$

$$= \frac{17189 \text{kg}}{h}$$

$$(5.24)$$

To this gas current of exit from the reactor, DRI, it is incorporated the quantity of water vapor consequence of the reduction process:

$$\frac{26579 \, \text{kg Fe}}{h} \cdot \frac{3 \cdot 18 \, \text{kg H}_2\text{O}}{2 \cdot 55.85 \, \text{kg Fe}} = 12849 \, \text{kg H}_2\text{O h}^{-1} \qquad (5.25)$$

Equally, to this current of exit from the reactor, it is necessary to subtract the quantity of hydrogen consumed by the reduction of the total quantity of hydrogen of the current, R; thus, the mass flow of hydrogen in G-DRI is:

$$\frac{5768 \, \text{kg gas R}}{h} \cdot 0.975 - \frac{1428 \, \text{kg H}_2}{h} = 4196 \, \text{kg H}_2 \, \text{h}^{-1} \qquad (5.26)$$

Consequently, the mass percentage composition of the G-DRI gas contains 74.75% H_2O, 24.41% H_2, and 0.84% CO_2. In a first stage, all the water vapor from the G-DRI condense: 12,849 kg H_2O h^{-1} to later subject the 4340 kg of dry gas h^{-1} to a partial discharge of CO_2 (milk of lime): 14 kg CO_2 h^{-1}. Finally, the 4326 kg of dry gas h^{-1} and purge of CO_2 are joined to the 1442 kg h^{-1} of the feeding of reductant gas, F, to provide the 5768 kg h^{-1} of recycled gas, R, to the reduction furnace.

Commentaries: With the flow diagram of Fig. 5.7 for a facility able to produce 233,000 tons of iron sponge per year (26,579 kg Fe h^{-1}), the following ironmaking operations might be pointed out:

- Recycling of all the gas that leaves the reduction unity.
- Requirement of purging of all the oxidizing gases (water vapor and carbon dioxide), as an increase in their concentration will cause a reduction of the driving force of the reaction and a diminishing of the reduction rate (Exercises 3.4 and 5.4).

Finally, it is necessary to point out that, habitually, all the poor gas, which is obtained in a DRI furnace, is used in a thermal power station located close to the iron and steelmaking installations. Consequently, the DRI plants are usually connected to a thermal power station.

Exercise 5.2 Calculate the coal consumption (coke) and the thermal energy that is required to obtain one ton of iron via processing hematite with 100% Fe_2O_3 and working under the following reduction mechanism:

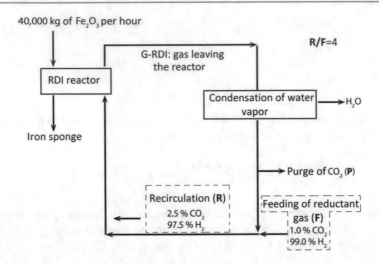

Fig. 5.7 Flow diagram of a plant for the production of iron sponge

- Composition and temperature of the materials that enter in the system (at 25 °C): Fe_2O_3 (100%), C (100%) and air (76.7% N_2 and 23.3% O_2 mass).
- Composition and temperature of the materials that leave the system: 1000 kg Fe and 45 kg C at 1500 °C. Residual reductant gas formed by CO, CO_2, and N_2.
- 35% of the iron is obtained by reduction with carbon:

$$FeO(s) + C(s) \rightarrow Fe(l) + CO(g) \tag{5.27}$$

- 65% of the remaining iron is obtained by the mechanism of reduction with gas:

$$FeO(s) + CO(g) \rightarrow Fe(s) + CO_2(g) \tag{5.28}$$

- All the FeO (stoichiometric wustite) that is required by the mechanisms of reduction with gas and coke is produced by the following reaction:

$$Fe_2O_3(s) + 2CO(g) \rightarrow 2FeO(s) + CO_2(g) + CO(g) \tag{5.29}$$

To calculate the coke consumption required for the mechanism of Fe_2O_3 reduction that is indicated in the enunciation, the reactions involved in the smelting reduction furnace are grouped in two blocks:

(a) Reactions at 1500 °C or higher temperatures:

$$FeO(s) + C(s) \rightarrow Fe(l) + CO(g) \tag{5.30}$$

$$C(s) + \frac{1}{2}O_2(g) \rightarrow CO(g) \tag{5.31}$$

(b) Reactions between 600 and 1150 °C:

$$FeO(s) + CO(g) \rightarrow Fe(s) + CO_2(g) \tag{5.32}$$

$$Fe_2O_3(s) + 2CO(g) \rightarrow 2FeO(s) + CO_2(g) + CO(g) \tag{5.33}$$

From the block of reactions at temperature $T \geq 1500$ °C, it is possible to calculate:

1. The quantity of carbon (coke) required to achieve the carbothermal reduction of the wustite, reaction (5.30), when only 35% of the iron is obtained by this reduction mechanism:

$$\frac{1\,mol\,C}{1\,mol\,Fe} \cdot \frac{1\,mol\,Fe}{55.85\,g\,Fe} \cdot \frac{12\,g\,C}{1\,mol\,C} \cdot \frac{1\,kg\,C}{1000\,g\,C} \cdot \frac{10^6\,g\,Fe}{1\,t\,Fe} \cdot 0.35 = 75\,kg\,C\,t^{-1}Fe \tag{5.34}$$

2. The quantity of CO produced in the reduction with coal and the Fe required:

$$\frac{1\,mol\,CO}{1\,mol\,Fe} \cdot \frac{28\,g\,CO}{1\,mol\,CO} \cdot \frac{1\,mol\,Fe}{55.85\,g\,Fe} \cdot \frac{1\,kg\,CO}{1000\,g\,CO} \cdot \frac{10^6\,g\,Fe}{1\,t\,Fe} \cdot 0.35 = 175\,kg\,CO\,t^{-1}Fe \tag{5.35}$$

$$\frac{1\,mol\,FeO}{1\,mol\,Fe} \cdot \frac{71.85\,g\,FeO}{1\,mol\,FeO} \cdot \frac{1\,mol\,Fe}{55.85\,g\,Fe} \cdot \frac{1\,kg\,FeO}{1000\,g\,FeO} \cdot \frac{10^6 g\,Fe}{1\,t\,Fe} \cdot 0.35 = 450\,kg\,FeO\,t^{-1}\,Fe \tag{5.36}$$

From the reactions in the range of temperatures 600–1150 °C, the quantity of carbon monoxide required will be:

1. To produce the required FeO, from the hematite, Fe_2O_3, of the feeding, two moles of CO are necessaries to obtain other two of FeO, reaction (5.33):

$$\frac{2\,mol\,CO}{2\,mol\,FeO} \cdot \frac{28\,g\,CO}{1\,mol\,CO} \cdot \frac{1\,mol\,FeO}{71.85\,g\,FeO} \cdot \frac{71.85\,g\,FeO}{55.85\,g\,Fe} \cdot \frac{1\,kg\,CO}{1000\,g\,CO} \cdot \frac{10^6\,g\,Fe}{1\,t\,Fe} = 501\,kg\,CO\,t^{-1}Fe \tag{5.37}$$

2. To reduce the FeO with CO and generate CO_2, reaction (5.32), and to produce 65% of the iron:

$$\frac{1\,mol\,CO}{1\,mol\,FeO} \cdot \frac{28\,g\,CO}{1\,mol\,CO} \cdot \frac{1\,mol\,FeO}{1\,mol\,Fe} \cdot \frac{1\,mol\,Fe}{55.85\,g\,Fe} \cdot \frac{1\,kg\,CO}{1000\,g\,CO} \cdot \frac{10^6\,g\,Fe}{1\,t\,Fe} \cdot 0.65$$
$$= 326\,kg\,CO\,t^{-1}Fe \tag{5.38}$$

However, it is necessary to take into account that, from this total quantity of carbon monoxide that is required to reduce the Fe_2O_3 into FeO and the FeO into Fe, the quantity of CO generated by the mechanism of direct reduction with carbon in the zone of high temperature must be subtracted, reaction (5.30). For that reason, the quantity of CO will be:

$$501 + 326 + 175 = 652\,kg\,CO\,t^{-1}\,Fe \tag{5.39}$$

To generate this quantity of carbon monoxide, the partial combustion of the carbon at temperatures equal or higher than 1500 °C should be taken into account, reaction (5.31):

$$\frac{1\,mol\,C}{1\,mol\,CO} \cdot \frac{12\,g\,C}{1\,mol\,C} \cdot \frac{1\,kg\,C}{1000\,g\,C} \cdot \frac{1\,mol\,CO}{28\,g\,CO} \cdot \frac{1000\,g\,CO}{1\,kg\,CO} \cdot \frac{652\,kg\,CO}{1\,t\,Fe} = 279\,kg\,C\,t^{-1}\,Fe$$
$$\tag{5.40}$$

Consequently, the coal quantity (ironmaking coke) required to reduce one ton of iron, under the conditions and mechanisms of reaction established in the enunciation, will be:

$$279 + 75 + 45 = 399\,kg\,C\,t^{-1}\,Fe \tag{5.41}$$

The calculation might be repeated if we consider that the mechanism of reduction with carbon (coke), described by the reaction (5.30), was responsible of the 15% or the 65% of the obtained iron. The results of the carbon (coke) requirements in the three situations that were studied (15, 35, and 65%) are shown in Table 5.2.

To analyze the reduction process from the energetic point of view, the percentage of direct reduction with carbon (coke) is of the 35%; in Appendix 2, we provide the standard enthalpy of four of the most relevant reactions, three exothermic and one endothermic:

Table 5.2 Consumption of carbon as a function of the percentage of reduction with carbon

Percentage of reduction with carbon (%)	15	35	65
Consumption of carbon in kg per ton of Fe	441	399	335

1. The partial combustion of the carbon at 2200 °C (flame temperature in the tuyeres of the blast furnace, Sects. 6.2 and 6.3), reaction (5.31):

$$\Delta H^0(\text{reaction } 5.31)(\text{kJ}) = -104.0 - 0.0075 \; T(\text{K})$$
$$= \Delta H^0(\text{reaction } 5.31)(2200 \; °C) = -122698 \, J \, (\text{mol C})^{-1}$$

$$(5.42)$$

2. Reduction of the FeO into Fe with CO gas, reaction (5.32) at 950 °C:

$$\Delta H^0(\text{reaction } 5.32)(\text{kJ}) = -18.0 - 0.0022 \; T(\text{K})$$
$$= \Delta H^0(\text{reaction } 5.32)(950°C) = -20691 \, J \, (\text{mol Fe})^{-1}$$

$$(5.43)$$

3. Reduction of the hematite, Fe_2O_3, into FeO by CO, reaction (5.33), at 600 °C:

$$\Delta H^0(\text{reaction } 5.33)(\text{kJ}) = -15.4 - 0.046 \; T(\text{K})$$
$$= \Delta H^0(\text{reaction } 5.33)(600 \; °C) = -55558 \, J \, (\text{mol Fe}_2\text{O}_3)^{-1}$$

$$(5.44)$$

4. Carbothermal reduction of the FeO at 1500 °C, reaction (5.30):

$$\Delta H^0(\text{reaction } 5.30)(\text{kJ}) = 167.0 - 0.0155 \; T(\text{K}) = \Delta H^0(\text{reaction } 5.30)(1500 \; °C)$$
$$= 139519 \, J \, (\text{mol Fe})^{-1}$$

$$(5.45)$$

In a first analysis, it could seem, without considering the sensible heat associated with the changes of state of the products and reagents, that the mechanism of reduction proposed for the hematite is energetically possible. Nevertheless, a better approximation to the problem is possible with the support of the HSC 5.1 (software for the analysis of both reactions and chemical equilibriums). In Table 5.3 we show, for the hypothesis of reduction of 35% of the hematite using the carbon (or that 35% of the iron is obtained by means of the reaction (5.30)), the thermal and mass characteristics of each one of the currents of entrance and exit to the smelting reduction model, as well as the energy in excess of the process, MJ t^{-1} of Fe.

Commentaries: When we talk about carbon consumption as the reductant required for the obtaining of iron from the hematite, it is not specified whether the carbon comes from the coke, the pulverized coal injection, any other combustible (fuel–oil or natural gas) or the recycled polymers. It is being introduced, with this purpose, the concept *RAR, Reduction Agent Rate* (Naito et al. 2015, p. 17).

Table 5.3 Thermal and mass characteristics of the currents of entrance and exit

Input	T (°C)	Mass (kg)	Enthalpy (MJ)
Fe_2O_3	25	1429.75	−7368.44
C	25	400.00	0.00
O_2 (g)	1200	372.40	461.21
N_2 (g)	1200	1228.00	1642.55
Output	T (°C)	Mass (kg)	Enthalpy (MJ)
Fe	1500	1000.00	1023.20
CO (g)	250	251.10	−931.60
CO_2 (g)	250	906.00	−7908.22
C	1500	45.05	112.09
N_2 (g)	250	1228.00	289.19
Balance	–	0.00	−2150.66

Other ideas that should be taken into account in the mass and energy balances, which were previously detailed and that allow putting into practice the proposed reduction mechanism with 35% of reduction with carbon are the performance in the process of solid–gas heat transfer and the carbon content of the ironmaking reductant (90% C), which lead to a value of the energy available equivalent to 60% of the -2150.66 MJ t^{-1} of Fe previously estimated using the HSC 5.1: -1291 MJ t^{-1} of Fe. However, it is necessary to subtract from this quantity the energy required for the following purposes:

(a) The mean Fe_2O_3 content in the ferric charge of a blast furnace (60% sinter, 30% pellets, and 10% grains of iron ore) is of 72%; that is, 28% of the 2151 MJ are 602 MJ.
(b) The heat losses through the walls and the hearth are estimated in 10% of the 2151 MJ: 215 MJ.
(c) The formation of slag during the process requires an energy consumption of around 276 MJ t^{-1} of pig iron (pig iron with 92% Fe means 300 MJ t^{-1} Fe).
(d) Other energy consumptions that were not considered: reduction of the Si, Mn, P, and S and H_2O of the injected air consume the remaining 174 MJ necessaries to close the balance.

Consequently, the addition of 602, 215, 300, and 174 MJ t^{-1} Fe leads to the 1291 MJ t^{-1} Fe calculated by means of the software HSC 5.1.

Exercise 5.3 The characteristics of the Fastmet process (Sect. 5.5) depend on the kinetics of the reaction between the small size iron ore particles and the carbon, for instance, between a mixture of iron ore fines and carbon, as happens with the powders collected in the blast furnace (30% C, 60% Fe_2O_3, 8% SiO_2, 1% ZnO, and 1% Na_2O). This phenomenon might be analyzed using the concepts developed by Ginstling–Brounshtein, Sect. 3.3 (Eq. 3.26). Calculate the time required for the total metallization of the iron if the reaction temperatures are 1200, 1250, or 1300 °C.

Data: Constant of Ginstling–Brounshtein, Eqs. (3.26) and (3.27), with the activation energy expressed in J:

$$k = 10^4 \cdot \exp\left(-\frac{143901}{R \cdot T}\right) \, \text{min}^{-1} \qquad (5.46)$$

We assume that the difference of CO pressure (driving force of the reduction) inside of the agglomerates (with the shape of brick or briquette, bars, nodules or pellets) is around 0.10 atm at 1300 °C. The mean global density of the agglomerates is 2.85 g cm^{-3}, and the mean size of the ensemble of particles that constitutes the agglomerates is 1.22 mm (1220 μm).

When the controlling step is the gaseous diffusion (in this case, carbon monoxide), the equation proposed by Ginstling–Brounshtein (Sect. 3.3, Eq. 3.26) is:

$$1 - \frac{2 \cdot X}{3} - (1 - X)^{\frac{2}{3}} = k \cdot t \qquad (5.47)$$

where X is the degree of reduction in a time t and k is the kinetics constant of the process. When the reduction is complete, the value of X is equal to one. In Table 5.4, we show the time, in minutes, required to achieve a reduction of the 95% and 100% at the temperatures of 1200, 1250, and 1300 °C.

One of the characteristics of the *Rotary Hearth Furnace (RHF)* (Naito et al. 2015, p. 32) is the speed of the reduction processes despite the diffusional impediments for the CO movement in the pores of the agglomerate. On the other hand, the value of the constant k, at 1300 °C, is equal to (Eq. 3.27):

$$k = 10^4 \cdot \exp\left(-\frac{143901 \, \text{J/mol}}{8.314 \frac{\text{J}}{\text{mol K}} \cdot 1573 \, \text{K}}\right) \text{min}^{-1} = 0.1665 \, \text{min}^{-1}$$

$$= \frac{2 \cdot D_{CO-e} \cdot \Delta C}{R_0^2 \cdot \rho_g} \qquad (5.48)$$

where the driving force of the reaction, ΔC, estimated by means of the pressure difference of the reductant gas with respect to the equilibrium, is a value that cannot hard be experimentally calculated:

Table 5.4 Degree of reduction as a function of the temperature

T (°C)	Degree of reduction	
	95%	100%
1200	19.9	28.7
1250	12.7	18.4
1300	8.4	12.1

$$\Delta C = \frac{\Delta P}{R \cdot T} = \frac{0.10\,\text{atm} \cdot 28\,\text{g/mol}}{0.082\,(\text{atm dm}^3/\text{mol K}) \cdot 1573\,\text{K}} = \frac{0.0217\,\text{g}}{\text{dm}^3} \cdot \frac{1\,\text{dm}^3}{10^3\,\text{cm}^3} \quad (5.49)$$
$$= 2.17 \cdot 10^{-5}\,\text{g cm}^{-3} \text{ of CO}$$

To achieve that the kinetics constant, k, of the Ginstling–Brounshtein equation could reach a value of $0.1665\,\text{min}^{-1}$, the weight that should be attributed to the CO effective diffusion coefficient, D_{CO-e}, inside of the agglomerate, would be:

$$0.1665\,\text{min}^{-1} = \frac{2 \cdot D_{CO-e} \cdot 2.17 \cdot 10^{-5}\,\text{g cm}^{-3}}{(0.061\,\text{cm})^2 \cdot 2.85\,\text{g cm}^{-3}} \rightarrow D_{CO-e} = 40.68\,\frac{\text{cm}^2}{\text{min}} \cdot \frac{1\,\text{min}}{60\,\text{s}}$$
$$= 0.678\,\text{cm}^2/\text{s}$$

$$(5.50)$$

which means correcting the diffusion of the gas in a N_2-CO-CO$_2$ atmosphere to a value of $3.835\,\text{cm}^2\,\text{s}^{-1}$ (Poirier and Geiger 1994, p. 456), with the restriction imposed by the transfer in a porous medium [open porosity of 25% and Eq. (2.46)]:

$$D_{CO-e} = 3.835\,\frac{\text{cm}^2}{\text{s}} \cdot \frac{0.25}{\sqrt{2}} = 0.678\,\text{cm}^2/\text{s} \quad (5.51)$$

Commentaries: The variation of the temperature with the coefficient of diffusion of the carbon monoxide in a gas, whose main phases are N_2-CO-CO$_2$, is:

$$D_{CO} = 9.76 \times 10^{-6}\,T^{1.75}\,\text{cm}^2/\text{s} \quad (5.52)$$

The novelty of the processes that use the *Rotary Hearth Furnace* is that as the reduction rate is fast and, on the other hand, the time of contact of the reaction gas with the formed metallic sponge is very short, the range of the kinetics mechanisms of the metallic product carburizing is very short, and there are not, for that reason, risks of liquid phases formation associated with the eutectic of the Fe-C system.

Exercise 5.4 One of the alternatives to the sponge iron, produced via solid–gas reactions (DRI), is the formation of the iron carbide, Fe$_3$C, according to the reaction (5.18). Calculate, at 600 °C, what would be the thermodynamic and kinetic variables that control the reaction.

Data: According to the HSC 5.1 database, the standard enthalpy and the standard free energy associated with the reaction (5.18) at 600 °C are:

$$\Delta_r H^0(\text{reaction 5.18}) = 434253\,\text{J}\,(2\,\text{mol Fe}_3\text{C})^{-1} \quad (5.53)$$

$$\Delta_r G^0(\text{reaction 5.18}) = -27457\,\text{J}\,(2\,\text{mol Fe}_3\text{C})^{-1} \quad (5.54)$$

Finally, the size of the iron ore, Fe_2O_3, which is treated in the fluidized bed, is 1000–3000 μm, and the total pressure of the gases in the reactor is 4.0 atm.

The enthalpy of the iron carbide synthesis at 600 °C is endothermic and, referred to the production of one ton of Fe_3C is:

$$\frac{434253 \, J}{2 \, mol \, Fe_3C} \cdot \frac{1 \, mol \, Fe_3C}{179.55 \, g \, Fe_3C} \cdot \frac{10^6 \, g}{1 \, t} = 1.21 \, GJ \, t^{-1} \, Fe_3C \qquad (5.55)$$

On the other hand, the standard free energy associated with the reaction (5.18), and its equilibrium constant at 600 °C is:

$$-\frac{27457 \, J}{2 \, mol \, Fe_3C} = -R \cdot T \cdot \ln K_{eq} \rightarrow \frac{27457 \, J}{2 \, mol \, Fe_3C} = 8.314 \frac{J}{mol \, K} \cdot 873 \, K \cdot \ln K_{eq}$$

$$(5.56)$$

$$\ln K_{eq} = 1.89 \rightarrow K_{eq} = 6.63 \qquad (5.57)$$

To know the partial pressures of the gases in the equilibrium (point of equilibrium), one of the pressures is fixed, for example, the partial pressure of the water in the gas, P_{H_2O}:

$$P_T = 4.0 \, atm = P_{H_2O} + P_{H_2} + P_{CH_4} \qquad (5.58)$$

On the other hand, we obtain the following relation if we take into account the equilibrium constant at 600 °C:

$$\frac{P_{H_2O}^9}{P_{H_2}^5 \cdot P_{CH_4}^2} = 6.63 \qquad (5.59)$$

Considering that the values of the partial pressure of equilibrium of the water in the gas $P_{H_2O}^=$ are: 0.1, 0.2, 0.8, 1.0, 1.2, and 1.75 atmospheres, it is possible to calculate, using the method of successive approximations, the methane partial pressure of equilibrium, $P_{CH_4}^=$, and also that of the hydrogen, $P_{H_2}^=$, in the gas, as it is described in Table 5.5 (atm).

Table 5.5 Equilibrium partial pressures	$P_{H_2O}^=$	$P_{CH_4}^=$	$P_{H_2}^=$
	0.1	3.894	0.006
	0.2	3.778	0.022
	0.8	2.901	0.299
	1.0	2.527	0.473
	1.2	2.092	0.708
	1.75	0	2.665

Table 5.6 Reaction times as a function of the particle radius

R (μm)	t_d (h) (diffusion)	t_q (h) (chemical reaction)	t_t (h) (total)
500	1.0×10^{-3}	1.99	2.00
1000	4.0×10^{-3}	3.98	4.01
1500	9.0×10^{-3}	5.97	6.03

It is possible to conclude that, out of the equilibrium, the partial pressures under which the process of iron carbide synthesis can take place are, approximately, 2.5 atm methane, 1.0 atm hydrogen, and 0.5 atm water vapor.

From the kinetics point of view, the formation of the iron carbide has two stages in series. The first stage is the reduction of the hematite ore using hydrogen, and the second stage is the diffusion of the carbon inside the iron sponge previously formed. It is assumed that the rate of recombination of the iron and carbon atoms in the carburized iron is very fast.

Using the unreacted core model (Sect. 3.3) and considering that the value of the hydrogen pressure at 600 °C in the gas is one atmosphere (the pressure of hydrogen in the equilibrium, equilibrium reaction (5.18), $P_{H_2}^=$, at 600 °C, is 0.15 atm), it is possible to obtain Table 5.6, where the reaction times (in hours) for particles of radius 500, 1000, and 1500 μm are collected.

On the other hand, the time required for the diffusion of the carbon in the alpha iron (ferrite) might be estimated using the following equation:

$$t = \frac{1}{3} \cdot \left(\frac{R_0^2}{D_{C-Fe-\alpha}} \right) \tag{5.60}$$

where the value of the coefficient of the carbon diffusion in the ferrite at 600 °C is equal to 1.016×10^{-7}–1.967×10^{-7} cm^2 s^{-1} (Van Vlack 1990, pp. 206–208; Pero-Sanz et al. 2018, p. 61). The times required for the diffusion of the carbon in the ferrite are summarized in Table 5.7.

However, during the formation (reduction) of the sponge iron, a considerable reduction in the particles size takes place (Exercise 3.2), and the values previously calculated might be smaller. Nevertheless, the times required for an optimal transformation of the hematite (16–20 h) are in accordance with the proposed kinetics model of reaction.

Commentaries: When the calculated energy consumption, 1.21 GJ t^{-1}, is compared with the real value, 13 GJ t^{-1}, it is possible to check that there is low energy efficiency in the technology. Concentrated solar energy might be an interesting technology for the future in order to improve the energy efficiency (Mochón et al. 2014; Fernández-González et al. 2018a, b, c, d, 2019a, b, c; Ruiz-Bustinza et al. 2013). Finally, the standard free energy associated with the reaction (5.18) at 600 °C is −30,314 J (2 mol Fe$_3$C)$^{-1}$ according to the data presented in Appendix 2.

Table 5.7 Reaction time as a function of the particle radius

R_0 (μm)	t_D (h) (Van Vlack)	t_D (h) (Pero-Sanz et al.)
500	2.28	1.18
1000	9.12	4.71
1500	20.51	10.59

$D_{C-Fe-\alpha}^{Van\,Vlack} = 1.016 \times 10^{-7}$ cm^2 s^{-1}; $D_{C-Fe-\alpha}^{Pero-Sanz\,etal.} = 1.967 \times 10^{-7}$ cm^2 s^{-1}

References

Babich A, Senk D, Gudenau HW, Mavrommatis TT (2008) Ironmaking. RWTH Aachen University, Aachen

Fernández-González D, Piñuela-Noval J, Verdeja LF (2018a) Iron ore agglomeration technologies in iron ores and iron oxide materials. In: Shatokha V (ed). Intechopen, Londres, pp. 61–80 (Chapter 4)

Fernández-González D, Ruiz-Bustinza I, González-Gasca C, Piñuela-Noval J, Mochón-Castaños J, Sancho-Gorostiaga J, Verdeja LF (2018b) Concentrated solar energy applications in materials science and metallurgy. Sol Energy 170(8):520–540

Fernández-González D, Prazuch J, Ruiz-Bustinza I, González-Gasca C, Piñuela-Noval J, Verdeja LF (2018c) Solar synthesis of calcium aluminates. Sol Energy 171(9):658–666

Fernández-González D, Prazuch J, Ruiz-Bustinza I, González-Gasca C, Piñuela-Noval J, Verdeja LF (2018d) Iron metallurgy via concentrated solar energy. Metals 8(11):873

Fernández-González D, Prazuch J, Ruiz-Bustinza I, González-Gasca C, Piñuela-Noval J, Verdeja LF (2019a) The treatment of basic oxygen furnace (BOF) slag with concentrated solar energy. Sol Energy 180(3):372–382

Fernández-González D, Prazuch J, Ruiz-Bustinza I, González-Gasca C, Piñuela-Noval J, Verdeja LF (2019b) Transformations in the Si-O-Ca system: silicon-calcium via solar energy. Sol Energy 181(3):414–423

Fernández-González D, Prazuch J, Ruiz-Bustinza I, González-Gasca C, Piñuela-Noval J, Verdeja LF (2019c) Transformations in the Mn-O-Ca system using concentrated solar energy. Sol Energy 184(5):148–152

Mochón J, Ruiz-Bustinza I, Vázquez A, Fernández D, Ayala JM, Barbés MF, Verdeja LF (2014) Transformations in the iron–manganese–oxygen–carbon system resulted from treatment of solar energy with high concentration. Steel Res Int 85(10):1469–1476

Naito M, Takeda K, Matsui Y (2015) Ironmaking technology for last 100 years: development to advanced technologies from introduction of technological know-how, and evolution to next-generation process. ISIJ Int 55(1):7–35

Pero-Sanz JA, Fernández-González D, Verdeja LF (2018) Physical metallurgy of cast irons. Springer, Cham

Poirier DR, Geiger GH (1994) Transport phenomena in materials processing. TMS, Warrendale, Pennsylvania

Ruiz-Bustinza I, Cañadas I, Rodríguez J, Mochón J, Verdeja LF, García-Carcedo F, Vázquez AJ (2013) Magnetite production from steel wastes with concentrated solar energy. Steel Res Int 84 (2):207–217

Van Vlack LH (1990) Elements of materials science and engineering, 6th edn. Addison-Wesley, New York

Production of Iron in the Blast Furnace

6

6.1 Introduction

Analyzing the history of the ironmaking allows better understanding the current production technologies and their future developments. Humans started to reduce iron ores in around 1500 BC in clay or stone recipients using charcoal as both reductant reagent and combustible. This technique was used in the Hittite Kingdom, in Asia Minor or Anatolia, which makes up most of the modern-day Turkey. There are data that confirms the presence of metallic iron before 1200 BC but coming from meteorites or as a result of localized reductions caused by forest fires.

Primitive civilizations found in the natural air flow an instrument to reach acceptable coal combustion rates. They succeeded their purpose locating the furnaces in the hillside of mountains or near to cliffs. For 28 centuries (1500 BC–1300 AC), the sponge iron was manufactured using this technique, making up the most relevant example of this period, known as Catalan Forge.

The sponge iron produced using this technique should be hit to segregate out the occluded slag and give the final shape to the part. In the Catalan Forge, forefather of the current processes of direct reduction (Sect. 5.1), the ore was located over the charcoal layer, and the ensemble was subjected to the heating process using the heat released in the charcoal partial combustion. Special attention should be paid to achieve both an incomplete combustion and a reductant environment in the furnace favoring the formation of the sponge iron, which after the forging process had the following approximate chemical composition: 0.10% C, 1.0% Si, 0.015% P, 0.020% S, and traces of Mn.

During the Fourteenth century, the decisive advance in the ironmaking appeared as the first blast furnaces were built in Central Europe. Metallurgists tried, successfully, to force the entrance of wet air in the furnace using bellows. Consequently, a noticeable improvement in the kinetics of the coal combustion, and an increase of the operating temperature were promoted. The capacity of production was increased, and the size of the furnace was redesigned again when the strength of the man was replaced in the movement of the bellows by the hydraulic energy.

J. I. Verdeja González et al., *Operations and Basic Processes in Ironmaking*,
Topics in Mining, Metallurgy and Materials Engineering,
https://doi.org/10.1007/978-3-030-54606-9_6

The sponge iron produced in the shaft furnace, by means of the iron ore reduction using carbon monoxide, can reach enough temperature to dissolve the carbon and form austenite (interstitial solid solution, carbon in the γ iron). If the quantity of carbon dissolved in the iron reached around 4.0%, Fe–C alloy is completely liquid at 1154 °C and a liquid iron alloyed with carbon that is known as pig iron is obtained as opposed that in the case of the Catalan Forge. The iron alloyed with carbon is casted into molds with different shapes (ingots), where the solidification of the liquid takes place. This way, the basis of one of the sectors at the first line of the technological development was established: cast irons (Pero-Sanz et al. 2018, pages ix–x).

The solidified product was brittle because the carbon in the pig iron appeared in the microstructure of the cast iron either as graphite or associated with the hard intermetallic composite known as cementite, Fe_3C. The form of the carbon in the cast iron, free, or associated with the Fe_3C, is related with the cooling rate or the presence of other metallic elements dissolved in the liquid iron, such as silicon or magnesium. Modern cast irons have reached a high degree of sophistication, and the brittle characteristic of the primitive irons has been surpassed, mainly due to the knowledge about the control of the microstructures by means of alloying elements and heat treatments (Pero-Sanz et al. 2019, Chap. 11; Pero-Sanz et al. 2018, pages 314–323).

Historically, the toughness of the cast irons could be only improved reducing the carbon content. The refinement process was achieved using horizontal furnaces where the metal was heated in contact with the air. They ensured that the thickness of the refined metals was not very big to allow the oxidation of the carbon into CO along the entire metallic mass, and to avoid a considerable oxidation of the iron. The time and the operations in the primitive process of a cast irons refining were long and tedious.

In the eighteenth century, another important milestone in the iron and steel-making industry was the substitution of the charcoal by the coke in the role of combustible and reductant agent in the shaft furnaces. This innovation resulted basic for the development of one of the fundamental pillars of the modern iron-making industry, the production of big quantities of unpurified liquid iron (*pig iron*), which is refined by oxidizing in the oxygen converter process. From the eighteenth century, the most important changes were focused on the technologies used to refine the pig iron to obtain steel. Meanwhile, the blast furnace has assimilated new technologies of process and production control, aimed at increasing the capacity and diminishing the operating costs. Blast furnaces able to produce 12,000 tons of pig iron per day have been built, and they had a hearth, in the zone of tuyeres, of 14 m (maximum at the beginning of the twenty-first century for the blast furnace).

In the last years, alternative processes to the sequence blast furnace–converter used to produce steel have been developed. These alternatives include the direct reduction of iron ore without using coking, sintering, and pelletizing plants. One of the factors that can guarantee the survival of the blast furnace, independently of the incorporation of all the advanced technologies in materials and control of processes, is the appearance of important reserves of coking coal. However, the environmental legislation in the industrialized countries is more and more restrictive, particularly that legislation related with the sulfur, nitrogen oxides, and cyanides emissions. Nowadays, pilot phase tests in big coke batteries are being carried out to control, in a more rational way, the pollution released through the numerous doors existing in the current coke batteries.

The production of sponge iron is other of the alternatives to the blast furnace (Chap. 5). The iron obtained by reduction of the ore at low temperature has characteristics very similar to those of the traditional Catalan Forge. Even when the process is similar, the technology is completely different due to the knowledge available about the physical–chemical changes that are produced in the iron ore when the temperature is changed in a reductant environment. However, as opposed to previous times, the sponge iron is not used to obtain bars or steel ingots by forge. It is used as raw material in electric power furnaces as replacement of the scrap. As single exception to this rule, it is possible to find the manufacture of iron or iron alloy powders (powder metallurgy) that are subsequently encapsulated and hot extruded as bars.

6.2 General Characteristics of the Blast Furnace

Modern blast furnaces might be considered the successors of the traditional shaft or cupola furnaces. They operate in countercurrent by means of a descending charge, which comprises a mixture of iron oxide, coke and fluxes, that is heated and reacts with the ascending flow of reductant gases. Figure 6.1 shows a general scheme of a blast furnace with all its auxiliary facilities, air heating stoves (wind), and industrial unit of casting, where the separation metal (hot metal, pig iron)–slag takes places as a result of the density difference.

In the blast furnace, the oxygen associated with the iron ore is eliminated using carbon as CO/CO_2. The metal tends to separate from: the most difficultly reducible oxides and the nonmetallic compounds associated with the gangue of the ferric charge and the coke ashes. To separate the metal–slag phases, it is necessary to melt the charge and achieve an oxidized liquid phase that contained the undesired impurities and this phase containing the impurities must be also insoluble in the high carbon liquid iron (pig iron).

To optimize the process, it operates at the minimum temperature at which both the smelting of the charge and the thermodynamics and kinetics of the reduction are compatibles. The iron contaminated with carbon, silicon, manganese, phosphorus, and sulfur, coming from the blast furnace, reaches a liquidus temperature close to 1475 K (1202 °C). The basic slags of the blast furnace (Fig. 6.2, SiO_2–Al_2O_3–CaO ternary diagram) start to soften at 1265 °C (anorthite–wollastonite–gehlenite ternary eutectic) while the acid slags begin to melt at 1170 °C (anorthite–wollastonite–tridymite ternary eutectic). However, in basic slags it is difficult to obtain liquidus

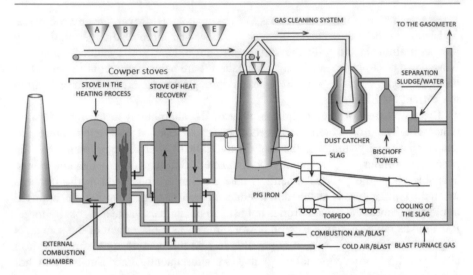

Fig. 6.1 General scheme of the blast furnace with the auxiliary facilities: A, sinter hopper; B, pellets hopper; C, iron ore hopper; D, coke hopper; E, supplementary raw materials hopper

temperatures lower than 1380 °C and in acid slags than 1300 °C. Moreover, the composition of the slag determines the minimum temperature of operation in the furnace.

Basically, a blast furnace is a shaft furnace (if it is analyzed as a chemical reactor, it is a plug flow reactor) with a refractory lining of variable thickness and quality, covered with a carbon steel sheet (15 mm in thickness) and where it is possible to distinguish the following elements (Fig. 6.3):

- **Throat**: zone for charging the burden [ferric charge + reductant (coke)]. The blast furnace gases circulate through the throat of the furnace toward the dry and wet cleaning systems. The temperature in the throat is 200–250 °C, the pressure is around 1.5 atm (there are blast furnaces in Russia and other countries that operate with bigger excesses of pressure), and the approximate composition of the gas is that indicated in Table 6.1 (Babich et al. 2008, page 153).
- **Stack**: upper conic trunk that has its maximum diameter in the lower part of the furnace stack, and it enlarges from the throat, allowing the burden to descent gradually at the same time that it expands due to the temperature increase caused by the reductant gases that flow in countercurrent.
- **Belt**: part with the biggest diameter of the furnace that allows joining the upper cone and the lower cone of the furnace (blast furnace bosh).
- **Bosh**: lower conic trunk that ends in the zone of tuyeres. In the bosh zone of the furnace, a reduction of the useful volume of the furnace (20%), to accompany the contraction derived from the partial melting of the burden, and to allow that the refractory lining could support the weight of the material located in upper layers, takes place. The tuyeres are located at the end of the bosh of the furnace.

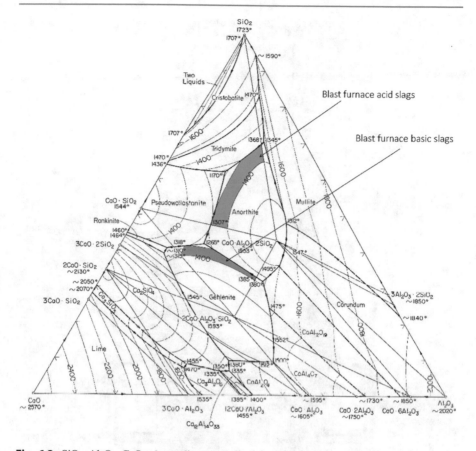

Fig. 6.2 SiO_2–Al_2O_3–CaO ternary diagram (edited from Levin et al. 1969)

Through them, preheated air (hot blast), pulverized coal or oxygen are injected
to produce the gasification of the coke.

- **Hearth**: cylindrical lower part of the furnace where the pig iron and the slag are
  located.

Reactions that take place inside of the furnace allow defining the following
zones:

- Top zone of heat transfer. Drying and heating of the burden.
- Chemical reserve zone. Reactions of solid–gas reduction or indirect reduction.
- Bottom productive zone or working zone. Direct reduction reactions, coke
  gasification (Boudouard reaction), slag formation, and dissolution of elements in
  the liquid iron.

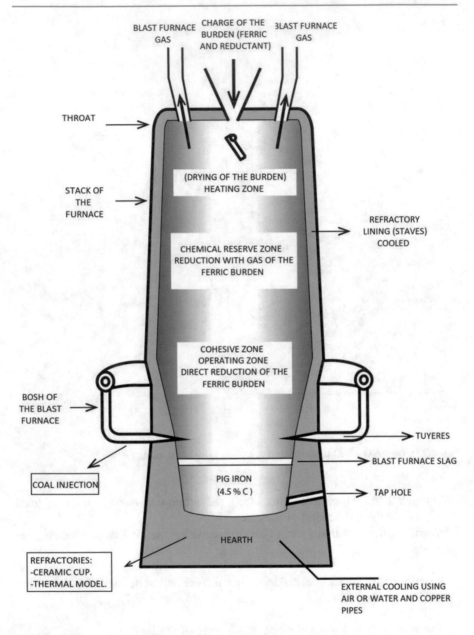

Fig. 6.3 Scheme of the elements and zones representatives of the blast furnace

Table 6.1 Composition of the blast furnace gas

Gaseous phase	% vol.
N_2	55–60
CO	17–25
CO_2	15–20
H_2	1–4
CH_4	0–0.5

Finally, if the physical characteristics of the burden as it advances from the top to the bottom are analyzed, the following zones can be defined:

- *Lumpy zone*: The burden is disposed in alternative layers of ferric material and coke (reductant) (Fig. 6.4).
- *Cohesive zone* or zone where the ferric charge starts to soften (iron ore, sinter, and pellets). The simulation of both the sinter heating process (Chap. 2) or pellet heating process (Chap. 3) is performed using the SiO_2–CaO–Fe_2O_3 ternary diagram.
- *Dripping zone*, where the ferric material reaches the liquidus temperature of the SiO_2–Al_2O_3–CaO ternary system (Fig. 6.2).
- *Heating zone* or zone of partial combustion of the coke (gasification).

The radial distribution of temperatures in the furnace that operates under ideal plug flow will be completely uniform. The different resistance to the thermal degradation of the iron ores, the coke characteristics, and the own dimensions of the stack cause preferential zones for the gas circulation in this zone of the furnace (stack). As a result, thermal gradients appear in the radial direction of the furnace stack. The thermal segregation problems were partially solved using a pretreated ferric burden (sinter and pellets).

The thermal profile of the charge, in the best situation, will be that shown in Fig. 6.5. In those zones close to the furnace walls, the temperature will be lower than in the zones close to the axis of the furnace. The furnace walls are manufactured in a refractory composite (SiC + plastic clay) embedded in a metallic matrix (steel shell), and they are cooled using copper pipes (*staves*) (Fig. 6.6).

On the other hand, if we analyze the horizontal section of the furnace, the thermal gradient of the burden is defined by the own segregation of the furnace burden (ferric and coke) in different granulometries (hot degradation tests of the ironmaking raw materials, Fig. 2.5). The most common types of segregation are *V*, inverse *V*, and *M*:

- *V-type segregation*: The particles of the biggest size are preferentially located in the center of the furnace, and thus, the gas circulation rate is higher in the center than in zones close to the furnace walls.
- *Inverse V-type segregation*: The biggest particles are preferentially close to the furnace walls, and thus, the reductant gas circulation rate is higher in this zone

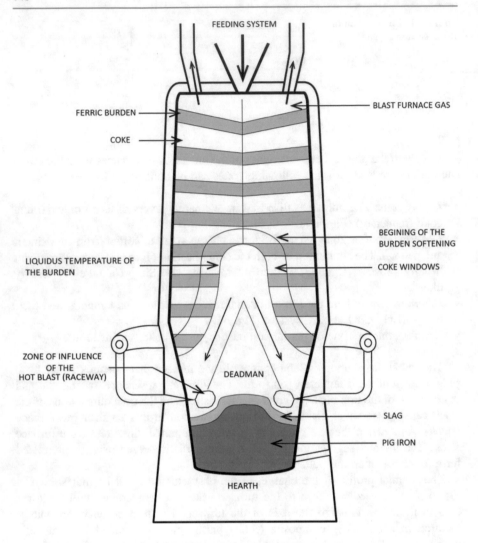

Fig. 6.4 Distribution of the ferric and reductant charges in the blast furnace

than in the center of the furnace (operating under these conditions would cause a bigger degradation of the furnace due to the abrasion mechanism).

- *M-type segregation*: It is a combination of the last two types of segregation; the flow of reductant gas goes through both the central zone and the zones close to the furnace walls.

The diameter of the blast furnace in the zone of tuyeres has been increased in the last decades, although the typical height of the furnace has not varied. One of the advantages that increasing the height of the furnace stack could mean is that the

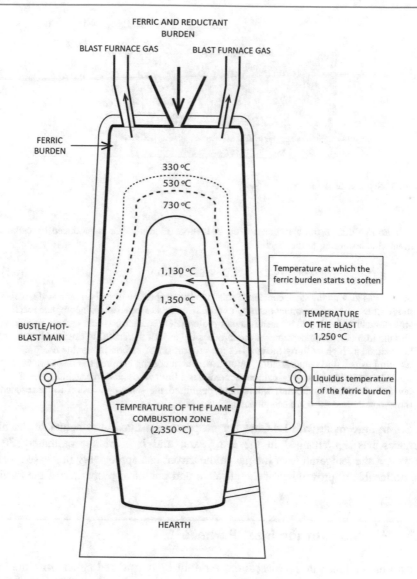

FERRIC AND REDUCTANT
BURDEN

BLAST FURNACE GAS   BLAST FURNACE GAS

FERRIC
BURDEN

330 °C

530 °C

730 °C

1,130 °C

Temperature at which the
ferric burden starts to soften

1,350 °C

BUSTLE/HOT-
BLAST MAIN

TEMPERATURE
OF THE BLAST
1,250 °C

Liquidus temperature
of the ferric burden

TEMPERATURE OF THE FLAME
COMBUSTION ZONE
(2,350 °C)

HEARTH

Fig. 6.5 Thermal heterogeneities in the burden of the blast furnace

solid–gas reactions in the chemical reserve zone might be favored and the use of the sensitive heat of the gases in the throat (top zone of heat transfer) might be also maximized. However, when the height of the furnace is increased, the compression stresses in the materials that form the charge is also increased, and the risks of breaking it, leading to the generation of fines, are increased. In 1888, Le Chatelier leveled the following criticism at the steel industry (Flinn 1963, page 284):

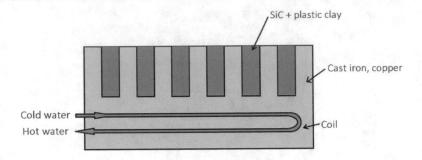

Fig. 6.6 Scheme of the staves

It is known that in the blast furnace the reduction of iron oxide is produced by carbon monoxide according to the reaction:

$$Fe_2O_3 + 3CO \rightleftharpoons 2Fe + 3CO_2 \qquad (6.1)$$

but the gas leaving the stack contains a considerable proportion of carbon monoxide, which thus carries away an important quantity of unutilized heat. Because this incomplete reaction was thought to be due to an insufficiently prolongated contact between carbon monoxide and the iron ore, the dimensions of the furnaces have been increased. In England they have been made as high as thirty meters. But the proportion of carbon monoxide escaping has not diminished, thus demonstrating, by an experiment costing hundred thousand francs, that the reduction of iron oxide by carbon monoxide is a limited reaction. Acquaintance with the laws of chemical equilibrium would have permitted the same conclusion to be reached more rapidly and far more economically.

So, in agreement with Le Chatelier proposal, the maximum height of the blast furnaces has not changed in the last years, and it is of approximately 30 m. However, the temperature of the gas in the throat has appreciably decreased due to the noticeable improvement of the physical and chemical properties of the burden.

6.3   Reactions in the Blast Furnace

Reactions produced in the blast furnace might be considered either from the thermodynamic or from the kinetics point of view. However, only the most noticeable thermodynamic aspects are quantified in the case of the iron ore solid–gas reduction reactions because the variables and mechanisms related with their kinetics were studied in previous chapters (Chaps. 2, 3, 4 and 5).

Even when examples of the typical reactions involved in the blast furnace are used, all the basic concepts developed for the kinetics of the metal–slag reactions are analyzed, and also those involved in the solids and liquids dissolution in the pig iron, which are applicable to all the reactions of the same nature that take place during the refining process in the converter, in the electric furnace, and during the secondary and tertiary metallurgies of the steel.

6.3.1   Iron Ores Reduction

Ballester et al. (2000) conceptually indicate the thermodynamic fundamentals of the iron oxides reduction using coke and gas (H_2, CO and CH_4). We also studied in the Chap. 5 the mechanisms involved in the reduction of iron oxides: the direct reduction with coke and the indirect reduction with CO gas (Exercise 5.2).

From the energy point of view, it is clearly demonstrated in the Chap. 5 that the mechanisms of direct reduction are clearly endothermic, while the reduction with carbon monoxide is slightly exothermic. However, from the perspective of the reductant gas consumption and as a function of the excesses that appear in the products of reaction, the following direct mechanism, when the CO/CO_2 ratio is 1, can be asses:

$$Fe_2O_3(s) + 2C(s) \rightarrow CO(g) + CO_2(g) + 2Fe(s,l) \qquad (6.2)$$

with a specific coke consumption of 215 kg of C per ton of Fe. Either, if the CO/CO_2 ratio is equal to 2, we will have:

$$4Fe_2O_3(s) + 9C(s) \rightarrow 6CO(g) + 3CO_2(g) + 8Fe(s,l) \qquad (6.3)$$

with a specific coke consumption of 242 kg of C per ton of Fe. Finally, if the CO/CO_2 ratio is 3:

$$5Fe_2O_3(s) + 12C(s) \rightarrow 9CO(g) + 3CO_2(g) + 10Fe(s,l) \qquad (6.4)$$

the specific coke consumption will be equal to 258 kg of C per ton of Fe. For the mechanism of indirect reduction with CO, if the CO/CO_2 ratio is 1:

$$Fe_2O_3(s) + 6CO(g) \rightarrow 3CO(g) + 3CO_2(g) + 2Fe(s) \qquad (6.5)$$

the specific consumption is 645 kg of C per ton of Fe. When the CO/CO_2 ratio is equal to 2:

$$Fe_2O_3(s) + 9CO(g) \rightarrow 6CO(g) + 3CO_2(g) + 2Fe(s) \qquad (6.6)$$

the specific consumption is 967 kg of C per ton of Fe. Finally, if the CO/CO_2 ratio is 3:

$$Fe_2O_3(s) + 12CO(g) \rightarrow 9CO(g) + 3CO_2(g) + 2Fe(s) \qquad (6.7)$$

the specific consumption is 1289 kg of C per ton of Fe. That is to say, if the direct mechanism is energetically unfavorable with respect to the reduction with gas, the low consumption of reductant gas makes attractive the reduction with coke (direct reduction) if compared with that using carbon monoxide (indirect reduction).

One of the secrets for the high energy efficiency and low reductant consumption is to know how to combine, in the suitable proportions, both mechanisms of reduction just like it is done in the blast furnace (Exercise 5.2). One of the problems of the direct reduction processes (*DRI*, Chap. 5) is the use of only one mechanism of reduction (indirect or with gas) to produce sponge iron. The heat associated with the process it is either favorable, using CO, or very slightly endothermic, using H_2 or CH_4, but the specific consumption of reductant gas per unit of produced iron is high.

This is one of the reasons that justify the failure of the Wiberg process, developed in Sweden at the beginning of the twentieth century (Fig. 6.7, Rosenqvist 1983, page 292). The alternative that the potential competitors in the reduction with gas (*Midrex* and *HYL*) propose to challenge the blast furnace is to operate with low energy consumptions even when they might have high specific consumptions of reductant gases. Meanwhile, the Wiberg process, even when it has a coke specific consumption very lower than that of the blast furnace for the pig iron (111 kg t^{-1} Fe), the electric power consumption grows up to 940 kWh t^{-1} Fe because it is necessary to regenerate the mix of reductant gas (CO + H_2) using coke and *fuel oil*. If the reductant gas of the Wiberg process would comprise only CO, the electric power consumption would be reduced a 30%, although the coke consumption would slightly grow.

The reduction of iron ore in the blast furnace mainly consists in the liberation of the oxygen associated with the metals and the combination of this oxygen with carbon. The carbon dioxide produced by indirect reduction reacts with the coke to regenerate the reductant gas (Boudouard reaction). Thus, the values of the potential of oxygen in the gas phase (POS, Potential of Oxygen in the System) change with the physical–chemical characteristics of the medium, the ferric layer or the coke. To

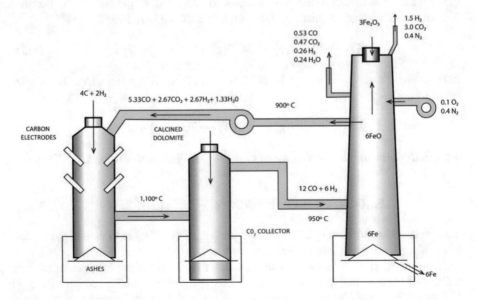

Fig. 6.7 Wiberg process. Mass balance in moles

illustrate this behavior in a hypothetical way, it is useful to imagine what could happen to a certain volume of gas (hot blast) that is injected by the tuyeres and ascends through the stack of the furnace. In Fig. 6.8, we show the changes that the potential of oxygen in the gas experiences over the trajectory of the furnace.

The air enters in the furnace after being compressed (around 3.5 atm) and preheated (1250 °C). The potential of oxygen in the system under these conditions (point A in Fig. 6.8) is equal to:

$$\Delta_r G^0(1253 \text{ K}) = R \cdot T \cdot \ln P_{O_2} \tag{6.8}$$

$$P_{O_2} \simeq 0.6 \text{ atm} \tag{6.9}$$

At the exit of the tuyeres, the oxygen of the air reacts with the coke of the burden. The temperature of the unit of volume of gas grows up to 2473 K (flame temperature estimated in this example). The value of the potential of oxygen in the system (POS) is:

$$\Delta_r G^0(2473 \text{ K}) = R \cdot T \cdot \ln P_{O_2} \tag{6.10}$$

where P_{O_2} is calculated considering the following equilibrium:

$$2C(s) + O_2(g) \rightleftharpoons 2CO(g) \tag{6.11}$$

which corresponds to the point B in Fig. 6.8. Before the moment when the volume of gas reaches the equilibrium with the coke layer, it enters in contact with the

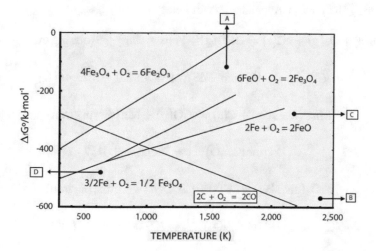

Fig. 6.8 Variation of the potential of oxygen in the blast furnace

stoichiometric wustite, FeO. The gas will reduce the wustite and the potential of oxygen in the system will grow and will be equal to:

$$\Delta_r G^0(2200 \text{ K}) = R \cdot T \cdot \ln P_{O_2} \tag{6.12}$$

where P_{O_2} is calculated considering the following equilibrium:

$$2Fe(l) + O_2(g) \rightleftharpoons 2FeO(l) \tag{6.13}$$

corresponding to the point C in Fig. 6.8. The alternative contacts between iron ore and coke layers produce changes in the potential of oxygen in the above-mentioned equilibriums. However, the temperature of the gas is lower, as a result of the heat transfer to the burden, and the coordinates of the hypothetic volume of gas are displaced toward the left side of Fig. 6.8.

Finally, the potential of oxygen in the gases of the throat at around 500 K is represented by the point D in Fig. 6.8.

6.3.2   Reduction of Metallic Oxides. Equilibrium Metal–Slag

Certain nonferric oxides that go with the iron in the burden of the furnace can be considered easily reducible, such as the oxidized combinations of manganese or phosphorus in the ores. Their reduction takes place in almost all the zones of the furnace in reactions such as:

$$P_2O_5(g) + 5C(s) \rightleftharpoons 5CO(g) + 2P(\text{dis}; \text{pig iron}) \tag{6.14}$$

The reduction of the sulfur (sulfide, sulfate, organic), manganese and silicon, considering the data of Appendix 2, will be:

$$FeS(s) + O_2(g) + 2C(s) \rightleftharpoons 2CO(g) + Fe(\text{dis}; \text{slag}) + S(\text{dis}; \text{pig iron}) \tag{6.15}$$

$$\Delta_r G^0_{(\text{reaction } 6.15)} \left[\text{kJ} \left(\text{mol FeS}\right)^{-1} \right] = -98.0 - 0.13 \cdot T(\text{K}) \tag{6.16}$$

$$MnO(\text{dis}; \text{slag}) + C(s) \rightleftharpoons CO(g) + Mn(\text{dis}; \text{pig iron}) \tag{6.17}$$

$$\Delta_r G^0_{(\text{reaction } 6.17)} \left[\text{kJ} \left(\text{mol MnO}\right)^{-1} \right] = 265.0 - 0.0157 \cdot T(\text{K}) \tag{6.18}$$

$$SiO_2(\text{dis}; \text{slag}) + 2C(s) \rightleftharpoons 2CO(g) + Si(\text{dis}; \text{pig iron}) \tag{6.19}$$

$$\Delta_r G^0_{(\text{reaction } 6.19)} \left[\text{kJ} \left(\text{mol SiO}_2\right)^{-1} \right] = 709.0 - 0.366 \cdot T(\text{K}) \tag{6.20}$$

These reduction reactions (reactions 6.17 and 6.19) are more complicated than those of the phosphorus and the sulfur, especially that of the SiO_2 (cristobalite, stable phase of the SiO_2 at high temperature) whose $\Delta_r G^0$ at 1773 K (1500 °C) has a very high positive value. As the operating variables of the blast furnace hearth do not coincide with the standard state of the reactions, it is necessary to calculate the $\Delta_r G$ of the reaction to check whether the process is thermodynamically possible.

In fact, all the $\Delta_r G$ of the above-mentioned reactions are possible (negative values), independently of the values of the $\Delta_r G^0$, because phosphorus, sulfur, manganese and silicon can appear as reaction products dissolved in the pig iron in very small concentrations (out of the equilibrium conditions).

Thus, it is more convenient to analyze all the mentioned reactions from the thermodynamic equilibrium point of view, that is to say, from the optical of the maximum development or extension reached for any spontaneous process.

The sulfur (reaction 6.15) is easily dissolved under weak reductant potentials. On the other hand, it is few probable that the operating conditions of the blast furnace might promote the simultaneous presence of the compounds of the reaction (6.15) [FeS(s), O_2 (g) and C(s)]. This way, to clearly know the dissolution and slagging of the sulfur mechanisms, it is more convenient to consider the reaction of oxygen and sulfur interchange between the slag of the blast furnace and the metal despite the equilibrium metal–slag considered by the reaction (6.15). The interchange reaction between metal and slag is the following:

$$CaO(dis; slag) + S(dis; pig\ iron) \rightleftharpoons CaS(dis; slag) + O(dis; pig\ iron) \quad (6.21)$$

$$k_{(reaction\ 6.21)} = \frac{a_O \cdot a_{CaS}}{a_S \cdot a_{CaO}} \quad (6.22)$$

The activity concentration of sulfur in the pig iron, a_S, will decrease when:

- The value of $k_{(reaction\ 6.21)}$ increases,
- Its equilibrium constant rises when the temperature grows, because the reaction is endothermic:

$$\Delta_r G^0_{(reaction\ 6.21)} \left[kJ\ (mol\ CaO)^{-1} \right] = 104.0 - 0.02714 \cdot T(K) \quad (6.23)$$

- The calcium oxide activity in the slag grows (basic slags),
- The activity of the oxygen in the metal is the lowest possible.

The analysis of the equilibrium between the metal and the slag in the case of the manganese, reaction (6.17), is:

$$k_{(reaction\ 6.17)} = \frac{P_{CO} \cdot a_{Mn}}{a_C \cdot a_{MnO}} \quad (6.24)$$

and this leads to the manganese activity increase in the pig iron when:

- The operating temperature is high $\left(k_{(\text{reaction }6.17)}\right)$,
- The basicity index of the slag increases, because the activity coefficient of the manganese oxide in the slag, γ_{MnO}, is directly proportional to the basicity,
- The activity of the carbon in the pig iron takes the biggest value.

Finally, in the case of the silicon, reaction (6.19):

$$k_{(\text{reaction }6.19)} = \frac{P_{\text{CO}}^2 \cdot a_{\text{Si}}}{a_{\text{SiO}_2} \cdot a_{\text{C}}^2} \tag{6.25}$$

And thus, the activity of the silicon in the pig iron will rise when:

- The operating temperature in the hearth grows.
- The basicity of the slag is low (acid slags).
- The activity of the carbon takes the biggest value.

The activity of the carbon in the pig iron can be calculated using the Fe–C (stable) diagram of Fig. 6.9, apart from using the interaction parameters of Henry's law in the liquid iron (Table 6.2). There is not clearly agreement about the critical lines and points of the Fe-C (stable) diagram. Figure 6.9 is the reference diagram proposed by the ASM International (Massalski et al. 1990); temperature and composition (in weight percentage) data are highlighted for certain critical points according to Pero-Sanz et al. (2018).

The carbon Raoult's activity in the pig iron, a_{C}, is equal to the product of the molar fraction of the carbon in the liquid phase, x_{C}, and the coefficient of activity, γ_{C} (Ballester et al. 2000, Chap. 20):

$$a_{\text{C}} = \gamma_{\text{C}} \cdot x_{\text{C}} \tag{6.26}$$

Particularly, throughout the liquidus line of the liquid–graphite equilibrium, the value of a_{C} is 1 (Fig. 6.9):

$$a_{\text{C}} = \gamma_{\text{C}} \cdot x_{\text{C}} = 1 \tag{6.27}$$

For a composition different to that of the liquid–graphite equilibrium at constant temperature:

$$\ln \gamma_{\text{C}} = \ln \gamma_{\text{C}}^* + \varepsilon_{\text{C}}^{\text{C}} \cdot x_{\text{C}} + \sum_{2}^{n} \varepsilon_{\text{C}}^j \cdot x_j \tag{6.28}$$

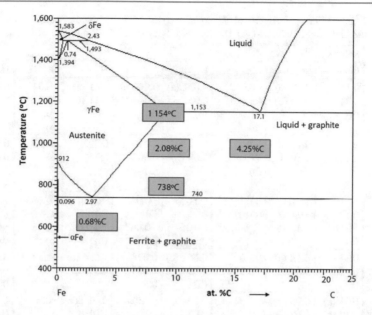

Fig. 6.9 Iron–carbon stable diagram (elaborated with data from Massalski et al. 1990; Pero-Sanz et al. 2018)

where γ_C^* is the coefficient of activity of the carbon at infinite dilution in the liquid iron, ε_C^C is the self-interaction parameter of the carbon in the liquid iron, and the addition represents the influence that the different solutes dissolved in the pig iron have on the carbon activity.

From the kinetics point of view, the metal–slag reaction of the sulfur, manganese, and silicon, in the hearth of the blast furnace, can be controlled by the diffusion, k_d, or by the chemical reaction of the chemical elements in the interfaces pig iron–slag, k_q, according to the equation:

$$\frac{1}{k_i} = \frac{1}{k_d} + \frac{1}{k_q} \tag{6.29}$$

where k_i is the kinetics coefficient of the equation of the reaction rate. The operating temperature can be sufficiently high to assume that the slow step of the process (controlling step) corresponds to the diffusion of the reagents between the reactive liquid phases, pig iron and slag (Fig. 6.10). As $k_i \simeq k_d$ when $k_q \gg k_d$, the reaction rate, v_r, expressed in percentage of sulfur, manganese or silicon incorporated to the pig iron per unit of time (mass transfer from the slag to the pig iron, $C_{i/m}^0 < C_{i/m}^=$), is equal to:

Table 6.2 Interaction parameters of Henry's law at 1600 °C in the liquid according to (a) Sigworth and Elliot, 1974, and (b) The Japanese Society for the Promotion of Science (1988)

Solutes i	Solutes j							
	C	Si	Mn	P	S	N	O	H
C	0.14 (a) 0.243 (b)	0.08 (a) 0.08 (b)	−0.012 (a) −0.008 (b)	0.051 (a) 0.051 (b)	0.046 (a) 0.044 (b)	0.11 (a) 0.11 (b)	−0.34 (a) −0.32 (b)	0.67 (a) 0.67 (b)
Si	0.18 (a) 0.18 (b)	0.11 (a) 0.10 (b)	0 (a) −0.015 (b)	0.11 (a) 0.09 (b)	0.056 (a) 0.066 (b)	0.092 (b)	−0.23 (a) −0.12 (b)	0 (a) 0.64 (b)
Mn	−0.07 (a) −0.054 (b)	0 (a) −0.033 (b)	0 (a) 0 (b)	−0.035 (a) −0.054 (b)	−0.048 (a) −0.047 (b)	−0.091 (a) −0.091 (b)	−0.083 (a) −0.083 (b)	−0.31 (a) −0.34 (b)
P	0.13 (a) 0.126 (b)	0.12 (a) 0.099 (b)	0 (a) −0.032 (b)	0.062 (a) 0.054 (b)	0.028 (a) 0.034 (b)	0.094 (a) 0.13 (b)	0.13 (a) 0.13 (b)	0.021 (a) 0.33 (b)
S	0.11 (a) 0.12 (b)	0.063 (a) 0.075 (b)	−0.026 (a) −0.027 (b)	0.24 (a) 0.035 (b)	−0.028 (a) −0.046 (b)	0.01 (a) 0.01 (b)	−0.27 (a) −0.27 (b)	0.12 (a) 0.41 (b)
O	−0.45 (a) −0.421 (b)	−0.131 (a) −0.066 (b)	−0.021 (a) −0.021 (b)	0.07 (a) 0.07 (b)	−0.133 (a) −0.133 (b)	0.057 (a) −0.14 (b)	−0.20 (a) −0.17 (b)	−3.1 (a) 0.73 (b)
H	0.06 (a) 0.06 (b)	0.027 (a) 0.027 (b)	−0.0014 (a) −0.002 (b)	0.011 (a) 0.015 (b)	0.008 (a) 0.017 (b)		−0.19 (a) 0.05 (b)	0 (a) 0 (b)

$$v_r = \frac{dC_{i/m}}{dt} = k_d \cdot \left(C_{i/m}^0 - C_{i/m}^= \right) \cdot \frac{A}{V} \qquad (6.30)$$

Or, if it was possible to know, by the boundary layer model, the thickness of this last one (Eq. 6.30) is:

$$\frac{dC_{i/m}}{dt} = \frac{D_i}{\delta_0} \cdot \left(C_{i/m}^0 - C_{i/m}^= \right) \cdot \frac{A}{V} \qquad (6.31)$$

where k_d is the coefficient of diffusional mass transfer, $C_{i/m}^0$ is the concentration of i expressed as percentage in the pig iron (an ideal behavior is assumed), $C_{i/m}^=$ is the concentration of i once reached the equilibrium (also in percentage), δ_0 is the thickness of the stationary layer through which the only possible mechanism of

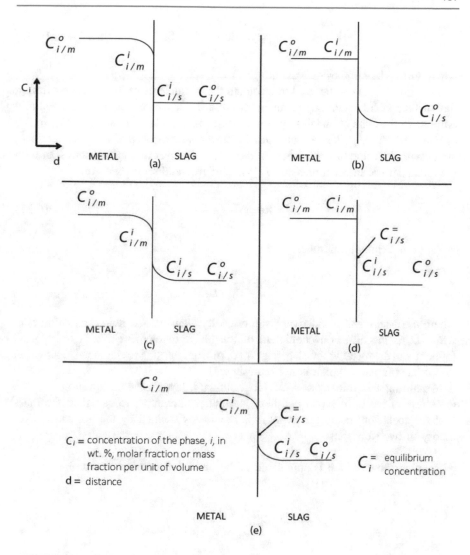

Fig. 6.10 Metal–slag reactions. Different concentration–penetration profiles as a function of the controlling step of the process. It is considered in all the cases that the mass transfer is performed from the metal to the slag (slag formation of impurities): **a** kinetical–diffusional control in the metallic phase; **b** kinetical–diffusional control in the slag; **c** mixed control of the diffusional transfer; **d** kinetical–chemical control; and, **e** mixed kinetical control, by interface reaction

mass transfer is the diffusion, and A/V is the mass–volume ratio in the reaction environment ($1/L$, L is the height of the hearth).

Under these conditions, the value of k_d, and thus that of k_i, might be estimated using the following correlation based on the analogies between the mass and the heat transfer processes:

$$\mathrm{Nu} = \frac{k_i \cdot 1}{D_i} = 0.332 \cdot \mathrm{Re}^{1/2} \cdot \mathrm{Sc}^{1/3} \tag{6.32}$$

where Nu is the number of Nusselt.

Equation (6.32) can be used to calculate the coefficients of mass transfer from a flat surface (solid or liquid) to an immiscible fluid, which circulates tangentially (similar to the contact metal–slag in the hearth of the blast furnace). In Eq. (6.32) (number of Nusselt), 1 is a characteristic linear dimension that can be identified, either with the value of δ, thickness of the boundary layer of concentrations, or with the characteristic linear dimension L (V/A) of Reynold's number, Re:

$$\mathrm{Re} = \frac{L \cdot \rho \cdot \upsilon}{\eta} \tag{6.33}$$

If, Sc is Schmidt's number:

$$\mathrm{Sc} = \frac{\eta}{\rho \cdot D_i} \tag{6.34}$$

where η, ρ and υ are, respectively, the viscosity, density, and fluid circulation rate, while D_i is the diffusion coefficient of the phase or element that is transported (either from the metal to the slag, Fig. 6.10, or from the slag to the metal, a situation in the blast furnace, which is not represented).

Assuming the thickness δ of the boundary layer of concentrations as the thickness of the tight layer, δ_0, there is another alternative to calculate the mass transfer coefficient using the hypothesis previously defined by the boundary layer theory in two situations:

- Laminar flow (which is applicable to the hearth of the blast furnace):

$$\delta = 1.5 \cdot D_i^{1/3} \cdot \left(\frac{\eta}{\rho}\right)^{1/6} \cdot \left(\frac{L}{\upsilon}\right)^{1/2} \tag{6.35}$$

- Turbulent flow:

$$\delta = 25.0 \cdot L \cdot \mathrm{Re}^{-0.9} \cdot \mathrm{Sc}^{-1.3} \tag{6.36}$$

6.3.3 Dissolution in the Pig Iron

The standard entropy and enthalpy of mixture for the dissolution of carbon, silicon, manganese, phosphorus, and sulfur in the liquid iron are calculated using the value of $\Delta_r G^0$ associated with the following reaction (Appendix 3):

Raoult's activity: $Me(s, l, g; a_i = 1) \leftrightarrow Me(dis; a_{H,i} = 1)$: Henry's activity

$$(6.37)$$

where a_i is the activity for a metal or element of Raoult's law and $a_{H,i}$ is the activity, for the same metal or element, of Henry's law of the diluted dissolutions. The previous reaction represents the standard free energy associated with the change of reference state of the elements dissolved in the liquid iron, which acts as solvent (Ballester et al. 2000, Chaps. 6 and 20). Knowing $\Delta_r G^0$, the value of $\Delta_r S^0$ is equal to:

$$\Delta_r S^0 = - \left(\frac{d\Delta_r G^0}{dT} \right)_P \qquad (6.38)$$

And finally, the enthalpy $\Delta_r H^0$ is calculated using the following equation:

$$\Delta_r H^0 = \Delta_r G^0 + T \cdot \Delta_r S^0 \qquad (6.39)$$

It is shown in Appendix 3 the variation of the free energy associated with the change of reference state (Eq. 6.38) as a function of the temperature. In the column $\pm J$ of Appendix 3, the maximum deviation for the linear regressions employed is indicated. In the following, two terminologies for the dissolution of one element (e.g., carbon) in the liquid iron will be used:

1. C(dis) or C(l) indicates the dissolution of the carbon in the liquid iron and its activity, or capacity of reaction, expressed by Raoult's activity, a_C.
2. $\underline{C}$ or C $(dis; a_{H,i} = 1)$ also represents that the carbon is dissolved in the liquid iron, although its activity, in this case, is expressed by Henry's activity, $a_{H,i}$.

6.3.4   Solved Exercises: Introductory Exercises

Exercise 6.1 A blast furnace produces pig iron (PI) with the following composition: 4.5% C, 1.0% Si, 1.0% Mn, 93.0% Fe, and 0.50% of other elements. A slag (S) with 48.0% CaO, 44.0% SiO_2, and 8% of other oxides is generated in the process. The ferric burden (FB) used to produce the pig iron, comprised of sinter, pellets and coarse iron ore, has the following chemical composition: 80% Fe_2O_3, 10% CaO, 9% SiO_2, and 1% of other oxides. The coke (C) used in the process has the following composition: 90% C, 6% SiO_2, 1% CaO, and 2% of other oxides. The blast furnace gas (BFG) generated in the process has the volumetric composition indicated as follows: 25% CO, 20% CO_2, and 55% N_2. Calculate per ton of pig iron produced in the blast furnace:

- The quantity of ferric burden that is required.
- The quantity of slag generated in the process.
- The coke specific consumption.
- The volume of blast furnace gas produced in the process.
- The quantity of air-blast (W) that is injected through the tuyeres of the furnace.

Data: Use the following relation of atomic weights: Fe, 55.85 g/mol; Si, 28.09 g/mol; Ca, 40.08 g/mol; C, 12.01 g/mol; O, 16.00 g/mol; N, 14.00 g/mol.

The scheme of the process is shown in Fig. 6.11.

To obtain the value of the ferric charge (FB) that is required to obtain the ton of pig iron, we make the balance to the iron in the system:

$$\{\text{Iron at the entrance}\} = \{\text{Iron at the exit}\} \tag{6.40}$$

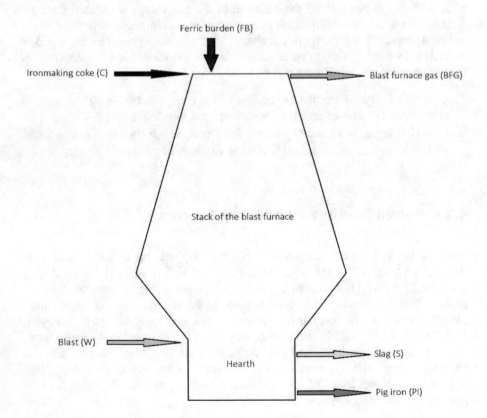

Fig. 6.11 Scheme of the blast furnace for the mass balance

$$FB(kg) \cdot \frac{80 \text{ kg Fe}_2O_3}{100 \text{ kg FB}} \cdot \frac{1 \text{ mol Fe}_2O_3}{159.70 \text{ kg Fe}_2O_3} \cdot \frac{2 \text{ mol Fe}}{1 \text{ mol Fe}_2O_3} \cdot \frac{55.85 \text{ kg Fe}}{1 \text{ mol Fe}}$$

$$= 1000 \text{ kg PI} \cdot \frac{93 \text{ kg Fe}}{100 \text{ kg PI}} \tag{6.41}$$

$$FB = 1662 \text{ kg t}^{-1}\text{pig iron} \tag{6.42}$$

To calculate the quantity of slag, S, and the coke consumption, C, the balances to both the calcium and the silicon are performed:

- Balance to the silicon:

$$FB(kg) \cdot \frac{9 \text{ kg SiO}_2}{100 \text{ kg FB}} \cdot \frac{1 \text{ mol SiO}_2}{60.09 \text{ kg SiO}_2} \cdot \frac{1 \text{ mol Si}}{1 \text{ mol SiO}_2} \cdot \frac{28.09 \text{ kg Si}}{1 \text{ mol Si}} + C(kg)$$

$$\cdot \frac{6 \text{ kg SiO}_2}{100 \text{ kg C}} \cdot \frac{1 \text{ mol SiO}_2}{60.09 \text{ kg SiO}_2} \cdot \frac{1 \text{ mol Si}}{1 \text{ mol SiO}_2} \cdot \frac{28.09 \text{ kg Si}}{1 \text{ mol Si}}$$

$$= S(kg) \cdot \frac{44 \text{ kg SiO}_2}{100 \text{ kg S}} \cdot \frac{1 \text{ mol SiO}_2}{60.09 \text{ kg SiO}_2} \cdot \frac{1 \text{ mol Si}}{1 \text{ mol SiO}_2} \cdot \frac{28.09 \text{ kg Si}}{1 \text{ mol Si}}$$

$$+ 1000 \text{ kg PI} \cdot \frac{1 \text{ kg Si}}{100 \text{ kg PI}} \tag{6.43}$$

- Balance to the calcium:

$$FB(kg) \cdot \frac{10 \text{ kg CaO}}{100 \text{ kg FB}} \cdot \frac{1 \text{ mol CaO}}{56.08 \text{ kg CaO}} \cdot \frac{1 \text{ mol Ca}}{1 \text{ mol CaO}} \cdot \frac{40.08 \text{ kg ca}}{1 \text{ mol Ca}} + C(kg)$$

$$\cdot \frac{1 \text{ kg CaO}}{100 \text{ kg C}} \cdot \frac{1 \text{ mol CaO}}{56.08 \text{ kg CaO}} \cdot \frac{1 \text{ mol Ca}}{1 \text{ mol CaO}} \cdot \frac{40.08 \text{ kg Ca}}{1 \text{ mol Ca}}$$

$$= S(kg) \cdot \frac{48 \text{ kg CaO}}{100 \text{ kg S}} \cdot \frac{1 \text{ mol CaO}}{56.08 \text{ kg CaO}} \cdot \frac{1 \text{ mol Ca}}{1 \text{ mol CaO}} \cdot \frac{40.08 \text{ kg Ca}}{1 \text{ mol Ca}} \tag{6.44}$$

In Eqs. (6.43) and (6.44), the values of the FB and PI are known, while on the contrary, the quantities of slag (S) and coke (C) to produce one ton of pig iron are not known. The values of S and C solving Eqs. (6.43) and (6.44) are:

$$S = 356 \text{ kg t}^{-1} \text{ pig iron} \tag{6.45}$$

$$C = 475 \text{ kg t}^{-1} \text{ pig iron} \tag{6.46}$$

The quantity of blast furnace gas that is generated in the process can be calculated by means of a balance to the carbon. To make this balance, it is necessary to consider that the volumetric composition of the gas (25% CO, 20% CO_2, and 55% N_2) can be transformed in mass composition (considering standard conditions):

$$25 \text{ mol CO} \cdot \frac{28.01 \text{ g CO}}{1 \text{ mol CO}} = 700.25 \text{ g CO}$$

$$\rightarrow \frac{700.25 \text{ g CO}}{700.25 \text{ g CO} + 880.2 \text{ g CO}_2 + 1540 \text{ g N}_2} = 22.44 \,\% \text{ CO}$$

$$(6.47)$$

$$20 \text{ mol CO}_2 \cdot \frac{44.01 \text{ g CO}_2}{1 \text{ mol CO}_2} = 880.2 \text{ g CO}_2$$

$$\rightarrow \frac{880.2 \text{ g CO}_2}{700.25 \text{ g CO} + 880.2 \text{ g CO}_2 + 1540 \text{ g N}_2} = 28.21 \% \text{CO}_2$$

$$(6.48)$$

$$55 \text{ mol N}_2 \cdot \frac{28 \text{ g N}_2}{1 \text{ mol N}_2} = 1540 \text{ g N}_2$$

$$\rightarrow \frac{1540 \text{ g N}_2}{700.25 \text{ g CO} + 880.2 \text{ g CO}_2 + 1540 \text{ g N}_2} = 49.35 \,\% \text{N}_2$$

$$(6.49)$$

Balance to the carbon:

$$C(\text{kg}) \cdot \frac{90 \text{ kg C}}{100 \text{ kg C}} = 1000 \text{ kg PI} \cdot \frac{4.5 \text{ kg C}}{100 \text{ kg PI}} + BFG(\text{kg})$$

$$\cdot \left[\frac{22.44 \text{ kg CO}}{100 \text{ kg BFG}} \cdot \frac{1 \text{ mol CO}}{28.01 \text{ kg CO}} \cdot \frac{1 \text{ mol C}}{1 \text{ mol CO}} \cdot \frac{12.01 \text{ kg C}}{1 \text{ mol C}} \right.$$

$$\left. + \frac{28.21 \text{ kg CO}_2}{100 \text{ kg BFG}} \cdot \frac{1 \text{ mol CO}_2}{44.01 \text{ kg CO}_2} \cdot \frac{1 \text{ mol C}}{1 \text{ mol CO}_2} \cdot \frac{12.01 \text{ kg C}}{1 \text{ mol C}} \right]$$

$$(6.50)$$

This way, as we know that the value of C is 475 kg C t^{-1} pig iron, we obtain a value for the blast furnace gas of 2210 kg BFG t^{-1} pig iron.

To know the volumetric flow (Nm^3) in standard conditions of the blast furnace gas (BFG), it is necessary to know that the weight of the blast furnace gas mol (25% CO, 20% CO_2, and 55% N_2) is 31.20 g. Consequently,

$$2210 \frac{\text{kg BFG}}{\text{t pig iron}} \cdot \frac{22.4 \text{ l BFG}}{31.20 \text{ g BFG}} \times \frac{10^3 \text{g BFG}}{1 \text{ kg BFG}} \cdot \frac{1 \text{ m}^3 \text{BFG}}{10^3 \text{l BFG}} \qquad (6.51)$$

$$= 1587 \text{ Nm}^3 \text{BFG } t^{-1} \text{pig iron}$$

Finally, to calculate the blast (W)-hot air that is injected via tuyeres of the blast furnace, it is necessary to take into account that the inert molecule of N_2 is not affected by the changes–transformations in the furnace. Consequently, the quantity of N_2 in the BFG is the same that in the hot blast that is injected (W):

$$2210 \frac{\text{kg BFG}}{\text{t pig iron}} \cdot \frac{49.35 \text{ kg N}_2}{100 \text{ kg BFG}} = 1091 \text{ kg N}_2 \text{ t}^{-1}\text{pig iron} \qquad (6.52)$$

The composition of the air is 79% N_2 and 21% O_2 in volumetric composition, so first we transform it in mass composition:

$$79 \text{ mol N}_2 \cdot \frac{28 \text{ g N}_2}{1 \text{ mol N}_2} = 2212 \text{ g N}_2 \rightarrow \frac{2212 \text{ g N}_2}{672 \text{ g O}_2 + 2212 \text{ g N}_2} = 76.70 \text{ %N}_2$$

$$(6.53)$$

$$21 \text{ mol O}_2 \cdot \frac{32 \text{ g O}_2}{1 \text{ mol O}_2} = 672 \text{ g O}_2 \rightarrow \frac{672 \text{ g O}_2}{672 \text{ g O}_2 + 2212 \text{ g N}_2} = 23.30 \text{ %O}_2$$

$$(6.54)$$

Thus, the quantity of air that is injected via tuyeres is:

$$1091 \frac{\text{kg N}_2}{\text{t pig iron}} \cdot \frac{100 \text{ kg air}}{76.70 \text{ kg N}_2} = 1422 \text{ kg air t}^{-1}\text{pig iron} \qquad (6.55)$$

The quantity of injected blast expressed in Nm^3 (standard conditions: 273 K and 1 atm), knowing that the weight of the air mol is 28.84 g, is:

$$1422 \frac{\text{kg air}}{\text{t pig iron}} \cdot \frac{22.4 \text{ l air}}{28.84 \text{ g air}} \times \frac{10^3 \text{g air}}{1 \text{ kg air}} \cdot \frac{1 \text{ m}^3\text{air}}{10^3 \text{l air}} = 1105 \text{ Nm}^3 \text{ blast t}^{-1}\text{pig iron}$$

$$(6.56)$$

Commentaries: A method to check, that the partial balances to the elements Fe, C, Ca, and Si, previously performed, are correct, is to verify the mass global balance in the furnace:

$$\text{Entrance: Ferric burden (FB)} + \text{Ironmaking coke (C)} + \text{Blast (W)} \qquad (6.57)$$
$$= 1662 + 475 + 1422 = 3559 \text{ kg}$$

$$\text{Exit: Pig iron (PI)} + \text{Slag (S)} + \text{Blast furnace gas (BFG)} = 1000 + 356 + 2210$$
$$= 3566 \text{ kg}$$

$$(6.58)$$

Even when there is a difference of 7 kg, the error involved in the calculations of the simplified balance to the blast furnace is smaller than 0.20%, which is reasonable.

Exercise 6.2 The conditions of the dry gas (without water vapor) in the zone of tuyeres of the blast furnace (Figs. 6.3 and 6.4) are: $T = 2200$ °C, total pressure of 3.0 atm, and 35% CO (in volume).

If the thermodynamic equilibrium conditions are reached, demonstrate that the system is defined and determine the corresponding partial pressures of the gases N_2, CO, CO_2, and O_2 in the zone of tuyeres.

If the thermodynamic equilibrium might be reached in the zone of tuyeres of the blast furnace under the conditions of the enunciation, there would be three components: C–O–N. The number of phases, F, would be two, the gas and the coke in excess (zone of inactive or low reactivity coke, around the tuyeres of the furnace). Consequently, the degrees of freedom of the system, L, will be:

$$L = 5 - F = 3 \tag{6.59}$$

The degrees of freedom are the temperature, the total pressure, and the composition of carbon monoxide in the gas (1.05 atm when we multiply the total pressure by the unitary volumetric composition).

The value of the standard free energy of the Boudouard equilibrium at 2473 K (Exercise 4.1, and Eq. 4.1) is:

$$(159,000 - 169 \cdot 2473)\frac{J}{mol\ CO_2} = 8.314\frac{J}{mol\ CO_2 \cdot K} \cdot 2473\ K \cdot \ln\left(\frac{P^2_{CO}}{P_{CO_2}}\right) \tag{6.60}$$

$$\ln\left(\frac{P^2_{CO}}{P_{CO_2}}\right) = 12.5939 \tag{6.61}$$

$$\frac{P^2_{CO}}{P_{CO_2}} = 294,753 \tag{6.62}$$

$$P_{CO} = P_T \cdot 0.35 = 3\ atm \cdot 0.35 = 1.05\ atm \tag{6.63}$$

$$\frac{P^2_{CO}}{294,753} = P_{CO_2} \rightarrow P_{CO_2} = 3.74 \times 10^{-6}\ atm \tag{6.64}$$

We estimate the standard free energy of formation of the carbon dioxide from the CO (Appendix 1):

$$CO(g) + 1/2O_2(g) \rightleftharpoons CO_2(g) \tag{6.65}$$

$$\Delta_r G^0(T) = -277.0 + 0.085 \cdot T(K)\ kJ\ mol^{-1}CO \tag{6.66}$$

The value of the equilibrium constant between the CO and the CO_2 at 2200 °C is:

$$\Lambda_r G^0(T) = -R \cdot T \cdot \ln(K_q) \rightarrow -66{,}795 \frac{J}{mol} = -8.314 \frac{J}{mol\ K} \cdot 2473\ K \cdot \ln(K_q)$$
$$\rightarrow K_q = 25.75$$

(6.67)

and, thus,

$$K_q = \frac{P_{CO_2}}{P_{CO} \cdot P_{O_2}^{1/2}} \rightarrow P_{O_2} = \left(\frac{P_{CO_2}}{P_{CO} \cdot K_q}\right)^2 = \left(\frac{3.74 \times 10^{-6}}{1.05 \cdot 25.75}\right)^2 = 1.91 \times 10^{-14}\ atm$$

(6.68)

This way, the partial pressure of the nitrogen in the gas, if the thermodynamic equilibrium was reached, would be:

$$P_T = P_{CO} + P_{CO_2} + P_{O_2} + P_{N_2} \rightarrow P_{N_2} = 3 - 1.05 - 3.74 \times 10^{-6} - 1.91 \cdot 10^{-14}$$
$$= 1.95\ atm$$

(6.69)

Commentaries: There are also water vapor and hydrogen in the atmosphere of the tuyeres zone of the blast furnace due to the moisture of the ambient air. However, the high temperatures of the flame front avoid the presence of undesirable cyclic hydrocarbons derived from the benzene and other complex hydrocarbons of high molecular weight.

Exercise 6.3 Calculate the temperature of the blast in the blast furnace (Fig. 6.5) to achieve that the temperature of the flame could reach 2300 °C. Equally, determine the effect on the temperature of the stabilized flame in 2300 °C of the following operations:

(a) Enrichment in oxygen of the hot blast injected by the tuyeres until reaching 24% O_2 (composition in volume).
(b) Injection, together with the hot blast, of 0.10 mol of pulverized coal.

Consider that the blast injected in the furnace does not contain water vapor, and that the coal–coke reaches the zone of tuyeres for its partial combustion at 1000 °C. It is assumed that the carbon combustion process is sufficiently fast to consider it adiabatic (it has not heat losses).

Data: Consulting the TAPP database, the average specific heats, at constant pressure, of the gases and the carbon, in kJ mol^{-1} K^{-1} are collected in Table 6.3.

Table 6.3 Average specific heats, at constant pressure of the gases and the carbon in kJ mol^{-1} K^{-1}

CO	N$_2$	O$_2$	Carbon (graphite)
0.03384	0.03347	0.03684	0.02334

The adiabatic reaction to be considered in this zone of the blast furnace (Fig. 6.11), where the smelting of the ferric charge, the direct reduction reactions (with coke) with the corresponding mass transfer from the slag to the metal, and the partial combustion of the carbon-coke to carbon monoxide take place, will be the next one (volumetric composition of the air: 79% N$_2$, 21% O$_2$; reaction (6.70) is adjusted to the mol of O$_2$):

$$2C(s) + \frac{79}{21}N_2(g) + O_2(g) \rightarrow 2CO(g) + \frac{79}{21}N_2(g) \tag{6.70}$$

The reference temperature for the entrance of solid (coke) or molten (ferric charge) raw materials to the zone of elaboration of the furnace is considered of 1000 °C. This way, all current of entrance, with temperature higher than 1000 °C, carries calorific energy, for instance, the current of hot blast (first term of Eq. 6.71). Other term that supplies energy to the elaboration zone is the partial combustion of coke (second term of Eq. 6.71). Using the data of Appendix 3, the heat of combustion of the coal at 1000 °C is 113.5 kJ mol^{-1} of C. The energy supplied is used to heat the products of the reaction (CO and N$_2$) up to the adiabatic temperature of the flame of 2300 °C from the reference temperature (Eq. 6.72).

$$\text{Heat supply (kJ)} = \left[\overline{c_p}(O_2) + \frac{79 \text{ mol N}_2}{21 \text{ mol O}_2} \cdot \overline{c_p}(N_2)\right] \cdot (T_{\text{blast}} - 1000 \text{ °C})$$
$$+ 2 \text{ mol C} \cdot 113.5 \frac{\text{kJ}}{\text{mol C}} \tag{6.71}$$

Heat consumption (kJ)

$$= \left[2 \text{ mol CO} \cdot \overline{c_p}(CO) + \frac{79 \text{ mol N}_2}{21 \text{ mol O}_2} \cdot \overline{c_p}(N_2)\right] \tag{6.72}$$
$$\cdot (2300 \text{ °C} - 1000 \text{ °C})$$

Equalizing both mathematic expressions, we obtain the temperature that the hot blast should reach, T_{blast}, to adjust the balance: 1152 °C.

$$\left[0.03684 \frac{kJ}{K \; mol \; O_2} + \frac{79 \; mol \; N_2}{21 \; mol \; O_2} \cdot 0.03347 \frac{kJ}{K \; mol \; N_2} \right] \cdot (T_{blast} - 1273 \; K)$$

$$+ 2 \; mol \; C \cdot 113.5 \frac{kJ}{mol \; C}$$

$$= \left[2 \; mol \; CO \cdot 0.03384 \frac{kJ}{K \; mol \; CO} + \frac{79 \; mol \; N_2}{21 \; mol \; O_2} \right.$$

$$\left. \cdot 0.03347 \frac{kJ}{K \; mol \; N_2} \right] \cdot (2573 \; K - 1273 \; K)$$

$$\tag{6.73}$$

$$0.1628 \cdot (T_{blast} - 1273 \; K) + 227 = (0.0677 + 0.1259) \cdot 1300 \tag{6.74}$$

$$T_{blast} = 1425 \; K \; (1152 \; °C) \tag{6.75}$$

Thus (at this temperature), the influence that the enrichment in oxygen (3%) of the gas might have over the adiabatic temperature of flame can be analyzed. In this case, the terms of the balance are:

$$\text{Heat supply (kJ)} = \left[0.03684 \frac{kJ}{K \; mol \; O_2} + \frac{76 \; mol \; N_2}{24 \; mol \; O_2} \cdot 0.03347 \frac{kJ}{K \; mol \; N_2} \right]$$

$$\cdot (1425 - 1273) K + 2 \; mol \; C \cdot 113.5 \frac{kJ}{mol \; C}$$

$$\tag{6.76}$$

Heat consumption (kJ)

$$= \left[2 \; mol \; CO \cdot 0.03384 \frac{kJ}{K \; mol \; CO} + \frac{76 \; mol \; N_2}{24 \; mol \; O_2} \right. \tag{6.77}$$

$$\left. \cdot 0.03347 \frac{kJ}{K \; mol \; N_2} \right] \cdot (T_{flame} - 1273 \; K)$$

where the value of the flame temperature, T_{flame}, that adjusts the balance is:

$$\left[0.03684 \frac{kJ}{K \; mol \; O_2} + \frac{76 \; mol \; N_2}{24 \; mol \; O_2} \cdot 0.03347 \frac{kJ}{K \; mol \; N_2} \right] \cdot (1425 - 1273) K$$

$$+ 2 \; mol \; C \cdot 113.5 \frac{kJ}{mol \; C}$$

$$= \left[2 \; mol \; CO \cdot 0.03384 \frac{kJ}{K \; mol \; CO} + \frac{76 \; mol \; N_2}{24 \; mol \; O_2} \right.$$

$$\left. \cdot 0.03347 \frac{kJ}{K \; mol \; N_2} \right] \cdot (T_{flame} - 1273 \; K)$$

$$\tag{6.78}$$

$$248.71 = 0.1737 \cdot (T_{\text{flame}} - 1273 \text{ K}) \tag{6.79}$$

$$T_{\text{flame}} = 2705 \text{ K}(2432 \,^{\circ}\text{C}) \tag{6.80}$$

Finally, keeping constant the temperature of the blast in 1152 °C, and analyzing the influence of the injection of the pulverized coal, at room temperature (25 °C), over the flame temperature, the reaction of partial combustion and the terms of supply and consumption of heat will be the next ones:

$$1.9\text{C(s)} + 0.1\text{C(s)} + \frac{79}{21}\text{N}_2(\text{g}) + \text{O}_2(\text{g}) \rightarrow 2\text{CO(g)} + \frac{79}{21}\text{N}_2(\text{g}) \tag{6.81}$$

$$\text{Heat supply (kJ)} = \left[0.03684 \frac{\text{kJ}}{\text{K mol O}_2} + \frac{79 \text{ mol N}_2}{21 \text{ mol O}_2} \cdot 0.03347 \frac{\text{kJ}}{\text{K mol N}_2} \right]$$
$$\cdot (1425 - 1273)\text{K} + 1.9 \text{ mol C} \cdot 113.5 \frac{\text{kJ}}{\text{mol C}}$$
$$\tag{6.82}$$

Heat consumption (kJ)

$$= \left[2 \text{ mol CO} \cdot 0.03384 \frac{\text{kJ}}{\text{K mol CO}} + \frac{79 \text{ mol N}_2}{21 \text{ mol O}_2} \right.$$
$$\left. \cdot 0.03347 \frac{\text{kJ}}{\text{K mol N}_2} \right] \cdot (T_{\text{flame}} - 1273 \text{ K}) + 0.1 \text{ mol C}$$
$$\cdot 0.02334 \frac{\text{kJ}}{\text{K mol C}} \cdot (1273 \text{ K} - 298 \text{ K})$$
$$\tag{6.83}$$

where the value of the flame temperature that adjusts the balance is:

$$\left[0.03684 \frac{\text{kJ}}{\text{K mol O}_2} + \frac{79 \text{ mol N}_2}{21 \text{ mol O}_2} \cdot 0.03347 \frac{\text{kJ}}{\text{K mol N}_2} \right] \cdot (1425 - 1273)\text{K}$$
$$+ 1.9 \text{ mol C} \cdot 113.5 \frac{\text{kJ}}{\text{mol C}}$$
$$= \left[2\text{mol CO} \cdot 0.03384 \frac{\text{kJ}}{\text{K mol CO}} + \frac{79 \text{ mol N}_2}{21 \text{ mol O}_2} \right.$$
$$\left. \cdot 0.03347 \frac{\text{kJ}}{\text{K mol N}_2} \right] \cdot (T_{\text{flame}} - 1273 \text{ K}) + 0.1 \text{ mol C}$$
$$\cdot 0.02334 \frac{\text{kJ}}{\text{K} \cdot \text{mol C}} \cdot (1273 \text{ K} - 298 \text{ K})$$
$$\tag{6.84}$$

$$240.39 = 0.1936 \cdot (T_{\text{flame}} - 1273 \text{ K}) + 2.2757 \tag{6.85}$$

$$T_{\text{flame}} = 2503 \text{ K}(2230 \,^{\circ}\text{C}) \tag{6.86}$$

That is to say, the injection of pulverized coal cools the elaboration zone (hearth and bosh). To adjust the losses, it is possible to speculate about the enrichment of the blast with oxygen; for instance, if it is reached a volumetric composition in the blast of 24% O_2, the temperature of the flame will rise up to 2353 °C when the temperature of the hot blast in the zone of tuyeres was of 1152 °C.

Commentaries: The calculations of the energy balance in the elaboration zone must be adapted to the conditions of design and operation of each furnace; for instance, for the blast furnace BF4 of Exercise 6.4, the equation of the flame front in the zone of tuyeres is:

$$T_{\text{flame}} = 1559 + 0.839 \cdot T_{\text{blast}} - 6.033 \cdot W(H_2O) - 4972 \cdot W(F) + 4972 \cdot W(O_2) \tag{6.87}$$

where $W(H_2O)$ is the moisture of the blast in g Nm^{-3} of blast, $W(F)$ is the quantity of fuel oil that can be injected via tuyeres expressed in kg Nm^{-3} of blast and $W(O_2)$ is the volume of oxygen injected in Nm^3 of O_2 per Nm^3 blast. In the particular case of the T_{blast} of 1200 °C, $W(H_2O)$ of 20 g Nm^{-3} of blast, $W(F)$ of 4.545×10^{-2} kg Nm^{-3} and without considering the enrichment in oxygen, we obtain a value of flame temperature of 2219 °C.

Exercise 6.4 The operation and geometric characteristics of eight blast furnaces are collected in Table 6.4.

Where P_{furnace} is the daily production of the furnace; D_{hearth} is the diameter of the hearth in the zone of tuyeres; V_{BF} is the operation useful volume of the blast furnace; H_{BF} is the height of the blast furnace; PCC are the heat losses in the hearth;

Table 6.4 Operation and geometric characteristics of eight blast furnaces

Blast furnace	BF1	BF2	BF3	BF4	BF5	BF6	BF7	BF8
Production (P_{furnace}) (t day^{-1})	450	1200	3000	4000	4500	6000	6700	10,000
Diameter (D_{hearth}) (m)	4.1	6.5	9.0	9.5	9.8	10.8	11.3	14.0
Volume (V_{BF}) (m^3)	425	90	1424	1624	1763	2255	2265	4100
Height (H_{BF}) (m)	18.5	24.0	26.0	29.0	29.4	30.4	31.4	36.0
PCC (GJ h^{-1})	29.05	41.95	55.23	60.08	63.60	68.58	72.96	87.70
D_{belly} (m)	5.40	7.58	10.69	10.50	10.83	12.22	12.80	15.50
AH (m^2)	66.80	108.74	179.54	249.95	289.08	320.19	370.10	417.83
AT (m^2)	360.29	604.95	936.59	1027.50	1075.26	1307.64	1362.96	1906.95
PCT (GJ h^{-1})	36.31	52.44	69.04	75.10	79.50	85.73	91.20	109.60

D_{belly} is the diameter of the furnace in the widest zone; AH is the cross area of the hearth; AT is the total area of the furnace; and, PCT are the total losses of the furnace.

Find the possible linear multivariate correlations between the capacity of production of the furnaces and the geometrical and operational variables above-indicated.

If we analyze the eight blast furnaces, we can establish two groups of linear multivariate correlations (Exercises 2.8 and 2.9):

(a) In a first group, we correlate the daily production, $P_{furnace}$, the diameter of the hearth, D_{hearth}, the useful volume of operation available, V_{BF}, the height of the furnace, H_{BF}, and the heat losses in the hearth PCC. The obtained result using EXCEL or MATLAB is:

$$P_{furnace}\left(t\ day^{-1}\right) = -814.8 - 1454.6 \cdot D_{hearth} + 1.5 \cdot V_{BF} - 263.9 \cdot H_{BF} \\ + 392.5 \cdot PCC \tag{6.88}$$

The values of the calculated daily production and the real daily production are collected in Table 6.5.

(b) In a second group, we correlate the daily production, $P_{furnace}$, the area of the hearth, AH, the total area of the furnace, AT, and the total heat losses of the furnace, PCT. The result is:

$$P_{furnace}\left(t\ day^{-1}\right) = 1152.22 + 14.30 \cdot AH + 9.75 \cdot AT - 143.70 \cdot PCT \tag{6.89}$$

The values of the calculated daily production and the real daily production are collected in Table 6.6.

Commentaries: It is possible to make more accurate correlations between the variables indicated in the exercise if the data of about hundreds of installed furnaces were known. However, it is also possible to consider that, from all the analyzed variables, there are some that suffer changes along the operation life of the furnace, production, heat losses, or operative volume of the furnace. Normally, the operative

Table 6.5 Calculated and real daily production of the blast furnaces (group (a) of variables)

Blast furnace	BF1	BF2	BF3	BF4	BF5	BF6	BF7	BF8
$P_{furnace}$ (real) (t day^{-1})	450	1200	3000	4000	4500	6000	6700	10,000
$P_{furnace}$ (calculated) (t day^{-1})	379	1212	3046	3731	4852	5753	6496	9893

Table 6.6 Calculated and real daily production of the blast furnaces (group (b) of variables)

Blast furnace	BF1	BF2	BF3	BF4	BF5	BF6	BF7	BF8
$P_{furnace}$ (real) (t day^{-1})	450	1200	3000	4000	4500	6000	6700	10,000
$P_{furnace}$ (calculated) (t day^{-1})	403	1070	2930	3953	4345	6161	6628	9970

volume of both the shaft and the hearth increases with the time due to the wear (corrosion of the refractory lining). In the last years of operation life of the blast furnace (limited by the duration of the hearth), the production tends to be bigger (Verdeja et al. 2014, pages 175–198).

Finally, with respect to the thermal losses through the furnace walls, the heat losses in the zone of the hearth (PCC) are around 80% of the total heat losses in the furnace (PCT).

Exercise 6.5 Calculate the volumetric flow rate, diffusional and convective, of the CO and the porosity of the charge in the shaft and the boshes (elaboration zone) for a blast furnace (characteristics of the BF4 in Exercise 6.4), considering further the following information:

(a) The pressure of the CO in the throat is 0.25 atm and in the hearth is 1.0 atm. The diffusion coefficient of the CO is 0.20 cm^2 s^{-1}.
(b) The volumetric flow of the gases in the furnace is 242,779 Nm^3 h^{-1}, and the carbon consumption is 500 kg per ton of pig iron.
(c) The total pressure in the tuyeres is 3.5 atm, and in the throat of the furnace is 1.5 atm. The drop of pressure, both in the elaboration zone and in the shaft of the furnace, is 1.0 atm (101,325 Pa). Assume (characteristics of the BF4 in Exercise 6.4) that the height of the furnace in the elaboration zone is 3.0 m while the corresponding to the shaft of the furnace is 20.0 m.
(d) The physical properties of the gas and the charge in the zone of elaboration–tuyeres–bosh of the furnace are: viscosity, 5.665×10^{-5} Pa s; density, 0.217 kg m^{-3}; and, average particle size of 2.50 cm. Similarly, in the shaft of the furnace: viscosity, 3.373×10^{-5} Pa s; density, 0.411 kg m^{-3}; and, average particle size of 2.50 cm.

First Fick's equation for the mass flow, J_m, of CO must be considered to obtain the volumetric flow rate, $\dot{V}$, derived from the diffusional transfer of CO between the hearth and the throat of the furnace (Figs. 6.4 and 6.5):

$$J_m = D_{CO} \cdot \left(\frac{dC}{dx}\right) = D_{CO} \cdot \left(\frac{\Delta C}{\Delta x}\right) \qquad (6.90)$$

where ΔC is the gradient of CO concentration, under standard conditions, between the hearth and the throat of the furnace, and Δx corresponds to 23.0 m (for the BF4 of Exercise 6.4), thus,

$$\Delta C = \frac{(1 - 0.25)\text{atm}}{0.082 \frac{\text{atm} \cdot \text{l}}{\text{mol K}} \cdot 273 \text{ K}} \cdot \frac{1\text{l}}{1\text{dm}^3} \cdot \frac{1\text{dm}^3}{10^3\text{cm}^3} = 3.35 \times 10^{-5}\text{mol CO cm}^{-3} \quad (6.91)$$

The diffusional mass flow of CO, J_m, taking into account that the coefficient of diffusion of the gas is 0.20 cm^2 s^{-1}, is:

$$J_m = D_{CO} \cdot \left(\frac{\Delta C}{\Delta x}\right) = 0.20 \frac{\text{cm}^2}{\text{s}} \cdot \left(\frac{3.35 \times 10^{-5}\text{mol CO cm}^{-3}}{2300 \text{ cm}}\right) \quad (6.92)$$
$$= 2.913 \times 10^{-9}\text{mol CO cm}^{-2}\text{s}^{-1}$$

The corresponding mass flow rate of CO, q_m, calculated for the biggest cross section of the furnace, 8.66×10^5 cm^2 (the diameter of the belly of the furnace BF4 is 10.5 m), is:

$$q_m = J_m \cdot A = 2.913 \times 10^{-9} \frac{\text{mol CO}}{\text{cm}^2 \text{ s}} \cdot 8.66 \times 10^5\text{cm}^2 \quad (6.93)$$
$$= 2.523 \times 10^{-3}\text{mol CO s}^{-1}$$

Finally, the diffusional volumetric flow rate of CO in the furnace, V_{dif}, is (we assume standard conditions):

$$V_{\text{dif}} = 2.523 \times 10^{-3} \frac{\text{mol CO}}{\text{s}} \cdot \frac{22.4\text{l CO}}{1\text{mol CO}} \cdot \frac{3600 \text{ s}}{1 \text{ h}} \quad (6.94)$$
$$= 203 \text{ litres of CO h}^{-1} \left(0.203 \text{ Nm}^3 \text{ h}^{-1}\right)$$

For the calculation of the convective volumetric flow rate, V_{con}, of CO in the BF4, it is precise to include, apart from the carbon consumption, the production of the furnace (Exercise 6.4): 4000 t pig iron day^{-1}. As one mol of carbon forms a mol of CO (we assume standard conditions):

$$V_{\text{con}} = \frac{500 \text{ kg C}}{1 \text{ t pig iron}} \cdot \frac{4000 \text{ t pig iron}}{\text{day}} \cdot \frac{1 \text{ day}}{24 \text{ h}} \cdot \frac{1000 \text{ g C}}{1 \text{ kg C}} \cdot \frac{1 \text{ mol C}}{12 \text{ g C}} \cdot$$
$$\frac{1 \text{ mol CO}}{1 \text{ mol C}} \cdot \frac{22.4 \text{ l CO}}{1 \text{ mol CO}} \cdot \frac{1 \text{ dm}^3\text{CO}}{1\text{l CO}} \cdot \frac{1 \text{ m}^3\text{CO}}{10^3 \text{ dm}^3\text{CO}} \quad (6.95)$$
$$= 155,556 \text{ Nm}^3\text{CO h}^{-1}$$

To determine the porosity of the bed of particles, both in the elaboration zone and in the shaft of the furnace, it is possible to use Ergun's equation (Ballester et al. 2000, pages 235–236):

$$\frac{\Delta P}{L} = \frac{150 \cdot v \cdot \eta}{D_p^2} \cdot \frac{(1 - \omega)^2}{\omega^3} + \frac{1.75 \cdot \rho \cdot v^2}{D_p} \cdot \frac{(1 - \omega)}{\omega^3} \tag{6.96}$$

where η is the viscosity in Pa s, ρ is the density of the gas, $\overline{D_p}$ is the average size of the particles of the static bed, and v is the gas rate along the channels and windows of the bed of particles. At the same time, the rate is function of the surface rate, v_0, of the gases in the furnace and the porosity of the bed ω:

$$v_0 = v \cdot \omega \tag{6.97}$$

The drop of pressure per unit of length of the bed of particles in the zone of elaboration (3 m of height over the level of tuyeres of the furnace) is (remember that 1 atm = 101,325 Pa):

$$\frac{\Delta P}{L} = \frac{1\,\text{atm}}{3\,\text{m}} \cdot \frac{101{,}325\,\text{Pa}}{1\,\text{atm}} = 33{,}775\,\text{Pa}\,\text{m}^{-1} \tag{6.98}$$

while in the shaft of the furnace, with a height of 20 m, is:

$$\frac{\Delta P}{L} = \frac{1\,\text{atm}}{20\,\text{m}} \cdot \frac{101{,}325\,\text{Pa}}{1\,\text{atm}} = 5066\,\text{Pa}\,\text{m}^{-1} \tag{6.99}$$

Inasmuch as the surface rate of the gases in the furnace (v_0) is calculated as a function of the cross section of either the belly (10.5 m in diameter) or the zone of tuyeres (9.5 m in diameter) of the furnace for a volumetric gas flow of 242,779 Nm^3 h^{-1}, the surface rate of the gas will be of 0.779 m s^{-1} in the belly and 0.951 m s^{-1} in the zone of tuyeres.

$$\text{Shaft: } v_0 = \frac{242{,}779\,\dfrac{\text{Nm}^3}{\text{h}}}{\pi \cdot \left(\frac{10.5\,\text{m}}{2}\right)^2} \cdot \frac{1\,\text{h}}{3600\,\text{s}} = 0.779\,\text{m}\,\text{s}^{-1} \tag{6.100}$$

$$\text{Tuyeres: } v_0 = \frac{242{,}779\,\dfrac{\text{Nm}^3}{\text{h}}}{\pi \cdot \left(\frac{9.5\,\text{m}}{2}\right)^2} \cdot \frac{1\,\text{h}}{3600\,\text{s}} = 0.951\,\text{m}\,\text{s}^{-1} \tag{6.101}$$

Finally, the porosity, analyzing the drop of pressure in the elaboration zone (33,775 Pa m^{-1}), or in the shaft of the furnace (5066 Pa m^{-1}), using Ergun's equation is:

(a) in the shaft of the furnace:

$$5066\frac{\text{Pa}}{\text{m}} = \frac{150 \cdot 3.373 \times 10^{-5}\ \text{Pa} \cdot \frac{0.779\ \text{m s}^{-1}}{\omega}}{(0.025\ \text{m})^2} \cdot \frac{(1-\omega)^2}{\omega^3}$$

$$+\frac{1.75 \cdot 0.411(\text{kg m}^{-3}) \cdot \left(\frac{0.779\ \text{m s}^{-1}}{\omega}\right)^2}{0.025\ \text{m}} \cdot \frac{(1-\omega)}{\omega^3} \rightarrow \omega \qquad (6.102)$$

$$= 0.3037\ (30.37\%)$$

(b) in the elaboration zone of the furnace:

$$33{,}775\frac{\text{Pa}}{\text{m}} = \frac{150 \cdot 5.665 \times 10^{-5}\ \text{Pa} \cdot \frac{0.951\ \text{m s}^{-1}}{\omega}}{(0.025\ \text{m})^2} \cdot \frac{(1-\omega)^2}{\omega^3}$$

$$+\frac{1.75 \cdot 0.217(\text{kg m}^{-3}) \cdot \left(\frac{0.951\ \text{m s}^{-1}}{\omega}\right)^2}{0.025\ \text{m}} \cdot \frac{(1-\omega)}{\omega^3} \rightarrow \omega$$

$$= 0.2062(20.62\%)$$

$$(6.103)$$

Commentaries: We appreciate that, apart from the temperature along the wall of the furnace as indication of the conditions at which the charge is found at different heights, the evaluation of the pressure is also an important indication.

The porosity of the bed in the elaboration zone is related with the quality of the coke and the ratio of pulverized coal that is injected via tuyeres. Equally, the porosity of the bed in the belly and upper zone of the shaft can be altered by the precipitation of metals and volatile compounds that are vaporized or volatilized in the hearth, and that condense in the shaft: sodium, potassium, zinc, and silicon monoxide (SiO, silica fume).

Exercise 6.6 In Exercise 6.2, we indicated the values of the thermodynamic equilibrium of the C–O system in the zone of tuyeres of a blast furnace. Determine, according to the thermodynamic equilibrium, the minimum concentration of atomic oxygen dissolved in a pig iron, liquid state, with 4.5% of carbon at 1450 and 1500 °C.

When we want to study the diluted solutions (liquid state or solid solution), it is convenient to use Henry's activity, $a_{H,i}$ (Ballester et al. 2000, pages 492–494), to calculate the capacity of reaction of the solutes. While in Raoult's activity, a_i, the state of reference for unitary activity is the compound or pure element, for Henry's activity, the state of reference or unitary activity is the dissolution at 1.0% in weight. It is logical, thus, that within these two different states of reference could exist noticeable difference of standard free energy, as it is shown in Appendix 3 for all the elements (solids, liquids or gases) that could be dissolved in the molten iron. For the molecular–gas oxygen, the energy of reaction for the change of state of reference is:

$$\frac{1}{2}O_2(g, 1.0 \text{ atm}) \leftrightarrow O\left(dis, a_{H,O} = 1\right) \tag{6.104}$$

$$\Delta_r G^0(T) = -117{,}000 - 3.14 \cdot T(\text{K}) \text{ J mol O}^{-1} \tag{6.105}$$

As it is a standard free energy, it is possible to assume that the value of the equilibrium constant of the reaction at 1500 °C is:

$$\Delta_r G^0(T) = -R \cdot T \cdot \ln K_{eq} \rightarrow (-117{,}000 - 3.14 \cdot 1773)\left(\frac{\text{J}}{\text{mol}}\right)$$
$$= -8.314\left(\frac{\text{J}}{\text{mol K}}\right) \cdot 1773 \text{ K} \cdot \ln K_{eq} \rightarrow K_{eq} = 4.084 \times 10^3 \tag{6.106}$$

and at 1450 °C is:

$$\Delta_r G^0(T) = -R \cdot T \cdot \ln K_{eq} \rightarrow (-117{,}000 - 3.14 \cdot 1723)\left(\frac{\text{J}}{\text{mol}}\right)$$
$$= -8.314\left(\frac{\text{J}}{\text{mol K}}\right) \cdot 1723 \text{K} \cdot \ln K_{eq} \rightarrow K_{eq} = 5.142 \times 10^3 \tag{6.107}$$

And, finally:

$$K_{eq} = \frac{a_{H,O}}{P_{O_2}^{1/2}} = \frac{(\%O) \cdot f_O}{P_{O_2}^{1/2}} \tag{6.108}$$

where $a_{H,O}$ is Henry's activity of the oxygen dissolved in the liquid iron in equilibrium with a gas where the partial pressure of oxygen (in atm) is P_{O_2}. If there is not a significant quantity of oxygen dissolved, the value of Henry's activity of the oxygen is equivalent to a concentration expressed in percentage:

$$a_{H,O} = (\%O) \cdot f_O = (\%O) \text{when} f_O = 1 \tag{6.109}$$

In this case, there are two possible interactions over the oxygen, that is to say, one of the own oxygen and the other of the carbon:

$$\log f_O = e_O^O \cdot (\%O) + e_O^C \cdot (\%C) \tag{6.110}$$

However, the contribution of the term corresponding to the interaction of the oxygen with itself is very small due to the small concentration of the oxygen compatible with the iron and the carbon at high temperature (Appendix 3), that is,

$$\log f_O \simeq e_O^C \cdot (\%C) = -0.4355 \cdot (\%C) \tag{6.111}$$

where the value of the interaction of the carbon with the oxygen corresponds to the mean of the results of Appendix 3. However, the influence or interaction of the carbon on the solubility of the oxygen at 1600 °C (Appendix 3) must be corrected at 1450 and 1500 °C. We assume, as general criterion, that the value of the interaction parameters grows when temperature decreases. In this case, as there is a negative interaction, when the temperature decreases, the value of e_O^C will be less negative:

- at 1500 °C:

$$e_O^C = -0.4355 \cdot \left(\frac{1773\ \text{K}}{1873\ \text{K}}\right) = -0.412 \tag{6.112}$$

- at 1450 °C:

$$e_O^C = -0.4355 \cdot \left(\frac{1723\ \text{K}}{1873\ \text{K}}\right) = -0.401 \tag{6.113}$$

Consequently, Henry's coefficient of activity will be:

- at 1500 °C:

$$\log f_O \simeq -0.412 \cdot (4.5\%) = -1.8540 \rightarrow f_O = 0.014 \tag{6.114}$$

- at 1450 °C:

$$\log f_O \simeq -0.401 \cdot (4.5\%) = -1.8045 \rightarrow f_O = 0.0157 \tag{6.115}$$

We use the reaction of formation of CO from the carbon (coke) and oxygen (Appendix 1) to estimate the partial pressure of the oxygen in the gas in equilibrium:

$$C(s) + \frac{1}{2}O_2(g) \leftrightarrow CO(g) \tag{6.116}$$

The standard free energy and the constant of equilibrium at 1500 and 1450 °C are:

$$\Delta_r G^0(T) = -118{,}000 - 84 \cdot T(K) \ J \ mol \ CO^{-1} \tag{6.117}$$

As it is a standard free energy, it is possible to assume that the value of the equilibrium constant of the reaction at 1500 °C is:

$$\Delta_r G^0(T) = -R \cdot T \cdot \ln K_{eq} \to (-118{,}000 - 84 \cdot 1773)\left(\frac{J}{mol}\right)$$

$$= -8.314\left(\frac{J}{mol \ K}\right) \cdot 1773K \cdot \ln K_{eq} \to K_{eq} = 7.3119 \times 10^7$$

$$\tag{6.118}$$

and at 1450 °C is:

$$\Delta_r G^0(T) = -R \cdot T \cdot \ln K_{eq} \to (-118{,}000 - 84 \cdot 1723)\left(\frac{J}{mol}\right)$$

$$= -8.314\left(\frac{J}{mol \ K}\right) \cdot 1723K \cdot \ln K_{eq} \to K_{eq} = 9.2329 \times 10^7 \tag{6.119}$$

When the pressure of the CO in tuyeres is 1.0 atm and the activity of the carbon is one (pure compound), the partial pressure of oxygen in equilibrium is at 1500 °C:

$$K_{eq} = \frac{P_{CO}}{P_{O_2}^{1/2} \cdot a_C} \to 7.3119 \times 10^7 = \frac{1 \ atm}{P_{O_2}^{1/2} \cdot 1} \to P_{O_2} = 1.8704 \times 10^{-16} \ atm$$

$$\tag{6.120}$$

and at 1450 °C is:

$$K_{eq} = \frac{P_{CO}}{P_{O_2}^{1/2} \cdot a_C} \to 9.2329 \times 10^7 = \frac{1 \ atm}{P_{O_2}^{1/2} \cdot 1} \to P_{O_2} = 1.1764 \times 10^{-16} \ atm$$

$$\tag{6.121}$$

at 2200 °C, in the flame front in the zone of tuyeres, the partial pressure of oxygen will be (Exercise 6.2, $P_{O_2} = 1.91 \times 10^{-14}$ atm, Eq. 6.68); or, in this case:

$$\Delta_r G^0(T) = -R \cdot T \cdot \ln K_{eq} \to (-118{,}000 - 84 \cdot 2473)\left(\frac{J}{mol}\right)$$

$$= -8.314\left(\frac{J}{mol \ K}\right) \cdot 2473K \cdot \ln K_{eq} \to K_{eq} = 7.5919 \times 10^6$$

$$\tag{6.122}$$

$$K_{eq} = \frac{P_{CO}}{P_{O_2}^{1/2} \cdot a_C} \rightarrow 7.5919 \times 10^6 = \frac{1 \text{ atm}}{P_{O_2}^{1/2} \cdot 1} \rightarrow P_{O_2}$$

$$= 1.735 \times 10^{-14} \text{ atm(very similar to the equation (6.68))}$$

$$(6.123)$$

Thus, the activity of Henry and the concentration of atomic oxygen in equilibrium in the molten iron will be:

• at 1500 °C:

$$a_{H,O} = K_{eq} \cdot P_{O_2}^{0.5} = 4.084 \times 10^3 \cdot \left(1.8704 \times 10^{-16}\right)^{0.5} = 5.5854 \times 10^{-5} \quad (6.124)$$

$$(\%O)^= = \frac{a_{H,O}}{f_O} = \left(\frac{5.5854 \times 10^{-5}}{0.014}\right) \times 10^4 = 40 \text{ ppm } (0.004\%) \qquad (6.125)$$

• at 1450 °C:

$$a_{H,O} = K_{eq} \cdot P_{O_2}^{0.5} = 5.142 \times 10^3 \cdot \left(1.1764 \times 10^{-16}\right)^{0.5} = 5.5771 \times 10^{-5} \quad (6.126)$$

$$(\%O)^= = \frac{a_{H,O}}{f_O} = \left(\frac{5.5771 \times 10^{-5}}{0.0157}\right) \times 10^4 = 35 \text{ ppm } (0.035\%) \qquad (6.127)$$

Commentaries: Using the standard free energy associated with the change of state of reference, it is possible to obtain the relation between Raoult's and Henry's activities and the coefficient of activity at infinite dilution, γ_i^*, of different solutes in the molten iron at a defined temperature (Ballester et al. 2000, page 495):

Raoult's activity: $M_e(s, 1, g; a_i = 1) \leftrightarrow M_e(\text{dis}; a_{H,i} = 1)$: Henry's activity

$$(6.128)$$

$$\Delta_r G^0 \left(a_i / a_{H,i}\right) = R \cdot T \cdot \ln\left(\frac{a_i}{a_{H,i}}\right) = R \cdot T \cdot \ln\left(\frac{\gamma_i^* \cdot 55.85}{100 \cdot PA_i}\right) \qquad (6.129)$$

Exercise 6.7 Estimate the carbon activity (Raoult's activity) in a binary solution iron–carbon with 4.5% of carbon at 1500 °C and compare it with that of a pig iron at the same temperature and the following composition: 4.5% C, 1.0% Mn, 0.5% Si, and 94.0% Fe.

Data: The maximum solubility of the carbon in the iron in the liquid state, $(\%C)^=$, as a function of the temperature in K, is:

$$(\%C)^= = 0.5656 + 2.58 \cdot 10^{-3} \cdot T \tag{6.130}$$

The variation of the interaction parameter carbon–carbon of Henry, e_C^C, with the temperature in K, is the following one:

$$e_C^C = \frac{158}{T} + 0.0581 \tag{6.131}$$

Finally, the relation between the parameter of interaction carbon–carbon of Henry with its equivalent in Roult, ε_C^C (Ballester et al. 2000, page 494) is:

$$\varepsilon_C^C = 49.48 \cdot e_C^C + 0.7851 \tag{6.132}$$

with results of applying the formula (6.129) listed above for C and Fe.

To calculate the value of the carbon activity in a Fe–C binary alloy, it is necessary to consider that these solutions in molten state show positive deviation from Raoult's law. That is to say, that the capacity of reaction of the carbon in the melt is greater than the expected by its concentration in molar fraction:

$$a_C = \gamma_C \cdot x_C \text{ where } \gamma_C > 1 \tag{6.133}$$

The value of the carbon molar fraction in a melt with 4.5 wt% C is:

$$x_C = \frac{\frac{4.5}{12.01}}{\frac{4.5}{12.01} + \frac{99.5}{55.85}} = 0.1797 \tag{6.134}$$

We must consider that, in non-ideal solutions, the coefficient of activity is function of the temperature and the concentration (Eq. 6.28) to calculate the value of the carbon coefficient of activity:

$$\ln \gamma_C = \ln \gamma_C^* + \varepsilon_C^C \cdot x_C \tag{6.135}$$

From the data of the enunciation (Eqs. 6.131 and 6.132), it is possible to estimate the value of ε_C^C at 1500 °C:

$$\varepsilon_C^C = 49.48 \cdot \left(\frac{158}{1773} + 0.0581 \right) + 0.7851 = 8.0693 \tag{6.136}$$

To obtain the coefficient of activity at infinite dilution of the carbon in the melt, γ_C^*, we consider that, for the maximum solubility of the carbon in the liquid iron at 1500 °C (5.14%C, Fig. 6.9), Raoult's activity of the carbon is equal to one:

$$1.0 = a_C = \gamma_C \cdot x_C = \gamma_C \cdot \frac{\frac{5.14}{12.01}}{\frac{5.14}{12.01} + \frac{94.86}{55.85}} \rightarrow \gamma_C = 4.9687 \qquad (6.137)$$

When the iron does not accept more carbon at 1500 °C, the value of the carbon coefficient of activity reaches 4.9687, and thus, the coefficient of activity at infinite dilution is:

$$\ln 4.9687 = \ln \gamma_C^* + 8.0693 \cdot \frac{\frac{5.14}{12.01}}{\frac{5.14}{12.01} + \frac{94.86}{55.85}} \rightarrow \gamma_C^* = 0.9793 \qquad (6.138)$$

At the same temperature, but for a concentration of 4.5% C, the carbon coefficient of activity will be:

$$\ln \gamma_C = \ln 0.9793 + 8.0693 \cdot 0.1797 \rightarrow \gamma_C = 4.1765 \qquad (6.139)$$

Thus, Raoult's activity of the carbon at 1500 °C, for a melt with 4.5% C, is:

$$a_C = \gamma_C \cdot x_C = 4.1765 \cdot 0.1797 = 0.7507 \qquad (6.140)$$

On the other hand, the influence of 1.0% Mn and 0.5% Si present in the iron melt on the carbon activity can be analyzed from the data of Appendix 3. The influence of the silicon and the manganese can be corrected at 1500 °C. When the interaction is positive:

$$e_i^j(1500\ °C) = \left(\frac{1873}{1773}\right) \cdot e_i^j(1600\ °C) \qquad (6.141)$$

and, when it is negative, using the following relation:

$$e_i^j(1500\ °C) = \left(\frac{1773}{1873}\right) \cdot e_i^j(1600\ °C) \qquad (6.142)$$

Moreover, it is possible to estimate Raoult's interaction parameters, ε_i^j from e_j^i through the formula (Ballester et al. 2000, page 494):

$$\varepsilon_i^j = \frac{230.3}{55.85} \cdot P \cdot A_j \cdot e_i^j + \frac{55.85 - P \cdot A_j}{55.85} \qquad (6.143)$$

Thus, the interaction parameters of the Mn and Si over the carbon, at 1500 °C, will be:

$$\varepsilon_C^{Mn} = -2.128 \text{ and } \varepsilon_C^{Si} = 6.003 \qquad (6.144)$$

The coefficient of activity and the carbon activity, in this pig iron with four elements at 1500 °C, is:

$$\ln \gamma_C = \ln \gamma_C^* + \varepsilon_C^C \cdot x_C + \varepsilon_C^{Si} \cdot x_{Si} + \varepsilon_C^{Mn} \cdot x_{Mn} \tag{6.145}$$

$$\ln \gamma_C = \ln 0.9793 + 8.0693 \cdot 0.1790 + 6.003 \cdot 0.0085 - 2.128 \cdot 0.0087 \tag{6.146}$$

$$\ln \gamma_C = 1.456 \rightarrow \gamma_C = 4.2888 \rightarrow a_C = \gamma_C \cdot x_C = 4.2888 \cdot 0.1790 = 0.7677 \tag{6.147}$$

Commentaries: While the manganese stabilizes the dissolution of the carbon in the melt, the silicon increases the carbon activity (Pero-Sanz et al. 2019). If the temperature is reduced and the quantity of silicon is increased, it is possible to propitiate the carbon precipitation in the melt before reaching the eutectic temperature of the iron–carbon binary system (Fig. 6.9), especially if the silicon interaction parameter over the carbon is made more positive.

6.4   Mass and Energy Balances. Operating Diagram

The entrance and exit flows that must be considered in the mass and energy balances of the blast furnace, in steady state, are (Fig. 6.3):

- Mass entrance:
  - Burden (ferric charge and reductant).
  - Hot blast.
  - Pulverized coal injected by the tuyeres.
- Mass exit:
  - Pig iron.
  - Slag.
  - Blast furnace gas.
- Energy entrance:
  - Hot blast.
  - Chemical energy of both the coke and pulverized coal.
  - Energy associated with the dissolution of elements in the liquid iron.
  - Energy associated with the formation of chemical phases in the slag.
- Energy exit:
  - Pig iron.
  - Slag.
  - Reduction reactions.
  - Blast furnace gas.
  - Heat losses.

The same as in the sintering process, it is possible using generic or specialized software (MATLAB, HSC, APL, or APL-J) to design a combined mass and energy balance whose entrance parameters are collected in Table 6.7. The cells filled in with an asterisk (*) indicate the variables (intensive or extensive) that are obtained as a consequence of the resolution of the system of equations set out by the balance (Tables 6.7 and 6.8).

Table 6.7 Mass and energy balances in the blast furnace (data of entrance to the software)

	Burden	Blast	Gas	Pig iron	Slag
$T(K)$	298	1523	473	1753	1753
Weight (kg)	*	*	*	1000	*
%N$_2$		76.7	*		
%O$_2$		23.3			
%Fe	8			95.1	
%C	*			4.5	
%Si				0.4	
%Fe$_2$O$_3$	*				
%FeO	4.0				
%CO$_2$			*		
%Ca$_2$SiO$_4$					*
%CO			*		
%H$_2$O	3.0		*		
%SiO$_2$	*				*
%Al$_2$O$_3$	2.0				*

Coke consumption per ton of pig iron: 440 kg; heat losses per ton of pig iron: 186,204 kJ; heat of reaction per ton of pig iron: $-712,447$ kJ; heat absorbed by the burden: 0; heat supplied by the hot blast: $-1,947,639$ kJ/t pig iron; heat absorbed by the gas: 466,788 kJ/t pig iron; heat absorbed by the pig iron: 1,382,373 kJ/t pig iron; heat absorbed by the slag: 624,717 kJ/t pig iron; adiabatic flame temperature: 2699 K.

We obtain the results collected in Table 6.8 using the following restrictions:

- The %CO in the gases of the throat, addition of CO and CO$_2$, is 70%, which means:

$$\frac{P_{CO}}{P_{CO} + P_{CO_2}} \cdot 100 = 70\% \tag{6.148}$$

- The basicity of the slag in % is (see Tables 6.7 and 6.8):

$$I_B = \frac{\%CaO}{\%SiO_2} \cdot 100 \tag{6.149}$$

$$Ca_2SiO_4 \rightarrow \%CaO = \frac{1 \text{ mol } Ca_2SiO_4}{172 \text{ g } Ca_2\,SiO_4} \cdot \frac{2 \text{ mol } CaO}{1 \text{ mol } Ca_2\,SiO_4} \cdot \frac{56 \text{ g } CaO}{1 \text{ mol } CaO} \cdot 100$$
$$= 65.10\% \tag{6.150}$$

Table 6.8 Mass and energy balances in the blast furnace (data of exit to the software)

	Burden	Blast	Gas	Slag
Weight (kg)	2181	1338	2128	391
%N$_2$			48.3	
%O$_2$				
%Fe				
%C	20.2			
%Si				
%Fe$_2$O$_3$	46.5			
%CaO	8.0			
%FeO				
%CO$_2$			14.6	
%Ca$_2$SiO$_4$				63.3
%CO			34.1	
%H$_2$O			3.0	
%SiO$_2$	8.4			20.6
%Al$_2$O$_3$				11.2

$$Ca_2SiO_4 \rightarrow \%SiO_2 = \frac{1 \text{ mol } Ca_2SiO_4}{172 \text{ g } Ca_2 \text{ SiO}_4} \cdot \frac{1 \text{ mol } SiO_2}{1 \text{ mol } Ca_2SiO_4} \cdot \frac{60.08 \text{ g } SiO_2}{1 \text{ mol } SiO_2} \cdot 100$$
$$- 34.90\%$$

$$(6.151)$$

$$I_B = \frac{391 \text{ kg} \cdot 0.633 \cdot 0.6510}{391 \text{ kg} \cdot 0.633 \cdot 0.3490 + 391 \text{ kg} \cdot 0.206} = 0.9653(96.53\%) \quad (6.152)$$

- Temperature of reaction is 298 K.
- Heat losses as fraction of the total energy if the system are 7%.

One of the restrictions (remember that the definition of "restrictions" is: Those variables previously determined to have a system of equations that can be solved) that can cause a certain surprise is considering that the temperature of reaction for those processes and chemical reactions involved in the furnace is 298 K. It is possible to demonstrate by means of thermodynamic calculations that for the majority of the reduction or oxidation reactions in the blast furnace, the variation of the enthalpy with the temperature is not excessively important. In the data of Appendix 2, it is indicated that the enthalpy of oxidation of the aluminum decreases 16 kJ each 1000 K, which represents only 0.1% of the energy released by the reaction. With the purpose of reaching a closer knowledge of the temperature of each and every one of the reactions that can take place in the blast furnace, it is possible that we could underestimate the importance of some other of the decisive

variables for the simulation of the blast furnace operation as; for instance, the
proper determination of the heat losses produced in the installation.

However, data collected in Table 6.8 acceptably (despite their limitations) represents the current operation of a blast furnace working without carbon injection.
Concretely, the coke consumption without injection of combustible is of approximately 450 kg t^{-1} of pig iron (the result obtained in the simulation in the balance of
Table 6.8 is 440 kg t^{-1} of pig iron).

In 1927, Reichardt proposed for the first time a representation of the enthalpies
of the solid and gaseous burden as a function of the temperature, for the rotating and
shaft furnaces (Reichardt diagrams). In the sixties, Rist developed the concept of the
operating line of the furnace to simulate and control in a graphical way the process
of oxygen transfer from the ferric burden to the gas (Rist diagrams, Figs. 6.12 and
6.13).

The operating line of the blast furnace can be expressed as follows:

$$Y = \mu \cdot X + b \qquad (6.153)$$

where the variable Y represents the quantity of oxygen in relation with the quantity
of oxygen in the ferric burden (solid) (moles of oxygen in relation with those of iron
in the ferric charge), X indicates the quantity of oxygen with respect to that of
carbon in the gas (moles of oxygen in relation with the moles of carbon in the gas),
μ is identified as the coke consumption per ton of pig iron (quantity of carbon with

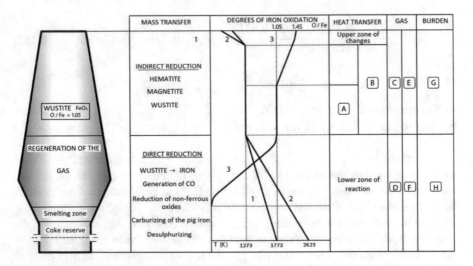

Fig. 6.12 Rist diagram of the general operating of the blast furnace: curve 1-temperature of the
solids; curve 2-temperature of the gas; curve 3-degree of oxidation of the iron; A-chemical
reservoir zone; B-thermal reservoir zone; C-utilization zone; D-zone of generation; E-CO$_2$ stable
zone; F-CO$_2$ unstable zone; G-zone of preparation of the burden; H-zone of elaboration of the pig
iron

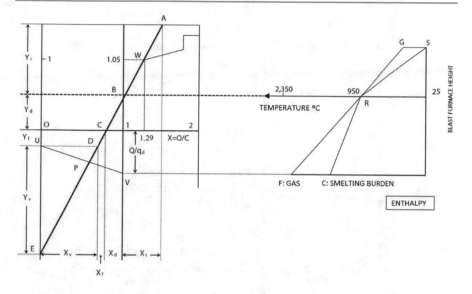

Fig. 6.13 Operating line. Rist and Reichardt diagrams

respect to the quantity of iron–pig iron produced), and b is the ordinate in the origin of the operating line of the furnace (identified with the production of the furnace, point E in Fig. 6.14, where the value of the ordinate $Y(E)$ is equal to $Y_v + Y_f$).

The coordinates of the operating line (left side of Fig. 6.14) can be calculated by means of the following equations:

- Point A-ferric burden and gas in the throat of the furnace:

$$X(A) = \frac{C_O}{C_C} = 1 + \frac{C_{CO_2}}{C_{CO} + C_{CO_2}} \tag{6.154}$$

$$Y(A) = \frac{C_O}{C_{Fe}} = \frac{\frac{3}{2} \cdot A + B}{A + B} \tag{6.155}$$

where C_O and C_C are the quantities of oxygen and carbon in the gas (expressed in mol), C_{CO} and C_{CO_2} are the volumetric compositions of the carbon monoxide and carbon dioxide, in percentage, in the throat of the furnace, C_{Fe} is the quantity of iron, A are the kilograms of iron 3+ (iron in the Fe_2O_3 burden of the blast furnace) per ton of pig iron in the iron and B are the kilograms of iron 2+ (iron in the FeO burden of the blast furnace). If the burden comprises only hematite, Fe_2O_3, $Y(A) = 1.5$. For instance, in a burden of 1253 kg of Fe_2O_3 per ton of pig iron, and 96 kg of FeO per ton of pig iron, the value of $Y(A)$ would be:

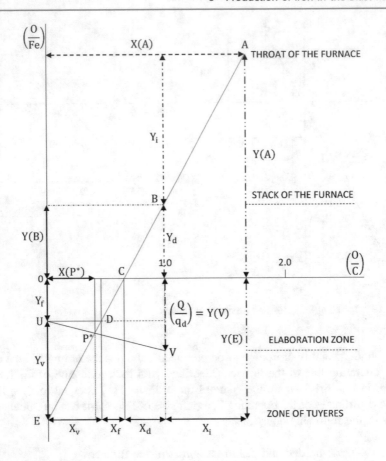

Fig. 6.14 Operating line

$$A = 253 \text{ kg Fe}_2\text{O}_3 \cdot \frac{1 \text{ kmol Fe}_2\text{O}_3}{159.69 \text{ kg Fe}_2\text{O}_3} \cdot \frac{2 \text{ kmol Fe}^{3+}}{1 \text{ kmol Fe}_2\text{O}_3} \cdot \frac{55 \text{ kg Fe}^{3+}}{1 \text{ kmol Fe}^{3+}}$$
$$= 863.11 \text{ kg Fe}^{3+} \tag{6.156}$$

$$B = 96 \text{ kg FeO} \cdot \frac{1 \text{ kmol FeO}}{71.84 \text{ kg FeO}} \cdot \frac{1 \text{ kmol Fe}^{2+}}{1 \text{ kmol FeO}} \cdot \frac{55 \text{ kg Fe}^{2+}}{1 \text{ kmol Fe}^{2+}} = 73.50 \text{ kg Fe}^{2+} \tag{6.157}$$

$$Y(A) = \frac{(3/2) \cdot 863.11 + 73.50}{863.11 + 73.50} = 1.461 \tag{6.158}$$

- Point D-charge in the zone of tuyeres:

$$X(D) = X_v - \frac{C_O}{C_C} = \left(\frac{2}{3.76}\right) \cdot \left(\frac{C_{N_2}}{100 - C_{N_2}}\right) \qquad (6.159)$$

$$Y(D) = Y_f = -[Y(Si) + Y(Mn) + Y(P) + Y(S)] \qquad (6.160)$$

where C_{N_2} is the volumetric composition of the nitrogen, in percentage, in the gases of the throat, $X(D) = X_v$ represents the fraction of reductant gas (CO) generated from the blast of tuyeres and, $Y(D)$ represents the fraction of oxygen atoms that are transferred to the gas by means of direct reduction mechanisms (with coke) of the nonferric phases (gangue-impurities) of the ore. Besides (see Exercise 6.11):

$$Y(Si) = \frac{C_{Si}}{100 \text{ g pig iron}} \cdot \frac{100 \text{ g pig iron}}{C_{Fe}} \cdot \frac{55.85 \text{ g Fe}}{\text{mol Fe}} \cdot \frac{1 \text{ mol Si}}{28.09 \text{ g Si}} \cdot \frac{2 \text{ mol O}}{1 \text{ mol Si}}$$
$$= 3.98 \cdot \left(\frac{C_{Si}}{C_{Fe}}\right)$$

$$(6.161)$$

$$Y(Mn) = \frac{C_{Mn}}{100 \text{ g pig iron}} \cdot \frac{100 \text{ g pig iron}}{C_{Fe}} \cdot \frac{55.85 \text{ g Fe}}{\text{mol Fe}} \cdot \frac{1 \text{ mol Mn}}{54.94 \text{ g Mn}} \cdot \frac{1 \text{ mol O}}{1 \text{ mol Mn}}$$
$$= 1.02 \cdot \left(\frac{C_{Mn}}{C_{Fe}}\right)$$

$$(6.162)$$

$$Y(P) = \frac{C_P}{100 \text{ g pig iron}} \cdot \frac{100 \text{ g pig iron}}{C_{Fe}} \cdot \frac{55.85 \text{ g Fe}}{\text{mol Fe}} \cdot \frac{1 \text{ mol P}}{30.97 \text{ g P}} \cdot \frac{5 \text{ mol O}}{2 \text{ mol P}} =$$
$$= 4.5 \cdot \left(\frac{C_P}{C_{Fe}}\right)$$

$$(6.163)$$

$$Y(S) = 1.75 \cdot M_{slag} \cdot \left(\frac{C_{S-slag}}{C_{Fe}}\right) \times 10^{-3} \qquad (6.164)$$

where C_{Si}, C_{Fe}, C_{Mn}, C_P, and C_{S-slag} are the quantities of the different chemical elements dissolved in the pig iron–slag, expressed in percentage and M_{slag} is the quantity of slag expressed in kilograms of slag produced per ton of pig iron.

- Point E-quantity of oxygen required in the production of pig iron:

$$Y(E) = \frac{C_O}{C_{Fe}} = -\left(Y_v + Y_f\right) \tag{6.165}$$

where Y_v is calculated from the gas flow blown per ton of pig iron. The ideal operating line of a furnace, for a certain capacity of production $Y(E)$ constant, is obtained by the union of the coordinates of the point $Y(E)$ with those of the non-stoichiometric wustite $(X(W) = 1.29, Y(W) = 1.05)$ in Fig. 6.13. From all the operating lines of a blast furnace that works with a defined value of $Y(E)$, the ideal operating line represents the working conditions that ensure a minimum coke consumption (slope of the line).

- Point B-position of the charge in the zone of chemical reserve of the furnace (Figs. 6.3, 6.12 and 6.13). As in the gas in contact with the ferric burden there is not or there is few CO_2, thus $X(B) = 1$. $Y(B) = Y_d$ represents the fraction of ferric burden that was reduced with coke by means of the direct mechanism, where the percentage of direct reduction is $100 \cdot Y_d/Y(A)$. On the other hand, the ordinate Y_i estimates the fraction of iron ore that is reduced with gas (indirect reduction mechanism), where the percentage of indirect reduction is $100 \cdot Y_i/Y(A)$.

Finally, the point V (see Fig. 6.14) is responsible of ensuring the thermal equilibrium that must exist in the zone where the pig iron is produced; we take 1000 °C (see Fig. 6.12, vertical line passing through 1) as the reference temperature for the thermal balance of the blast furnace. The coordinates of the point V ($X(V)$ and $Y(V)$) are:

$$X(V) = 1 \tag{6.166}$$

$$Y(V) = \frac{C_O}{C_{Fe}} = \frac{Q}{q_d} \tag{6.167}$$

where q_d is the heat associated with the direct reduction of the FeO (with coke) where the Boudouard mechanism is not yet involved and Q corresponds to an addition of terms that express the different types of calorific energy exchanges in the smelting and coke reserve zones of the blast furnace (Fig. 6.12). The addends that form Q have the following signs:

- The positive addends refer to:
  - Sensible heats of the pig iron, the slag and the blast (because they are at temperature >1000 °C);
  - Partial combustion of the coke;
  - Reduction of the FeO with gas;

- Addends corresponding to the heat associated with the dissolution of the elements in the pig iron (Si, Mn, P, and S).
- The negative addends refer to:
  - Carbothermal reduction of both the FeO, q_d, and the impurities associated with the ore (Mn, Si, P, S).
  - The heat losses in this zone of the furnace, which are close to the 80% of the total heat losses produced in the furnace.

As it is possible to see in the data of Figs. 6.12, 6.13 and Table 6.7, it is difficult to provide accurate values for the temperature of the gases in the zone of tuyeres, although the systems used to control the temperature have noticeably evolved. If we want to express the value of the adiabatic flame temperature, we should take into account that this one is the maximum temperature that can be reached in a point of the flame front if all the heat of the coke partial combustion is used to increase the temperature of the CO and the N_2 up to the above-indicated value. However, it is possible to correlate the flame temperature in each particular furnace, by optical pyrometry, with variables as important for the furnace operation as the hot blast temperature (air) of the stove, the moisture of the blast, the quantity of pulverized coal injected by the tuyeres (or any other liquid or gaseous combustible) or the oxygen enrichment of the blast, as well.

6.4.1 Solved Exercises: Operation Line

Exercise 6.8 The equation of the operation line of the blast furnace (Figs. 6.13 and 6.14) has different points of interest highlighted with the letters A, B, C, D, and E. Demonstrate that the abscissa of the point A, $X(A)$ is obtained using Eq. (6.154).

In the case of the dry gas of the throat of the blast furnace, free of water vapor and that comprises CO, CO_2, N_2 and small quantities of O_2, the balance of the carbon and the oxygen, expressed in moles, will be:

$$n(C) = n(CO) + n(CO_2) \qquad (6.168)$$

$$n(O) = n(CO) + 2 \cdot n(CO_2) + 2 \cdot n(O_2) \simeq n(CO) + 2 \cdot n(CO_2) \qquad (6.169)$$

Thus, the relation between the moles of oxygen, $n(O)$, and the moles of carbon, $n(C)$, in the point A is:

$$\frac{n(O)}{n(C)} = \frac{n(CO) + 2 \cdot n(CO_2)}{n(CO) + n(CO_2)} = \frac{n(CO) + n(CO_2)}{n(CO) + n(CO_2)} + \frac{n(CO)}{n(CO) + n(CO_2)} \qquad (6.170)$$

$$X(A) = \frac{n(O)}{n(C)} = 1 + \frac{n(CO)}{n(CO) + n(CO_2)} \qquad (6.171)$$

Multiplying and dividing the second addend of the equality by the factor $100/n_T$, where n_T is the total number of moles of the gas in throat of the furnace, we obtain:

$$X(A) = 1 + \frac{\%CO_2}{\%CO + \%CO_2} = 1 + \frac{C_{CO_2}}{C_{CO} + C_{CO_2}} \tag{6.172}$$

Commentaries: It is usual, as control instrument (Exercise 6.3), adding to the blast that is injected by the tuyeres of the furnace (environmental air) a certain fixed quantity of water. With the excess of carbon (coke) that is found in the hearth, the water vapor is transformed into CO and H_2, producing H_2 and H_2O in the gases of the throat. The presence of H_2O is also consequence of either the reduction of the iron ore with hydrogen or the moisture of the raw materials. In this case, the mass balance in the gases of the throat will be:

$$n(O) = n(CO) + 2 \cdot n(CO_2) + n(H_2O) \tag{6.173}$$

For that reason, the balance with the carbon will be equal to the atomic relation between the oxygen and this element:

$$\frac{n(O)}{n(C)} = 1 + \frac{n(CO_2)}{n(CO) + n(CO_2)} + \frac{n(H_2O)}{n(CO) + n(CO_2)} \tag{6.174}$$

$$X(A) = 1 + \frac{\%CO_2}{\%CO + \%CO_2} + \frac{\%H_2O}{\%CO + \%CO_2} \tag{6.175}$$

The value of the abscissa of the point A ($X(A)$) is function of the volumetric percentage of the gases that leave the blast furnace.

Exercise 6.9 Develop a generalized equation to estimate the value of the ordinate Y (A) of the point A of the blast furnace operation line (Exercise 6.8), as a function of the different oxidized species of the iron in the feeding of the furnace (degree of oxidation of the charge).

While the value of the abscissa of the point A of the operation line represents the composition of the gas that leaves the furnace (Exercises 6.8), the ordinate ($Y(A)$) values the quality or level of reduction of the iron feed that counteracts with the gas. From the chemical point of view, it is easy to determine the quantity of Fe^{3+} or Fe^{2+} that is available in the sample. A mass balance with the moles of oxygen, $n(O)$, in the ore charge of three-valence iron, at a rate of a kg t^{-1} of pig iron, turns into:

$$n(O)_{Fe^{3+}} = a \cdot \left[\frac{kg\ Fe(3+)}{t_{pig\ iron}}\right] \cdot \left(\frac{3\ mol\ O}{2\ mol\ Fe}\right) \cdot \left(\frac{1\ mol\ Fe}{55.85\ g\ Fe}\right) \times \left(\frac{10^3 g\ Fe}{1\ kg\ Fe}\right)$$

$$\tag{6.176}$$

$$n(O)_{Fe^{3+}} = \frac{3}{2} \cdot a \times \left(\frac{10^3}{55.85}\right) \cdot \left(\frac{mol\ O}{t_{pigiron}}\right) \tag{6.177}$$

The corresponding balance of oxygen moles, for the two-valence iron in the charge, at a rate of b kg t^{-1} of pig iron, is:

$$n(O)_{Fe^{2+}} = b \cdot \left[\frac{kg\ Fe(2+)}{t_{pig\ iron}}\right] \cdot \left(\frac{1\ mol\ O}{1\ mol\ Fe}\right) \cdot \left(\frac{1\ mol\ Fe}{55.85\ g\ Fe}\right) \times \left(\frac{10^3 g\ Fe}{1\ kg\ Fe}\right) \tag{6.178}$$

$$n(O)_{Fe^{2+}} = b \times \left(\frac{10^3}{55.85}\right) \cdot \left(\frac{mol\ O}{t_{pig\ iron}}\right) \tag{6.179}$$

$$n(O) = n(O)_{Fe^{3+}} + n(O)_{Fe^{2+}}$$
$$= \frac{3}{2} \cdot a \times \left(\frac{10^3}{55.85}\right) \cdot \left(\frac{mol\ O}{t_{pig\ iron}}\right) + b \times \left(\frac{10^3}{55.85}\right) \cdot \left(\frac{mol\ O}{t_{pig\ iron}}\right)$$
$$= \left[\frac{3}{2} \cdot a + b\right] \times \left(\frac{10^3}{55.85}\right) \cdot \left(\frac{mol\ O}{t_{pig\ iron}}\right) \tag{6.180}$$

The balance in moles of three-valence, two-valence, and total iron in the charge that feeds the furnace is:

$$n(Fe)_{Fe^{3+}} = a \cdot \left[\frac{kg\ Fe(3+)}{t_{pig\ iron}}\right] \cdot \left(\frac{1\ mol\ Fe}{55.85\ g\ Fe}\right) \times \left(\frac{10^3 g\ Fe}{1\ kg\ Fe}\right) \tag{6.181}$$

$$n(Fe)_{Fe^{3+}} = a \times \left(\frac{10^3}{55.85}\right) \cdot \left(\frac{mol\ Fe}{t_{pig\ iron}}\right) \tag{6.182}$$

$$n(Fe)_{Fe^{2+}} = b \cdot \left[\frac{kg\ Fe(2+)}{t_{pig\ iron}}\right] \cdot \left(\frac{1\ mol\ Fe}{55.85\ g\ Fe}\right) \times \left(\frac{10^3 g\ Fe}{1\ kg\ Fe}\right) \tag{6.183}$$

$$n(Fe)_{Fe^{2+}} = b \times \left(\frac{10^3}{55.85}\right) \cdot \left(\frac{mol\ Fe}{t_{pig\ iron}}\right) \tag{6.184}$$

$$n(Fe) = n(Fe)_{Fe^{3+}} + n(Fe)_{Fe^{2+}}$$
$$= a \times \left(\frac{10^3}{55.85}\right) \cdot \left(\frac{mol\ Fe}{t_{pig\ iron}}\right) + b \times \left(\frac{10^3}{55.85}\right) \cdot \left(\frac{mol\ Fe}{t_{pig\ iron}}\right)$$
$$= (a+b) \times \left(\frac{10^3}{55.85}\right) \cdot \left(\frac{mol\ Fe}{t_{pig\ iron}}\right) \tag{6.185}$$

The relation between the moles of oxygen and iron per ton of pig iron in the charge leads to Eq. (6.155):

$$Y(A) = \frac{C_O}{C_{Fe}} = \frac{n(O)}{n(Fe)} = \frac{\left[\frac{3}{2} \cdot a + b\right] \times \left(\frac{10^3}{55.85}\right) \cdot \left(\frac{mol\ O}{t_{pig\ iron}}\right)}{(a+b) \times \left(\frac{10^3}{55.85}\right) \cdot \left(\frac{mol\ Fe}{t_{pig\ iron}}\right)} = \frac{(3/2) \cdot a + b}{a+b} \quad (6.186)$$

Commentaries: If the moles of Fe_2O_3 (*a*), FeO (*b*), and Fe_3O_4 (*c*) are known, coming from the recycling of the mill scale, per unit of matter in the charge that feeds the furnace or in the produced pig iron, the value of the ordinate $Y(A)$ will be:

$$n(O)_{Fe^{2+},^{3+}} = c \cdot \left[\frac{kg\ Fe(2+,3+)}{t_{pig\ iron}}\right] \cdot \left(\frac{4\ mol\ O}{3\ mol\ Fe}\right) \cdot \left(\frac{1\ mol\ Fe}{55.85\ g\ Fe}\right) \times \left(\frac{10^3 g\ Fe}{1\ kg\ Fe}\right)$$
$$(6.187)$$

$$n(O)_{Fe^{2+},^{3+}} = \frac{4}{3} \cdot c \times \left(\frac{10^3}{55.85}\right) \cdot \left(\frac{mol\ O}{t_{pig\ iron}}\right) \quad (6.188)$$

$$n(O) = n(O)_{Fe^{3+}} + n(O)_{Fe^{2+}} + n(O)_{Fe^{2+},^{3+}}$$
$$= \frac{3}{2} \cdot a \times \left(\frac{10^3}{55.85}\right) \cdot \left(\frac{mol\ O}{t_{pigiron}}\right) + b \times \left(\frac{10^3}{55.85}\right) \cdot \left(\frac{mol\ O}{t_{pigiron}}\right) + \frac{4}{3}$$
$$\cdot c \times \left(\frac{10^3}{55.85}\right) \cdot \left(\frac{mol\ O}{t_{pigiron}}\right)$$
$$= \left[\frac{3}{2} \cdot a + b + \frac{4}{3} \cdot c\right] \times \left(\frac{10^3}{55.85}\right) \cdot \left(\frac{mol\ O}{t_{pigiron}}\right)$$
$$(6.189)$$

$$n(Fe)_{Fe^{2+},^{3+}} = c \cdot \left[\frac{kg\ Fe(2+,3+)}{t_{pig\ iron}}\right] \cdot \left(\frac{1\ mol\ Fe}{55.85\ g\ Fe}\right) \times \left(\frac{10^3 g\ Fe}{1kg\ Fe}\right) \quad (6.190)$$

$$n(Fe)_{Fe^{2+},^{3+}} = c \times \left(\frac{10^3}{55.85}\right) \cdot \left(\frac{mol\ Fe}{t_{pig\ iron}}\right) \quad (6.191)$$

$$n(Fe) = n(Fe)_{Fe^{3+}} + n(Fe)_{Fe^{2+}} + n(Fe)_{Fe^{2+},^{3+}}$$
$$= a \times \left(\frac{10^3}{55.85}\right) \cdot \left(\frac{mol\ Fe}{t_{pigiron}}\right) + b \times \left(\frac{10^3}{55.85}\right) \cdot \left(\frac{mol\ Fe}{t_{pigiron}}\right) + c$$
$$\times \left(\frac{10^3}{55.85}\right) \cdot \left(\frac{mol\ Fe}{t_{pigiron}}\right) = (a+b+c) \times \left(\frac{10^3}{55.85}\right) \cdot \left(\frac{mol\ Fe}{t_{pigiron}}\right)$$
$$(6.192)$$

And finally, Y(A) will be:

$$Y(A) = \frac{C_O}{C_{Fe}} = \frac{n(O)}{n(Fe)} = \frac{\left[\frac{3}{2} \cdot a + b + \frac{4}{3} \cdot c\right] \times \left(\frac{10^3}{55.85}\right) \cdot \left(\frac{mol\ O}{t_{pig\ iron}}\right)}{(a+b+c) \times \left(\frac{10^3}{55.85}\right) \cdot \left(\frac{mol\ Fe}{t_{pig\ iron}}\right)}$$

$$= \frac{\frac{3}{2} \cdot a + b + \frac{4}{3} \cdot c}{a+b+c} = \frac{9 \cdot a + 6 \cdot b + 8 \cdot c}{6 \cdot (a+b+c)}$$

(6.193)

Exercise 6.10 Demonstrate that the ideal operation line of a blast furnace that passes through the point $X = 1.29$, where the gas is in equilibrium with the $Fe_{0.95}O$ ($FeO_{1.05}$) or non-stoichiometric wustite (point W of Fig. 6.13).

Data: The temperature of the solid–gas equilibrium is 1000 °C (chemical reserve zone of the furnace, Fig. 6.12), the total pressure in the tuyeres is 4.81 atm and the pressure corresponding to the equilibrium between the wustite, $Fe_{0.95}O$, and the gas in $X = 1.29$, is 2.0 atm, the pressure of CO in tuyeres is 1.01 atm, and the standard free energy associated with the reduction of the $Fe_{0.95}O$ by the CO is:

$$\Delta_r G^0(T) = -15,700 + 20 \cdot T(K)(J/mol\ Fe_{0.95}O)$$

(6.194)

The reaction and the standard free energy of the wustite reduction with CO are:

$$Fe_{0.95}O(s) + CO(g) \leftrightarrow CO_2(g) + 0.95Fe(s)$$

(6.195)

$$\Delta_{Fe_{0.95}O}G^0(T = 1273\ K) = 9760\ J/mol\ Fe_{0.95}O$$

(6.196)

The value corresponding to the equilibrium constant, $K_{Fe_{0.95}O}$, at 1273 K, is:

$$\Delta_{Fe_{0.95}O}G^0(T = 1273\ K) = -R \cdot T \cdot \ln K_{Fe_{0.95}O} \rightarrow K_{Fe_{0.95}O}$$

$$= \exp\left(-\frac{9760\ \frac{J}{mol}}{8.314\ \frac{J}{mol\ K} \cdot 1273\ K}\right) = 0.3976$$

(6.197)

$$K_{Fe_{0.95}O} = 0.3976 = \frac{P_{CO_2}}{P_{CO}}$$

(6.198)

On the other hand, the partial pressure of CO in the tuyeres of furnace is of 1.01 atm and the volumetric fraction will be:

$$\%CO = \frac{P_{CO}}{P_T} \cdot 100 = \frac{1.01\ atm}{4.81\ atm} \cdot 100 = 21\%$$

(6.199)

In the ascendant path of the gas along the shaft of the furnace, from the point E to the point B (Fig. 6.14), the direct reduction (with coke) takes place, and from the point B to the point A, the indirect reduction (with gas) of the ferric charge takes place. The volumetric fraction of CO in the gas diminishes, while the fraction of CO_2 grows (in the zone of tuyeres, the volumetric fraction of the CO_2 in the gas is negligible, Exercise 6.2). In the central zone of the furnace, the thermal equilibrium between the descending charge and the reductant gas at 1000 °C is reached (chemical reserve zone, Fig. 6.12). Thus, the addition of partial pressures of CO and CO_2 in the belly of the furnace (central part of the furnace) will be:

$$P_{CO} + P_{CO_2} = 1.01 \text{ atm} \qquad (6.200)$$

Once reached the thermodynamic equilibrium, the relation of partial pressures indicated by the equilibrium constant provides an equation to estimate the pressures of CO and CO_2:

$$K_{Fe_{0.95}O} = \frac{P_{CO_2}}{P_{CO}} \rightarrow K_{Fe_{0.95}O} = \frac{1.01 - P_{CO}}{P_{CO}} \rightarrow K_{Fe_{0.95}O} \cdot P_{CO} + P_{CO} = 1.01 \rightarrow P_{CO}$$

$$= \frac{1.01}{1 + K_{Fe_{0.95}O}} = \frac{1.01}{1 + 0.3976} = 0.72 \text{ atm} \rightarrow P_{CO_2}$$

$$= 1.01 \text{ atm} - 0.72 \text{ atm} = 0.29 \text{ atm}$$

$$(6.201)$$

On the other hand, as the total pressure in the chemical reserve zone is 2.0 atm, the volumetric fractions of CO and CO_2 in the gas are:

$$\%CO = \frac{P_{CO}}{P_T} \cdot 100 = \frac{0.72 \text{ atm}}{2.0 \text{ atm}} \cdot 100 = 36.0\% \qquad (6.202)$$

$$\%CO_2 = \frac{P_{CO_2}}{P_T} \cdot 100 = \frac{0.29 \text{ atm}}{2.0 \text{ atm}} \cdot 100 = 14.5\% \qquad (6.203)$$

Thus, the relation between the moles of oxygen and the moles of carbon in the gases of the belly of the furnace (Exercise 6.8 and Eq. 6.154) will be:

$$X(W) = \frac{n(O)}{n(C)} = 1 + \frac{n(CO_2)}{n(CO) + n(CO_2)} = 1 + \frac{14.5}{36.0 + 14.5} = 1.29 \qquad (6.204)$$

Commentaries: We used in the exercise a value of the standard free energy for the non-stoichiometric wustite, $Fe_{0.95}O$, of:

$$\Delta_r G^0(Fe_{0.95}O - Fe)(T) = -261.3 + 0.065 \cdot T(K)(kJ/mol \, Fe_{0.95}O) \qquad (6.205)$$

$$\Delta_r G^0 (Fe_{0.95}O - Fe)(1273 \ K) = -178.56 \ kJ/mol \ Fe_{0.95}O \qquad (6.206)$$

which is comparable with the value of -182.26 kJ/FeO of the stoichiometric wustite, FeO (Appendix 1). The reducibility test of the iron raw materials, to check the greater or smaller oxygen loss in the iron ores in front of the reductant gases, depends on the nature of the wustite formed during the process. A wustite with the maximum standard energy of formation (close to the equilibrium, ΔG^0 is less negative, the kinetics play a relevant role, Eq. 6.31) will be faster reduced in the shaft of the furnace (Fig. 2.6 and Exercises 2.6 and 2.7). In the approach to evaluate the operating line of the furnace, it is not considered the existence of wustite that could be better or worse reduced than that considered in the exercise. In the specialized bibliography, the equilibrium between the gas and the non-stoichiometric wustite is usually represented, at 1000 °C, in the center of the shaft of the furnace, using the letter W, that is, with coordinates in Fig. 6.13 of (1.29, 1.05). Thus, in some way, the coordinates of W depend on the reducibility of the wustite.

Exercise 6.11 From the negative ordinates of the operation line, the first point of interest is D (Fig. 6.13). Check the equations to estimate the abscissa X_v (Eq. 6.159), and the ordinate Y_f (Eq. 6.160) associated with this point. Determine also the coordinates of the point D for a blast furnace that produces: 6000 t day^{-1} of pig iron (4.50% C, 1.00% Mn, 0.50% Si, 0.04% P, 0.04% S, and 93.92% Fe), 250 kg of slag per ton of pig iron with 0.02% S, and gas with 23% CO, 20% CO_2, and 57% N_2.

The composition in nitrogen of the gases that leave the furnace is related with the quantity of oxygen that, by means of the air (blast), is injected via the tuyeres. If we assume a volumetric composition for the air of 79% N_2 and 21% O_2, the number of moles of oxygen, $n(O)$, in the gas of tuyeres is directly proportional to that of the nitrogen in the gas of the throat:

$$n(O) = \frac{21 \ mol \ O_2}{79\% \ N_2} \cdot \frac{2 \ mol \ O}{1 \ mol \ O_2} \cdot \%N_2 = \frac{2 \ mol \ O}{3.76\% \ N_2} \cdot \% \ N_2 \qquad (6.207)$$

On the other hand, the quantity of CO in the gas that can be produced by partial combustion of the oxygen with excess of coke stored in the hearth of the furnace, and the number of moles of carbon associated with its formation, $n(C)$, also allows estimating the composition of the gas of the throat without including the nitrogen:

$$n(C) = 100 - \%N_2 \qquad (6.208)$$

Thus, the relation between the moles of oxygen and the moles of carbon (Eq. 6.159) is:

$$X(D) = \frac{n(O)}{n(C)} = X_v = \frac{\dfrac{2 \text{ mol O}}{3.76\% \text{ N}_2} \cdot \% \text{ N}_2}{(100 - \%\text{N}_2) \text{ mol C}} = \frac{2}{3.76} \cdot \left(\frac{\%\text{N}_2}{100 - \%\text{N}_2}\right)\left(\frac{\text{mol O}}{\text{mol C}}\right)$$

$$= \frac{2}{3.76} \cdot \left(\frac{57}{100 - 57}\right)\left(\frac{\text{mol O}}{\text{mol C}}\right) = 0.7051 \frac{\text{mol O}}{\text{mol C}}$$

$$(6.209)$$

As the values of the abscissas of the operation line move in positive values, 0.00–2.00 mol O per mol C, if all the gas that leaves the furnace was CO_2, the ordinates would vary within 1.50 (point A) and -1.50 (point E) moles O per mol Fe, depending on the blast injected via tuyeres. As positive ordinates, we have the points A (Exercise 6.9) and W (Exercise 6.10), and as negatives from the point C to the E (Fig. 6.14), and these last two points indicate the concentration of moles of O per mol of Fe generated by:

(a)   the blast injected via tuyeres,
(b)   the direct reduction with coke of the nonferrous elements that accompany the iron.

According to Eq. (6.160), Y_f represents the addition of the contribution of the moles of O per mol of Fe due to: the silicon $Y(Si)$, the manganese $Y(Mn)$, the phosphorus $Y(P)$, and the sulfur $Y(S)$. To obtain the contribution of the silicon, considering the reaction (6.19):

$$Y(Si) = \left(\frac{\%Si}{100 \text{ g pig iron}}\right) \cdot \left(\frac{100 \text{ g pig iron}}{\%Fe}\right) \cdot \left(\frac{55.85 \text{ g Fe}}{1 \text{ mol Fe}}\right) \cdot \left(\frac{1 \text{ mol Si}}{28.09 \text{ g Si}}\right)$$

$$\cdot \left(\frac{2 \text{ mol O}}{1 \text{ mol Si}}\right) = 3.98 \cdot \frac{\%Si}{\%Fe} = 3.98 \cdot \frac{0.50}{93.92}\left(\frac{\text{mol O}}{\text{mol Fe}}\right)$$

$$= 0.0212 \frac{\text{mol O}}{\text{mol Fe}}$$

$$(6.210)$$

To calculate the contribution of the manganese, according to the carbothermal reduction (reaction 6.17):

$$Y(Mn) = \left(\frac{\%Mn}{100 \ g \ pig \ iron}\right) \cdot \left(\frac{100 \ g \ pig \ iron}{\%Fe}\right) \cdot \left(\frac{55.85 \ g \ Fe}{1 \ mol \ Fe}\right) \cdot \left(\frac{1 \ mol \ Mn}{54.94 \ g \ Mn}\right)$$

$$\cdot \left(\frac{1 \ mol \ O}{1 \ mol \ Mn}\right) = 1.02 \cdot \frac{\%Mn}{\%Fe} = 1.02 \cdot \frac{1.00}{93.92} \left(\frac{mol \ O}{mol \ Fe}\right)$$

$$= 0.0109 \ \frac{molO}{molFe}$$

$$(6.211)$$

The contribution of the phosphorus, (reaction 6.14), is:

$$Y(P) = \left(\frac{\%P}{100 \ g \ pig \ iron}\right) \cdot \left(\frac{100 \ g \ pig \ iron}{\%Fe}\right) \cdot \left(\frac{55.85 \ g \ Fe}{1 \ mol \ Fe}\right) \cdot \left(\frac{1 \ mol \ P}{30.97 \ g \ P}\right)$$

$$\cdot \left(\frac{5 \ mol \ O}{2 \ mol \ P}\right) = 4.51 \cdot \frac{\%P}{\%Fe} = 4.51 \cdot \frac{0.04}{93.92} \left(\frac{mol \ O}{mol \ Fe}\right)$$

$$= 0.0019 \ \frac{mol \ O}{mol \ Fe}$$

$$(6.212)$$

Finally, to estimate the quantity of oxygen provided by the sulfur, we consider the exchange reaction (6.21) and the percentage of sulfur in the slag, $\%S_{slag}$, as well as the quantity of slag (kg) generated per ton of pig iron, M_{slag}:

$$Y(S) = \left(\frac{\%S_{slag}}{100 \ g \ slag}\right) \cdot \left(\frac{1 \ mol \ S}{32.06 \ g \ S}\right) \cdot \left(\frac{1 \ mol \ O}{1 \ mol \ S}\right) \cdot \left(\frac{M_{slag}}{t \ pig \ iron}\right) \times \left(\frac{10^3 \ g \ slag}{1 \ kg \ slag}\right)$$

$$\cdot \left(\frac{1 \ t \ pig \ iron}{10^6 \ g \ pig \ iron}\right) \cdot \left(\frac{100 \ g \ pig \ iron}{\%Fe}\right) \cdot \left(\frac{55.85 \ g \ Fe}{1 \ mol \ Fe}\right)$$

$$= 1.74 \times 10^{-3} \cdot \left(\frac{\%S_{slag} \cdot M_{slag}}{\%Fe}\right) \cdot \left(\frac{mol \ O}{mol \ Fe}\right)$$

$$= 9.263 \times 10^{-5} \ \frac{mol \ O}{mol \ Fe}$$

$$(6.213)$$

So, the value of Y_f $(Y(D))$ is:

$$Y_f = Y(D) = -[Y(Si) + Y(Mn) + Y(P) + Y(S)]$$

$$= -(0.0212 + 0.0109 + 0.0019 + 9.263 \cdot 10^{-5}) \frac{mol \ O}{mol \ Fe} \quad (6.214)$$

$$= -0.0341 \ \frac{mol \ O}{mol \ Fe}$$

Accordingly, the coordinates of the point D are $X(D) = 0.7051$ and $Y(D) = -0.0341$.

Commentaries: In the ironmaking, it is usually understood as "yield" of the CO in the blast furnace the ratio CO/CO_2 (% vol.) of the gases in the throat of the furnace, η, and thus, it is possible to estimate the calculation of the abscissa of the point A (Eq. 6.154):

$$X(A) = 1 + \frac{1}{1 + \eta} \tag{6.215}$$

Exercise 6.12 Other of the critical points of the operation line of the blast furnace with negative coordinate is the point E (Fig. 6.13). Determine its coordinates for a blast furnace with the following characteristics (data also indicated in Exercise 6.11):

- Daily production: 6000 tons of pig iron,
- Blowing flow rate: 225,500 Nm^3 h^{-1},
- Composition of the produced pig iron: 4.50% C, 1.00% Mn, 0.50% Si, 0.04% P, 0.04% S, and 93.92% Fe.
- Production of 250 kg of slag per ton of pig iron with 0.02% S.
- Composition of the gases in the throat: 23% CO, 20% CO_2, and 57% N_2.

The position of the point E is determined by the value of the abscissa, $X(E)$, which takes the value of zero, while the value of the ordinate, $Y(E)$, is:

$$Y(E) = Y_f + Y_v \tag{6.216}$$

The value of Y_f was calculated in Exercise 6.11 ($Y_f = -0.0341$), and the value of Y_v is estimated by calculating the specific consumption of blast in the blast furnace, CEV, in Nm^3 of air per ton of pig iron:

$$Y_v = CEV = \left(\frac{225{,}500\,Nm^3}{h}\right) \cdot \left(\frac{day}{6000\,t\,pig\,iron}\right) \cdot \left(\frac{24\,h}{1\,day}\right) \tag{6.217}$$
$$= 902\,Nm^3/t\,pig\,iron$$

On the other hand, the moles of O per ton of pig iron, obtained from the calculation of the CEV, are:

$$Y_v = \frac{CEV\,Nm^3}{t\,pig\,iron} \cdot \frac{21\,m^3\,O_2}{100\,m^3\,air} \times \frac{10^3\,l\,O_2}{1\,m^3 O_2} \cdot \frac{1\,mol\,O_2}{22.41\,O_2} \cdot \frac{2\,mol\,O}{1\,mol\,O_2} \tag{6.218}$$
$$= 18.75 \cdot CEV \left(\frac{mol\,O}{t\,pig\,iron}\right)$$

So, the value of Y_v in moles of O per mol of Fe is:

$$Y_v = \frac{18.75 \cdot \text{CEV mol O}}{\text{t pig iron}} \cdot \frac{\text{t pig iron}}{10^6 \text{ g pig iron}} \cdot \frac{100 \text{ g pig iron}}{\%\text{Fe}} \cdot \frac{55.85 \text{ g Fe}}{1 \text{ mol Fe}}$$

$$= 0.1047 \cdot \left(\frac{\text{CEV}}{\%\text{Fe}}\right) \left(\frac{\text{mol O}}{\text{mol Fe}}\right) = 0.1047 \cdot \left(\frac{902}{93.92}\right) \left(\frac{\text{mol O}}{\text{mol Fe}}\right)$$

$$= -1.0055 \frac{\text{mol O}}{\text{mol Fe}}$$

$$(6.219)$$

And $Y(E)$ is:

$$Y(E) = -0.0341 - 1.0055 = -1.0396 \frac{\text{mol O}}{\text{mol Fe}} \qquad (6.220)$$

Consequently, the coordinates of the point E are $X(E) = 0$ and $Y(E) = -1.0396$.

Commentaries: The relation between the specific consumption of blast, CEV, in Nm^3 per ton of pig iron, the blast rate, CS, in $\text{Nm}^3 \text{ h}^{-1}$, and the daily production of pig iron, P, in t day^{-1}, will be:

$$\text{CEV} = \frac{24 \cdot CS}{P} \qquad (6.221)$$

Exercise 6.13 The operation line for a certain operation of the blast furnace is characterized by:

$$Y = 1.82 \cdot X - 1.15 \qquad (6.222)$$

The coke that feeds the furnace has 94%C (balance ashes), and the pig iron obtained has 94.0% Fe, 4.0%, and 2% of other elements dissolved in the pig iron (that do not have influence in the problem). On the other hand, the volumetric analysis of the gases of the throat is: 31% CO, 19% CO_2, and 50% N_2. If the blast that is injected via tuyeres does not contain water vapor (dry air with composition: 79% N_2 and 21% O_2 in volume), calculate:

(a) the degree of oxidation of the charge,
(b) the percentage of direct reduction (with coke),
(c) the specific coke consumption, expressed in kilograms per ton of pig iron,
(d) the specific air (blast) consumption injected via tuyeres, expressed in Nm^3 per ton of pig iron.

The ordinate of the point $X(A)$ in Fig. 6.13 (Eq. 6.154) is:

$$X(A) = 1 + \frac{C_{CO_2}}{C_{CO} + C_{CO_2}} = 1 + \frac{19}{31 + 19} = 1.38 \qquad (6.223)$$

where the yield of the gas in the furnace, η, is:

$$X(A) = 1 + \frac{C_{CO_2}}{C_{CO} + C_{CO_2}} = 1 + \frac{1}{1 + \eta} \rightarrow \eta = \frac{\%CO}{\%CO_2} = \frac{31}{19} = 1.63 \qquad (6.224)$$

Thus, the value of $Y(A)$, degree of oxidation of the charge, would be the relation of moles of oxygen and moles of iron (Eq. 6.222):

$$Y(A) = \frac{n(O)}{n(Fe)} = 1.82 \cdot 1.38 - 1.15 = 1.36 \; \frac{mol\,O}{mol\,Fe} \qquad (6.225)$$

The contribution of the direct reduction mechanism (with coke, when $X(B) = 1$) in the obtaining of pig iron will be (Fig. 6.13):

$$\%\,direct\,reduction = \frac{Y(B)}{Y(A)} \cdot 100 \overset{X(B)=1}{\Rightarrow} \%\,direct\,reduction$$

$$= \frac{1.82 \cdot 1 - 1.15}{1.36} \cdot 100 = 49\% \qquad (6.226)$$

To calculate the coke consumption, firstly, we consider the coke used in the reduction of iron and its impurities, μ:

$$\mu = \frac{1.82\,mol\,C}{mol\,Fe} \cdot \frac{12.01\,g\,C}{1\,mol\,C} \cdot \frac{100\,g\,coke}{94\,g\,C} \cdot \frac{1\,mol\,Fe}{55.85\,g\,Fe} \cdot \frac{940\,g\,Fe}{1\,kg\,pig\,iron}$$

$$\times \frac{10^3\,kg\,pig\,iron}{1\,t\,pig\,iron} \cdot \frac{1\,kg\,coke}{10^3\,g\,coke} = 391 \; \frac{kg\,coke}{t\,pig\,iron} \qquad (6.227)$$

Secondly, we consider the dissolution of the carbon in the pig iron (pig iron that leaves the furnace), γ:

$$\gamma = \frac{40\,g\,C}{kg\,pig\,iron} \cdot \left(\frac{100\,g\,coke}{94\,g\,C} \right) \times \frac{10^3\,kg\,pig\,iron}{1\,t\,pig\,iron} \cdot \frac{1\,kg\,coke}{10^3\,g\,coke}$$

$$= 42 \; \frac{kg\,coke}{t\,pig\,iron} \qquad (6.228)$$

The addition of both contributions, μ and γ, is equivalent to the total coke in the blast furnace:

$$\mu + \gamma = \frac{n(C)}{n(Fe)} = (391 + 42)\,\frac{g\;coke}{t\;pig\;iron} = 433\,\frac{g\;coke}{t\;pig\;iron} \tag{6.229}$$

Moreover, the specific consumption of blast, Y_v, can be calculated using Eq. (6.159) and Fig. 6.13:

$$X(D) = \frac{n(O)}{n(C)} = X_v = \frac{\dfrac{2\;mol\;O}{3.76\%\;N_2}\cdot\%\,N_2}{(100 - \%N_2)\,mol\,C} = \frac{2}{3.76}\cdot\left(\frac{\%N_2}{100 - \%N_2}\right)\left(\frac{mol\,O}{mol\,C}\right)$$
$$= \frac{2}{3.76}\cdot\left(\frac{50}{100 - 50}\right)\left(\frac{mol\,O}{mol\,C}\right) = 0.53\,\frac{mol\,O}{mol\,C} \tag{6.230}$$

$$Y(D) = 1.82 \cdot 0.53 - 1.15 = -0.18\,\frac{mol\,O}{mol\,Fe} \tag{6.231}$$

$$Y_v = Y(E) - Y(U) = 0.97\,\frac{mol\,O}{mol\,Fe} = \frac{n(O)}{n(Fe)} \tag{6.232}$$

$$Y_v = \frac{0.97\;mol\,O}{mol\,Fe}\cdot\frac{1\;mol\,O_2}{2\;mol\,O}\cdot\frac{22.41O_2}{1\;mol\,O_2}\cdot\frac{100\,l\;air}{21\,1O_2}\cdot\frac{1\;mol\,Fe}{55.85\,g\;Fe}\cdot\frac{940\,g\;Fe}{1\,kg\;pig\;iron}$$
$$= 871\,\frac{1\;air}{kg\;pig\;iron}\cdot\frac{1\,m^3air}{10^3\,l\;air}\times\frac{10^3\,kg\;pig\;iron}{1\,t\;pig\;iron}$$
$$= 871\,\frac{Nm^3air}{t\;pig\;iron} \tag{6.233}$$

Consequently, the specific consumption of blast, CEV, is 871 Nm^3 per ton of pig iron.

Comment: The blast rate (air) blown via tuyeres is also usually expressed in Nm^3 per unit of time (day or hour). In this case, it would be necessary to know the production of the installation in tons of pig iron per day (Exercise 6.12).

Exercise 6.14 Calculate the equation of the ideal operation line of a blast furnace whose ferric charge comprised only Fe_2O_3 (hematite), being the composition of the gases that leave the furnace: 20% CO, 20% CO_2, 59% N_2, and 1% H_2. On the other hand, the produced pig iron has 5.0% C and 95% Fe. Find the following:

(a) Calculate the specific coke consumption.
(b) Estimate the production of pig iron of the furnace, under ideal operation conditions, being the blowing flow rate of 150,000 Nm^3 h^{-1}.

Assume that the quantity of oxygen that is transferred to the gas by direct reduction (with coke) of the nonferrous species or elements of the charge is negligible if compared with the quantity of oxygen supplied by the blast (as it is possible to deduce from the chemical analysis of the produced pig iron).

As the ordinates of the points A and W are known (Fig. 6.13), it is possible to obtain the equation of the operation line of the furnace:

- For the point A:

$$Y(A) = \frac{n(O)}{n(Fe)} = \frac{\frac{3}{2} \cdot A + B}{A + B} \tag{6.234}$$

$B = 0$ because the charge comprised only Fe_2O_3, and in this way:

$$Y(A) = \frac{\frac{3}{2} \cdot A}{A} = 1.5 \, \frac{mol \, O}{mol \, Fe} \tag{6.235}$$

$$X(A) = 1 + \frac{\%CO_2}{\%CO_2 + \%CO} = 1 + \frac{20}{20 + 20} = 1.5 \, \frac{mol \, O}{mol \, C} \tag{6.236}$$

- For the point W (Exercise 6.10), considering $X(W) = 1.29$ and $Y(W) = 1.05$:

$$1.05 = 1.29 \cdot \mu + b \tag{6.237}$$

$$1.5 = 1.5 \cdot \mu + b \tag{6.238}$$

The result of solving the two previous equations is:

$$\mu = 2.14 \, and \, b = -1.71 \tag{6.239}$$

That is, the equation of the operation line of the furnace is:

$$Y = 2.14 \cdot X - 1.71 \tag{6.240}$$

The participation of the direct (with coke) and indirect (with gas) reduction mechanisms is:

$$\% \, direct \, reduction = \frac{Y(B)}{Y(A)} \stackrel{X(B)=1}{\Rightarrow} \% \, direct \, reduction$$
$$= \frac{2.14 \cdot 1 - 1.71}{1.5} \cdot 100 = 29\% \tag{6.241}$$

$$\% \, indirect \, reduction = 100 - \% \, direct \, reduction = 71\% \tag{6.242}$$

The contribution of the carbon dissolved in the iron to the specific coke consumption, γ, is:

$$\gamma = \frac{5\,\text{g C}}{95\,\text{g Fe}} \cdot \frac{1\,\text{mol C}}{12.01\,\text{g C}} \cdot \frac{55.85\,\text{g Fe}}{1\,\text{mol Fe}} = 0.245 \, \frac{\text{mol C}}{\text{mol Fe}} \tag{6.243}$$

In total, the addition of both:

$$\gamma + \mu = \frac{n(\text{C})}{n(\text{Fe})} = 0.245 + 2.140 = 2.385 \, \frac{\text{mol C}}{\text{mol Fe}} \tag{6.244}$$

Thus, the specific coke consumption is:

$$2.385 \, \frac{\text{mol C}}{\text{mol Fe}} \cdot \frac{12.01\,\text{g C}}{1\,\text{mol C}} \cdot \frac{1\,\text{kg C}}{10^3\,\text{g C}} \cdot \frac{1\,\text{mol Fe}}{55.85\,\text{g Fe}} \cdot \frac{950\,\text{g Fe}}{1\,\text{kg pig iron}}$$
$$\times \frac{10^3\,\text{kg pig iron}}{1\,\text{t pig iron}} = 487 \, \frac{\text{kg C}}{\text{t pig iron}} \tag{6.245}$$

We calculate now the consumption of air per ton of pig iron:

$$Y(E) = 1.71 \, \frac{\text{mol O}}{\text{mol Fe}} \cdot \frac{1\,\text{mol O}_2}{2\,\text{mol O}} \cdot \frac{1\,\text{mol air}}{0.21\,\text{mol O}_2} \cdot \frac{22.41\,\text{air}}{1\,\text{mol air}} \cdot \frac{1\,\text{Nm}^3\,\text{air}}{10^3\,\text{l air}}$$
$$\cdot \frac{1\,\text{mol Fe}}{55.85\,\text{g Fe}} \cdot \frac{950\,\text{g Fe}}{1\,\text{kg pig iron}} \times \frac{10^3\,\text{kg pig iron}}{1\,\text{t pig iron}} \tag{6.246}$$
$$= 1551.3 \, \frac{\text{Nm}^3\,\text{air}}{\text{t pig iron}}$$

We consider that the blowing flow rate is $150{,}000\,\text{Nm}^3\,\text{h}^{-1}$ to calculate the production of the furnace:

$$150{,}000 \, \frac{\text{Nm}^3\,\text{air}}{\text{h}} \cdot \frac{24\,\text{h}}{1\,\text{day}} \cdot \frac{\text{t pig iron}}{1551.3\,\text{Nm}^3\,\text{air}} = 2320.6 \, \frac{\text{t pig iron}}{\text{day}}$$
$$= 847{,}032 \, \frac{\text{t pig iron}}{\text{year}} \tag{6.247}$$

Commentaries: The capacity of production of a blast furnace depends on the blowing flow rate (quantity of hot blast that can be injected to the hearth per hour). Besides, this capacity of production is conditioned by the hearth's size and the duration of the campaign (lifetime) of the furnace. Habitually, the worn-out hearths have a bigger capacity of storing pig iron and their production is greater. However, the magnitude of the wear of the lining and bottom of the hearth must not endanger the security of the installation.

Conceptually, the ideal operation line passes through the points P and W (Exercise 6.17), although if the points A and W or E and W are known, it is possible to adjust a pseudo-ideal operation line for these conditions.

Exercise 6.15 To estimate the coordinates of the point P of the operation line, we must value the terms involved in the energy balance of the blast furnace in the elaboration zone (zone H, Fig. 6.12): variable Q, Eq. (6.167), which corresponds to a summation of terms that express the different types of calorific energy exchanged in the smelting zone (H) and in the chemical reserve zone (A) of the blast furnace, Figs. 6.12 and 6.13.

However, to estimate the numerical values, we must take into account that the operation conditions of the blast furnace are those indicated in Exercise 6.11. Apart from considering the above-indicated variables, it is necessary to add:

- The temperature of the liquid phases that leave the furnace (pig iron and slag) is 1500 °C.
- The reference temperature for the energy balance in the elaboration zone is 1000 °C (1273 K; see Fig. 6.12). Equally, the temperature of the injected hot blast is 1200 °C.
- 25% of the iron in the pig iron comes from the direct reduction mechanism (with coke) of the FeO. The rest of the iron is obtained by reduction with CO gas. However, from the 75% of the iron reduced with gas, 50% of the total iron is reduced in the zone of preparation of the burden (zone G, Fig. 6.12) and the 25% of the rest of the total iron is reduced in the elaboration zone of the furnace.

Most of the data about the standard enthalpy of the reactions are found in Appendix 2. We only indicate the source of the thermodynamic data when these data have a different origin.

From the terms involved in the calculation of Q (Eq. 6.167), some of them are positive (supply of heat) and others are negatives (they require heat supply). We should remember that the enthalpy and the heat have different signs (exothermic reaction, $\Delta H < 0$, heat released, $q < 0$, and vice versa). We have within the positive terms:

(A)   Sensitive heat supplied by the pig iron, q_{sa}:

The variation of the specific heat, at constant pressure, of the pig iron with the temperature (Verdeja et al. 2014, page 196) is:

$$c_p = 1260 - 1.06 \cdot T + 4.66 \cdot 10^{-4} \cdot T^2 \; \left(\text{J kg}^{-1} \, \text{K}^{-1} \right) \tag{6.248}$$

At 1250 °C, the value of the specific heat of the pig iron is:

$$c_p = 1260 - 1.06 \cdot 1523 + 4.66 \times 10^{-4} \cdot 1523^2 = 726.52 \, \frac{\text{J}}{\text{kg K}} \tag{6.249}$$

Taking 1000 °C as reference temperature of the balance:

$$q_{sa} = 726.52 \frac{J}{kg\,K} \cdot (1773 - 1273)K = 363{,}260 \frac{J}{kg\,pig\,iron} \qquad (6.250)$$

$$q_{sa} = 363{,}260 \frac{J}{kg\,pig\,iron} \cdot \frac{1\,kg\,pig\,iron}{10^3\,g\,pig\,iron} \cdot \frac{1\,g\,pig\,iron}{0.9392\,g\,Fe} \cdot \frac{55.85\,g\,Fe}{1\,mol\,Fe} \cdot \frac{1\,kJ}{10^3\,J}$$

$$= 21.60 \frac{kJ}{mol\,Fe}$$

$$(6.251)$$

(B)   Sensitive heat supplied by the slag, q_{se}:

We take the average value of the three slags of Exercise 6.33, 1445.2 J kg^{-1} K^{-1}, as value specific heat of the slag. The sensitive heat supplied by this melt will be:

$$q_{se} = 1445.2 \frac{J}{kg\,K} \cdot (1773 - 1273)K = 722{,}600 \frac{J}{kg\,slag} \qquad (6.252)$$

$$q_{se} = 722{,}600 \frac{J}{kg\,slag} \cdot \frac{250\,kg\,slag}{10^3\,kg\,pig\,iron} \cdot \frac{1\,kg\,pig\,iron}{939.2\,g\,Fe} \cdot \frac{55.85\,g\,Fe}{1\,mol\,Fe} \cdot \frac{1\,kJ}{10^3\,J}$$

$$= 10.74 \frac{kJ}{mol\,Fe}$$

$$(6.253)$$

(C)   Heat of the partial combustion of the carbon (coke), q_c:

The variation with the temperature of the standard enthalpy associated with the reaction (6.254):

$$C(s) + \frac{1}{2}O_2(g) \leftrightarrow CO(g) \qquad (6.254)$$

is:

$$\Delta_r H^0(T) = -104.0 - 0.0075 \cdot T(K)\ \left(kJ\,mol\,O^{-1}\right) \qquad (6.255)$$

Considering the temperature of reaction equal to that used as reference for the thermal balance, that is 1000 °C; the standard enthalpy corresponding to the reaction is:

$$\Delta_r H^0(1273\,K) = -104.0 - 0.0075 \cdot 1273 = -113.55 \frac{kJ}{mol\,O} \qquad (6.256)$$

To express the supply in kJ mol Fe^{-1}, it is necessary to remember that the quantity of oxygen generated in the blast furnace per unit of iron, $|Y_v|$, was calculated in Exercise 6.12: 1.0055 mol O per mol Fe. Thus, the value of q_c is:

$$q_c = Y_v \cdot \Delta_r H^0 = - \left[1.0055 \frac{\text{mol O}}{\text{mol Fe}} \cdot \left(-113.55 \frac{\text{kJ}}{\text{mol O}} \right) \right] = 114.17 \frac{\text{kJ}}{\text{mol Fe}}$$

(6.257)

(D) Sensitive heat supplied by the hot blast (blast free of moisture) injected via tuyeres of the furnace, q_{sv}:

From the database of the software HSC 5.1, we indicate, as a function of the temperature, the values of the specific heat at constant pressure of the air whose composition is 79% N_2 and 21% O_2 (Table 6.9).

The sensitive heat supplied by the blast free of moisture (contribution of the hot blast) is calculated considering the value of the ordinate $|Y_v|$ (Exercise 6.12):

$$q_{sv} = 0.08341 \frac{\text{kJ}}{\text{mol O} \cdot \text{K}} \cdot (1473 - 1273)\text{K} \cdot 1.0055 \frac{\text{mol O}}{\text{mol Fe}} = 16.77 \frac{\text{kJ}}{\text{mol Fe}}$$

(6.258)

(E) Reduction with gas in the elaboration zone, q_{rg}:

As part of the FeO reduction process takes place in the elaboration zone, the variation of the standard enthalpy with the temperature of reaction is:

$$FeO(s) + CO(g) \leftrightarrow CO_2(g) + Fe(s)$$

(6.259)

$$\Delta_r H^0(T) = -18.0 - 0.0022 \cdot T(K) \frac{\text{kJ}}{\text{mol Fe}}$$

(6.260)

Table 6.9 Specific heat as a function of the temperature

T (°C)	c_p N_2 (kJ mol^{-1} K^{-1})	c_p O_2 (kJ mol^{-1} K^{-1})	c_p air (kJ mol^{-1} K^{-1})	c_p air (kJ mol O^{-1} K^{-1})
1000	0.03401	0.03565	0.03436	0.08180
1100	0.03436	0.03603	0.03471	0.08265
1200	0.03467	0.03637	0.03503	0.08341
1300	0.03496	0.03669	0.03532	0.08409

Considering the temperature of reaction equal to that used as reference for the thermal balance, that is 1000 °C; the standard enthalpy corresponding to the reaction is:

$$\Delta_r H^0(1273\ \text{K}) = -18.0 - 0.0022 \cdot 1273 = -20.8\ \frac{\text{kJ}}{\text{mol Fe}} \tag{6.261}$$

And consequently, the value of q_{rg} when 25% of the iron is reduced with CO in the elaboration zone is:

$$q_{rg} = 0.25 \cdot 20.8\ \frac{\text{kJ}}{\text{mol Fe}} = 5.20\ \frac{\text{kJ}}{\text{mol Fe}} \tag{6.262}$$

From the group of terms that require heat for their development, we can consider the next ones:

(F) The carbothermal reduction (direct reduction with coke) of the FeO, q_{rd-Fe}:

$$\text{FeO(s)} + \text{C(s)} \leftrightarrow \text{CO(g)} + \text{Fe(s)} \tag{6.263}$$

$$\Delta_r H^0(T) = 167.0 - 0.0155 \cdot T(\text{K})\frac{\text{kJ}}{\text{mol Fe}} \tag{6.264}$$

The variation with the temperature of the standard enthalpy of this reaction is:

$$\Delta_r H^0(1273\ \text{K}) = 167.0 - 0.0155 \cdot 1273 = 147.269\ \frac{\text{kJ}}{\text{mol Fe}} \tag{6.265}$$

Finally, to calculate the energy required for the carbothermal reduction of 25% of the iron:

$$q_{rd-Fe} = -0.25 \cdot 147.269\ \frac{\text{kJ}}{\text{mol Fe}} = -36.82\ \frac{\text{kJ}}{\text{mol Fe}} \tag{6.266}$$

(G) Carbothermal reduction of other elements, q_{rd-Me}:

The standard enthalpy of the reduction of the elements Mn, Si, and P that can be found in the composition of the pig iron in the quantities indicated in Exercise 6.11 is:

$$\Delta_r H^0(T)_{\text{eq. (6.17)}} = 290.0 - 8.66 \times 10^{-4} \cdot T(\text{K})\ (\text{kJ mol Mn}^{-1}\ \text{or kJ mol O}^{-1}) \tag{6.267}$$

$$\Delta_r H^0(T)_{eq.\,(6.19)} = 754.0 - 0.0168 \cdot T(K) \; (kJ\,mol\,Si^{-1}\,or\,kJ\,2\,mol\,O^{-1}) \quad (6.268)$$

$$\Delta_r H^0(T)_{eq.\,(6.14)} = 1094.0 - 0.0384 \cdot T(K) \left(kJ\,mol\,P^{-1}\,or\,kJ\left(\frac{5}{2}\right)mol\,O^{-1} \right)$$
$$(6.269)$$

Finally, the reduction of the SO_2 (g) with carbon is:

$$SO_2(g) + 2C(s) \leftrightarrow 2CO(g) + \frac{1}{2}S_2(g) \quad\quad (6.270)$$

$$\Delta_r H^0(T)_{eq.\,(6.270)} = 155.0 - 0.0159 \cdot T(K)\left(kJ\,mol\,S^{-1}\,or\,kJ\,2\,mol\,O^{-1}\right) \quad (6.271)$$

Considering that the temperature of the carbothermal reduction for the silicon and the manganese is 1773 K, while in the case of the sulfur and the phosphorus is 1273 K, the heat supply in $kJ\,mol\,Fe^{-1}$ in the zone of elaboration zone are (see Exercise 6.11):
The variation with the temperature of the standard enthalpy of this reaction is:

$$\Delta_r H^0(1773\,K)_{eq.\,(6.17)} = 290.0 - 8.66 \times 10^{-4} \cdot 1773 = 288.46\,\frac{kJ}{mol\,O} \quad (6.272)$$

$$q_{rd-Mn} = -288.46\,\frac{kJ}{mol\,O} \cdot 0.0109\,\frac{mol\,O}{mol\,Fe} = -3.14\,\frac{kJ}{mol\,Fe} \quad (6.273)$$

$$\Delta_r H^0(T)_{eq.\,(6.19)} = 754.0 - 0.0168 \cdot 1773 = 724.21\,\frac{kJ}{2\,mol\,O} \quad (6.274)$$

$$q_{rd-Si} = -724.21\,\frac{kJ}{2\,mol\,O} \cdot 0.0212\,\frac{mol\,O}{mol\,Fe} = -7.67\,\frac{kJ}{mol\,Fe} \quad (6.275)$$

$$\Delta_r H^0(T)_{eq.\,(6.14)} = 1094.0 - 0.0384 \cdot 1273\,\frac{kJ}{\frac{5}{2}mol\,O} \quad (6.276)$$

$$q_{rd-P} = -1045.12\,\frac{kJ}{\frac{5}{2}mol\,O} \cdot 0.0019\,\frac{mol\,O}{mol\,Fe} = -0.79\,\frac{kJ}{mol\,Fe} \quad (6.277)$$

$$\Delta_r H^0(T)_{eq.\,(6.270)} = 155.0 - 0.0159 \cdot 1273\,\frac{kJ}{2\,mol\,O} \quad (6.278)$$

$$q_{rd-S} = -134.76\,\frac{kJ}{2\,mol\,O} \cdot 9.263 \times 10^{-5}\,\frac{mol\,O}{mol\,Fe} = -0.0062\,\frac{kJ}{mol\,Fe} \quad (6.279)$$

Consequently, the value of q_{rd-Me} is:

$$q_{rd-Me} = q_{rd-Mn} + q_{rd-Si} + q_{rd-S} + q_{rd-P}$$

$$= (-3.14 - 7.67 - 0.79 - 0.0062)\left(\frac{kJ}{mol\,Fe}\right) \qquad (6.280)$$

$$= -11.61\frac{kJ}{mol\,Fe}$$

(H) The dissolution of the elements C, Mn, Si, P, and S dissolved in the pig iron, q_{dis-Me}:

As the elements Mn, Si, P, and S are obtained by carbothermal reduction of their oxides, we must take into account the possible energy exchange by their dissolution in the iron. Apart from these four elements, we must also consider the dissolution of the reductant (carbon) in the iron. To make this estimation, we identify the origin ordinate (independent term) of the variation with the temperature of the standard free energies of Appendix 3 (column of A) with the value of the dissolution enthalpy of the C, Mn, Si, P, and S. The obtained values, in kJ mol Fe^{-1}, are (considering also the data of the enunciation of Exercise 6.11):

$$q_{dis-C} = -19,000\frac{J}{mol\,C} \cdot \frac{1\,mol\,C}{12.01\,g\,C} \cdot \frac{4.5\,g\,C}{93.92\,g\,Fe} \cdot \frac{55.85\,g\,Fe}{1\,mol\,Fe} \cdot \frac{1\,kJ}{10^3\,J}$$
$$= -4.43\frac{kJ}{mol\,Fe} \qquad (6.281)$$

$$q_{dis-Mn} = -4084\frac{J}{mol\,Mn} \cdot \frac{1\,mol\,Mn}{54.94\,g\,Mn} \cdot \frac{1\,g\,Mn}{93.92\,g\,Fe} \cdot \frac{55.85\,g\,Fe}{1\,mol\,Fe} \cdot \frac{1\,kJ}{10^3\,J}$$
$$= -0.04\frac{kJ}{mol\,Fe} \qquad (6.282)$$

$$q_{dis-P} = 140,000\frac{J}{mol\,P} \cdot \frac{1\,mol\,P}{30.97\,g\,P} \cdot \frac{0.04\,g\,P}{93.92\,g\,Fe} \cdot \frac{55.85\,g\,Fe}{1\,mol\,Fe} \cdot \frac{1\,kJ}{10^3\,J}$$
$$= 0.11\frac{kJ}{mol\,Fe} \qquad (6.283)$$

$$q_{dis-S} = 130,000\frac{J}{mol\,S} \cdot \frac{1\,mol\,S}{32.06\,g\,S} \cdot \frac{0.04\,g\,S}{93.92\,g\,Fe} \cdot \frac{55.85\,g\,Fe}{1\,mol\,Fe} \cdot \frac{1\,kJ}{10^3\,J}$$
$$= 0.10\frac{kJ}{mol\,Fe} \qquad (6.284)$$

$$q_{dis-Si} = 132,000 \frac{J}{mol\,Si} \cdot \frac{1\,mol\,Si}{28.09\,g\,Si} \cdot \frac{0.5\,g\,Si}{93.92\,g\,Fe} \cdot \frac{55.85\,g\,Fe}{1\,mol\,Fe} \cdot \frac{1\,kJ}{10^3\,J}$$

$$= 1.40 \frac{kJ}{mol\,Fe} \tag{6.285}$$

Summing up, the contribution by dissolution of these five elements in the iron is (in total):

$$q_{dis-Me} = q_{dis-C} + q_{dis-Mn} + q_{dis-P} + q_{dis-S} + q_{dis-Si}$$

$$= -4.43 - 0.04 + 0.11 + 0.10 + 1.40 = -2.86 \frac{kJ}{mol\,Fe} \tag{6.286}$$

(I) Heat losses in the elaboration zone, q_{per}:

According to the results of Exercise 6.4, a blast furnace with a daily production of 6000 tons leads to a total thermal loss of 9.120×10^7 kJ h^{-1}. If 80% of the total losses are produced in the elaboration zone, the contribution will be:

$$q_{per} = 9.120 \times 10^7 \frac{kJ}{h} \cdot \frac{24\,h}{1\,day} \cdot \frac{day}{6000\,t\,pig\,iron} \cdot \frac{1\,t\,pig\,iron}{10^6\,g\,pig\,iron}$$

$$\cdot \frac{100\,g\,pig\,iron}{93.92\,g\,Fe} \cdot \frac{55.85\,g\,Fe}{1\,mol\,Fe} = -17.35 \frac{kJ}{mol\,Fe} \tag{6.287}$$

The total addition of the positive terms is:

$$q_{sa} + q_{se} + q_c + q_{sv} + q_{rg} = 21.60 + 10.74 + 114.17 + 16.77 + 5.20$$

$$= 168.48 \frac{kJ}{mol\,Fe} \tag{6.288}$$

The total addition of negative terms of the heat balance is:

$$q_{rd} + q_{rd-Me} + q_{dis-Me} + q_{per} = -36.82 - 11.61 - 2.86 - 17.35$$

$$= -68.64 \frac{kJ}{mol\,Fe} \tag{6.289}$$

Thus, the value of Q (Eq. 6.167) will be 99.84 kJ mol Fe^{-1}, and the corresponding value of $Y(V)$:

$$Y(V) = \frac{Q}{q_d} = -\frac{99.84}{147.269} = -0.6779 \tag{6.290}$$

where q_d represents the heat of direct reduction whether all FeO was directly reduced by the carbon.

Commentaries: We emphasize that the estimations to calculate the value of the ordinate in the point V of Fig. 6.13 depend, apart from on the operation variables of the furnace, on the agreement of the used thermodynamic data to the reality. For instance, in all the reactions where carbon participates, the state of matter assigned to the carbon by the thermodynamics corresponds to the carbon-graphite. Naturally, the structure of the coke, although similar, is different from that of the graphite and, consequently, all the reactions in which this element is involved should be reconsidered.

Whatever, the positive energy contributions derived from the possible formation of either intermetallic compounds in the pig iron or metallic silicates in the slag were not considered. A first approximation to the calculation of $Y(V)$ reaches values within -0.3500 and -0.7500.

Exercise 6.16 The last point of interest upon the operation line of the blast furnace is P^* (Fig. 6.13) that indicates the energetic equilibrium that is reached for the production of pig iron and slag with certain physical, chemical, and thermal characteristics. Calculate the values of $X(P^*)$ and $Y(P^*)$ assuming the operation conditions of Exercises 6.11 and 6.15.

The value of the abscissa of the point P^* of the operation line is estimated with the equation:

$$X(P^*) = \frac{n(O)}{n(C)} = \frac{|q_g|}{|q_g| + |q_c| + |q_{sv}|} \tag{6.291}$$

where q_g is the standard enthalpy of the carbon gasification (coke)(absolute value of the heat) and q_c and q_{sv} (calculated in Exercise 6.15) represent the heat or energy supplied by the partial combustion of the carbon and the hot blast injected via tuyeres (1200 °C):

$$q_c = -113.55 \frac{kJ}{mol\,O} \tag{6.292}$$

$$q_{sv} = 16.77 \frac{kJ}{mol\,O} \tag{6.293}$$

If we analyze the enthalpy of the carbon gasification in the form of coke (Boudouard reaction) using the data of Appendix 2, it is possible to state that the variation with the temperature of the standard enthalpy of the reaction:

$$C(s) + CO_2(g) \rightarrow 2CO(c) \tag{6.294}$$

is:

$$\Delta_r H^0(T) = 185.0 - 0.0133 \cdot T(K)\,(kJ\,mol\,CO_2^{-1}) \tag{6.295}$$

At 1250 °C (1523 K), the value of the standard enthalpy can be expressed as follows:

$$\Delta_r H^0(1523\,K) = 185.0 - 0.0133 \cdot 1523 = 164.74\,\frac{kJ}{mol\,CO_2}$$

$$= 164.74\,\frac{kJ}{2\,mol\,CO} = 82.37\,\frac{kJ}{mol\,O} = 82.37\,\frac{kJ}{mol\,C} \qquad (6.296)$$

(in this zone of the furnace, both the direct reduction with coke and the indirect reduction with gas (CO) takes place, which always implies a stoichiometric relation mol of oxygen per mol of iron). Thus, the value of the abscissa will be:

$$X(P^*) = \frac{82.37\,\frac{kJ}{mol\,C}}{82.37\,\frac{kJ}{mol\,O} + 113.55\,\frac{kJ}{mol\,O} + 16.77\,\frac{kJ}{mol\,O}} = 0.3873\,\frac{mol\,O}{mol\,C} \quad (6.297)$$

To determine the ordinate $Y(P^*)$, as it is possible to check in Fig. 6.13, triangles are drawn between the points U, P^* and V, which leads to the following equality:

$$\frac{Y(U) - Y(V)}{(1)} = \frac{Y(U) - Y(P^*)}{X(P^*)} \qquad (6.298)$$

$$Y(P^*) = Y(U) - X(P^*) \cdot [Y(U) - Y(V)] \qquad (6.299)$$

As the value of $Y(U) = Y_f$ and as $Y(V)$ is obtained using Eq. (6.167), it is possible to calculate $Y(P^*)$ using the value of the ordinates of U and V and the abscissa of the point P^*. From Exercise 6.11, the ordinate of U (Y_f) is -0.0341 and the value of the ordinate of the point V, $Y(V)$ is -0.6779 (Exercise 6.15), thus,

$$Y(P^*) = -0.0341 - 0.3873 \cdot [-0.0341 + 0.6779] = -0.2834 \qquad (6.300)$$

Commentaries: The point P^* represents the locus where the operation line intersects with that line that passes through U and V. All the operation lines of a furnace that could keep constant the composition of the pig iron and the slag can rotate to optimize the productivity (CEV, Exercise 6.12) with the minimum coke consumption. If the operation conditions are modified, the ordinates of the points U and V are also modified, but the position of the point P^* will be over the tie-line that joins them.

As a function of the specific operation characteristics of each furnace, it will be important to define with the maximum accurateness as possible, the coordinates of the point P^*, which have the values $X(P^*) = 0.26 \sim 0.65$ and $Y(P^*) = -0.10 \sim -0.45$.

Exercise 6.17 A blast furnace has operation conditions with coordinates of the point P^*: $Y(P^*) = -0.34$ and $X(P^*) = 0.52$. Define the equation of the ideal operation line and determine the following operation parameters of the furnace: daily production, coke consumption, and degree of oxidation of the charge. Data:

(a)   The characteristics of the pig iron are the same of Exercise 6.11.
(b)   The maximum flow that the tuyeres can supply each hour has a value of 250,000 Nm^3 of air.
(c)   The carbon content of the coke is 90%.

The slope of the ideal operation line is (Exercise 6.10):

$$\mu = \left(\frac{1.05 + 0.34}{1.29 - 0.52}\right) - \frac{1.39}{0.77} = 1.81 \; \frac{\text{mol C}}{\text{mol Fe}} \qquad (6.301)$$

The value of the origin ordinate, $Y(E)$, Eq. (6.153), is:

$$Y(E) = b = 1.05 - 1.81 \cdot 1.29 = -1.28 \; \frac{\text{mol O}}{\text{mol Fe}} \qquad (6.302)$$

Consequently, the equation of the ideal operation line will be:

$$Y = 1.81 \cdot X - 1.28 \qquad (6.303)$$

The production of the furnace can be estimated from the value of the ordinate Y_v, remembering that for the ensemble pig iron–slag of Exercise 6.11, $Y_f = -0.0341$:

$$Y(E) = Y_v + Y_f \qquad (6.304)$$

$$Y_v = Y(E) - Y_f = -1.28 - (-0.0341) = -1.2459 \; \frac{\text{mol O}}{\text{mol Fe}} \qquad (6.305)$$

$$1.2459 \; \frac{\text{mol O}}{\text{mol Fe}} \cdot \frac{1 \, \text{mol O}_2}{2 \, \text{mol O}} \cdot \frac{1 \, \text{mol air}}{0.21 \, \text{mol O}_2} \cdot \frac{22.4 \, \text{l air}}{1 \, \text{mol air}} \cdot \frac{1 \, \text{m}^3 \text{air}}{10^3 \, \text{l air}} \cdot \frac{1 \, \text{mol Fe}}{55.85 \, \text{g Fe}}$$

$$\cdot \frac{939.2 \, \text{g Fe}}{1 \, \text{kg pig iron}} \cdot \frac{10^3 \, \text{kg pig iron}}{1 \, \text{t pig iron}} = 1117.4 \frac{\text{Nm}^3 \text{air}}{\text{t pig iron}}$$

$$(6.306)$$

The daily production of pig iron, P, will be (Exercise 6.12):

$$P = \frac{250,000 \, \text{Nm}^3 \, \text{air}}{\text{h}} \cdot \frac{24 \, \text{h}}{\text{day}} \cdot \frac{\text{t pig iron}}{1117.4 \, \text{Nm}^3 \, \text{air}} = 5369.6 \; \frac{\text{t pig iron}}{\text{day}} \qquad (6.307)$$

The coke consumption is estimated by adding the contributions of μ and γ (Exercise 6.13):

$$\mu = 1.81 \frac{\text{mol C}}{\text{mol Fe}} \cdot \frac{12.01 \text{ g C}}{1 \text{ mol C}} \cdot \frac{100 \text{ g coke}}{90 \text{ g C}} \cdot \frac{1 \text{ kg coke}}{10^3 \text{ g coke}} \cdot \frac{1 \text{ mol Fe}}{55.85 \text{ g Fe}}$$
$$\cdot \frac{939.2 \text{ g Fe}}{1 \text{kg pig iron}} \cdot \frac{10^3 \text{ kg pig iron}}{1 \text{ t pig iron}} = 406.2 \frac{\text{kg coke}}{\text{t pig iron}} \qquad (6.308)$$

$$\gamma = \frac{45 \text{ g C}}{\text{kg pig iron}} \cdot \frac{10^3 \text{ kg pig iron}}{1 \text{ t pig iron}} \cdot \frac{100 \text{ g coke}}{90 \text{ g C}} \cdot \frac{1 \text{kg coke}}{10^3 \text{ g coke}} = 50 \frac{\text{kg coke}}{\text{t pig iron}}$$
$$(6.309)$$

$$\mu + \gamma = (406 + 50)\left(\frac{\text{kg coke}}{\text{t pig iron}}\right) = 456 \frac{\text{kg coke}}{\text{t pig iron}} \qquad (6.310)$$

The degree of oxidation of the charge, $Y(A)$, is calculated using the value of $X(A)$ (Eq. 6.154) and the composition of the gases in the throat: 23% CO, 20% CO_2, and 57% N_2. Thus,

$$X(A) = 1 + \frac{20}{20 + 23} = 1.4651 \qquad (6.311)$$

$$Y(A) = 1.81 \cdot 1.4651 - 1.28 = 1.3661 \qquad (6.312)$$

Commentaries: The coordinates of the point P^* are indispensable to define the equation of the ideal operation line, which indicates the best conditions of operation of the blast furnace and can be a suitable complement for the automatization and control of the furnace operation. Moreover, it is a parameter similar to the equilibrium constant in a system, toward which all the reactions should be directed, although it is never reached.

Finally, to assess the importance of reaching the highest level of accurateness possible in the estimation of P^*, for the values of $X(P^*) = 0.52$ and $Y(P^*) = -0.29$, the equation of the operation line is $Y = 1.74 \cdot X - 1.19$, the daily production would be 5820 tons of pig iron, the coke consumption would be of 441 kg per ton of pig iron, and the degree of oxidation of the charge would be 1.3545 mol of O per mol of iron.

6.5 Pretreatment of the Pig Iron

The introduction of new operations, which could control in a more demanding way the P and S concentrations in the finished product (steel), has been one of the main arguments of the Japanese ironmaking industry for the installation, between the steelworks and the blast furnace, of plants for the pretreatment of the pig iron.

The European ironmaking industry has been more prudent in the development of the pretreatment processes due to, mainly, the search of a simplification in the metallurgical operations as a fundamental hypothesis of its research activities. While in the Japanese ironmaking industry the desulfurizing and dephosphorization installations are usual, most of the pretreatment plants existing in the European industry have as objective the slagging of the sulfur.

The blast furnace does not eliminate the phosphorus, and almost all this element, available in the raw materials, is dissolved in the pig iron. Moreover, the blast furnace desulfurizes bad, as in the best situation only 25–30% of the sulfur, is collected in the furnace slag (Eq. 6.21) when using basic slags of gehlenite (Fig. 6.2).

6.5.1 Sulfur Removal

We shown in Sect. 6.3.2 the main equation (Eq. 6.21) that regulates the equilibrium of the sulfur between the metallic phase and the slag. The carbothermal reduction of the sulfur in reductant environments is a process thermodynamically favorable, Eq. (6.15). Equally, we highlight the importance of the activities of certain basic oxides in the slag as they increase the entrapment of the sulfur. However, the sulfur refinement in the ironmaking is not completed until the secondary metallurgy.

On the other hand, the pig iron is a suitable medium for the reaction (6.21); especially if it has some calcium compound that does not increase the O_2 potential in the medium and the formation of calcium sulfides is promoted to produce the slagging of the sulfur dissolved in the pig iron.

A calcium compound that has the above-indicated characteristics is the calcium carbide, CaC_2. In a desulfurizing installation, calcium carbide powders in an argon or dry air flow are injected to the pig iron (transported in ladle or torpedo car). Depending on the type of injection, the possible reactions are (Pero-Sanz et al. 2018):

- Under inert conditions (argon):

$$CaC_2(s) + S(dis; pig\ iron) \leftrightarrow CaS(dis; slag) + 2C(dis; pig\ iron) \qquad (6.313)$$

- Under oxidizing conditions (air) with partial oxidation of the carbon into carbon monoxide:

$$CaC_2(s) + S(dis; pig\ iron) + O(dis; slag)$$
$$\leftrightarrow CaS(dis; slag) + CO(g) + C(dis; pig\ iron) \qquad (6.314)$$

- Under oxidizing conditions (air) with total oxidation of the carbon into carbon monoxide:

$$CaC_2(s) + 2O(dis; pig\ iron) + S(dis; pig\ iron) \leftrightarrow CaS(dis; slag) + 2CO(g)$$

$$(6.315)$$

Equation (6.315) is the result of the addition of the following reactions:

$$CaC_2(s) + 3O(dis; pig\ iron) \leftrightarrow CaO(dis; slag) + 2CO(g) \qquad (6.316)$$

$$CaO(dis; slag) + S(dis; pig\ iron) \leftrightarrow CaS(dis; slag) + O(dis; pig\ iron) \quad (6.317)$$

6.5.2   Phosphorus Removal

Most of the phosphorus of the raw materials is assimilated to the pig iron (reaction 6.14)($\simeq 75\%$), and the rest finishes in the slag. If we want to produce a pig iron with low phosphorus content, the only alternative is restricting the presence of phosphorus in the burden of the furnace. The steelworks of the USA and Canada operate with burden (ferric + coke) whose phosphorus content is around 0.02%. Considering the data of the mass and energy balances of Tables 6.7 and 6.8, the quantity of phosphorus in the pig iron is:

$$P(dis; pig\ iron) = \frac{Burden\ (kg)}{Slag\ (kg) + Pig\ Iron\ (kg)} \cdot P(burden)$$
$$= \frac{2181\ kg}{391\ kg + 1000\ kg} \cdot P(burden) \simeq 1.6 \cdot 0.02\% = 0.032\%$$

$$(6.318)$$

while in the steelworks of Europe and Japan, where ores from Brazil, Venezuela, and Australia are used, the phosphorus content reaches (and many times exceed) the 0.05%:

$$P(dis; pig\ iron) \simeq 1.6 \cdot 0.05\% = 0.08\% \qquad (6.319)$$

If iron ores with low phosphorus are used, it is possible to recycle into the blast furnace part of the slag generated in the conversion process, and still producing low phosphorus steels (<0.015%) in steels (pearlitic) with 0.70% C. Formally, the elimination of the phosphorus dissolved in the liquid iron can be performed via slag or gaseous phase. The combinations of the oxygen with the phosphorus, PO and P_2O_5 (Appendix 1), are compounds that, at the average temperature of operation in the ironmaking processes, are found in gaseous state. Thus, we could think that, with the suitable oxygen potential, the phosphorus dissolved in the liquid iron can be removed in gaseous state.

At 1500 °C and with a total pressure of one atmosphere, it is possible to dephosphorize the liquid iron when the composition of products and reagents is:

- Average concentration of phosphorus dissolved in the liquid iron: 0.05%.
- POS (potential of oxygen of the system) of 1000 ppm.
- $P_{P_2O_5} = P_{PO} = 10^{-3}$ atm

The reactions of the phosphorus oxidation are:

$$2P(dis; pig\ iron) + 5O(dis; pig\ iron) \leftrightarrow P_2O_5(g) \qquad (6.320)$$

$$P(dis; pig\ iron) + O(dis; pig\ iron) \leftrightarrow PO(g) \qquad (6.321)$$

The values of free energy for the above-mentioned reactions are (Appendix 1 and Appendix 3):

$$\Delta_r G^0_{(reaction6.320)}(1500\ °C) = 366{,}000\ J\ (mol\ P_2O_5)^{-1} \qquad (6.322)$$

$$\Delta_r G^0_{(reaction6.321)}(1500\ °C) = 158{,}000\ J\ (mol\ PO)^{-1} \qquad (6.323)$$

These free energies indicate that it is not possible to remove the phosphorus from the iron in gaseous state. The only possibility of dephosphorizing is using a slag. The affinity of the alkaline (Li_2O, Na_2O, and K_2O) and alkaline earth (CaO, MgO, BaO y SrO) oxides available in the slag by the phosphorus oxides to form phosphates is a decisive factor to have a $\Delta_r G^0$ negative in the process. That is to say, the addition of the processes of phosphorus oxidation (Eq. 6.320) and the reaction of the P_2O_5 (g) with the basic oxides of the slag (Eq. 6.324):

$$P_2O_5(g) + 3CaO(dis; slag) \leftrightarrow Ca_3P_2O_8(dis; slag) \qquad (6.324)$$

lead to the following global reaction of dephosphorization:

$$2P(dis; pig\ iron) + 5O(dis; pig\ iron) + 3CaO(dis; slag) \leftrightarrow Ca_3P_2O_8(dis; slag) \qquad (6.325)$$

The value of the free energy of the reaction (6.325), at 1500 °C and at a total pressure of one atmosphere under the following conditions: phosphorus concentration of 0.05% dissolved in the iron, POS equals to 1000 ppm, a_{CaO} in the slag is equal to 1 and $a_{Ca_3P_2O_8}$ is equal to 1, is:

$$\Delta_r G_{(6.325)}(1500\ °C) = -166{,}000\ J\ (mol\ Ca_3P_2O_8)^{-1} \qquad (6.326)$$

Equally, the slagging of the phosphorus can be produced by combination of the P_2O_5 with other alkaline and alkaline earth oxides (for instance, sodium and barium):

$$2P(\text{dis; pig iron}) + 5O(\text{dis; pig iron}) + 3Na_2O(\text{dis; slag}) \leftrightarrow Na_6P_2O_8(\text{dis; slag})$$
$$(6.327)$$

$$2P(\text{dis; pig iron}) + 5O(\text{dis; pig iron}) + 3BaO(\text{dis; slag}) \leftrightarrow Ba_3P_2O_8(\text{dis; slag})$$
$$(6.328)$$

However, the most used slag-forming reagent in the ironmaking is the calcium oxide, although small additions of Na_2O favor the thermodynamics and kinetics of the process.

According to the reaction (6.325), the thermodynamic variables involved in the slagging of the phosphorus are:

- the activity of the calcium oxide.
- the activity of the phosphate in the slag.
- the potential of oxygen of the system (POS).
- the quantity of phosphorus dissolved in the iron.
- the temperature (the equilibrium constant diminishes when the temperature is increased).

The objective of any dephosphorizing process is calculating the distribution of the phosphorus between the slag and the metal under defined conditions of pressure (around atmospheric conditions) and temperature. With this purpose, models are developed to try to predict the values that the partition coefficient between the slag and the metal reach out supported by thermodynamic models and empiric equations.

From the thermodynamic models that were developed to predict the partition coefficient, that proposed by Flood, Forland, and Grjotheim (FFG) is considered. The FFG model considers that a basic slag comprises a series of ionic species (anions and cations) of ideal behavior that reaches the behavior of the ideal solutions: The formation of the slag is achieved with a mixing enthalpy of zero and an entropy of ideal mixture (Coudurier et al. 1985):

$$a_{O^{2-}} = x_{O^{2-}} \tag{6.329}$$

$$a_{Ca^{2+}} = x_{Ca^{2+}} \tag{6.330}$$

$$a_{CaO} = N_{e,Ca^{2+}} \cdot N_{e,O^{2-}} \tag{6.331}$$

where $N_{e,Ca^{2+}}$ and $N_{e,O^{2-}}$ are, respectively, the equivalent fractions of calcium cations and oxygen anions, in the ideal mixture of anions and cations that form the slag and $x_{O^{2-}}$ and $x_{Ca^{2+}}$ are the molar fraction of the calcium and oxygen ions.

The partition coefficient of the phosphorus between the metal and the slag is function of the greater or smaller participation of the so-called *elemental reactions of interchange* in the global process.

That is to say, the reaction of the phosphorus slagging in the ionic form is:

$$2P(\text{dis}; \text{pig iron}) + 5O(\text{dis}; \text{pig iron}) + 3O^{2-}(\text{dis}; \text{slag}) \leftrightarrow P_2O_8^{6-}(\text{dis}; \text{slag})$$

$$(6.332)$$

if we consider the participation of the different cations, coming from the dissociation of the basic oxides, such as Ca^{2+}, Mg^{2+}, Mn^{2+}, Fe^{2+}, and Na^+, it is possible to write all the possible simultaneous reactions of the phosphorus slagging with the corresponding equilibrium constants:

$$2P(\text{dis}; \text{pig iron}) + 5O(\text{dis}; \text{pig iron}) + 3CaO(\text{dis}; \text{slag}) \leftrightarrow Ca_3P_2O_8(\text{dis}; \text{slag})$$
$$\rightarrow K_{Ca^{2+}}$$

$$(6.333)$$

$$2P(\text{dis}; \text{pig iron}) + 5O(\text{dis}; \text{pig iron}) + 3MgO(\text{dis}; \text{slag}) \leftrightarrow Mg_3P_2O_8(\text{dis}; \text{slag})$$
$$\rightarrow K_{Mg^{2+}}$$

$$(6.334)$$

$$2P(\text{dis}; \text{pig iron}) + 5O(\text{dis}; \text{pig iron}) + 3FeO(\text{dis}; \text{slag}) \leftrightarrow Fe_3P_2O_8(\text{dis}; \text{slag})$$
$$\rightarrow K_{Fe^{2+}}$$

$$(6.335)$$

$$2P(\text{dis}; \text{pig iron}) + 5O(\text{dis}; \text{pig iron}) + 3Na_2O(\text{dis}; \text{slag}) \leftrightarrow Na_6P_2O_8(\text{dis}; \text{slag})$$
$$\rightarrow K_{Na^+}$$

$$(6.336)$$

According to Flood, the calculation of the equilibrium constant of the reaction (6.332) can be expressed as follows:

$$K_P = \frac{x_{P_2O_8^{6-}}}{x_{O^{2-}}^3 \cdot a_P^2 \cdot a_O^5} \qquad (6.337)$$

and leads to:

$$\log K_P = N_{e,Ca^{2+}} \cdot \log K_{Ca^{2+}} + N_{e,Mg^{2+}} \cdot \log K_{Mg^{2+}} + \cdots \qquad (6.338)$$

Using as reference the model of Flood, Forland, and Grjotheim, the IRSID has considered and proposed the following equation for the calculation of K_P (Gaye 1982):

$$\log K_P = N_{e,Ca^{2+}} \cdot \log K_{Ca^{2+}} + N_{e,Mg^{2+}} \cdot \log K_{Mg^{2+}} + \cdots + g_P \qquad (6.339)$$

where the term g_P is added to take into account the possible deviations from ideality of mixed cations in the slag:

$$g_P = -8.18 + 1.41 \cdot S + \frac{13,200}{T} + D \cdot x_{O^{2-}} + E \cdot x_{O^{2-}}^2 \qquad (6.340)$$

$$D = 40.55 - 6.95 \cdot S - \frac{64,500}{T} \qquad (6.341)$$

$$E = -13.10 + 5.91 \cdot S + \frac{22,100}{T} \qquad (6.342)$$

$$S = \frac{x_{SiO_4^{4-}}}{x_{SiO_4^{4-}} + x_{PO_4^{3-}}} \qquad (6.343)$$

Finally, the values calculated using the IRSID model for the possible elemental reactions of interchange in the dephosphorizing process are:

$$\log K_{Ca^{2+}} = \frac{82,200}{T} - 34.2 \qquad (6.344)$$

$$\log K_{Mn^{2+}} = \frac{64,370}{T} - 31.4 \qquad (6.345)$$

$$\log K_{Na^+} = \frac{100,600}{T} - 31.7 \qquad (6.346)$$

$$\log K_{Mg^{2+}} = \frac{75,200}{T} - 34.2 \qquad (6.347)$$

$$\log K_{Fe^{2+}} = \frac{59,300}{T} - 33.9 \qquad (6.348)$$

Moreover, to calculate the activity of the ferrous oxide of the slag, a_{FeO}, the IRSID uses the following equation:

$$a_{FeO} = \gamma_{FeO} \cdot x_{O^{2-}} \cdot x_{Fe^{2+}} \rightarrow \log a_{FeO} = \log(x_{O^{2-}} \cdot x_{Fe^{2+}}) + \log \gamma_{FeO} \qquad (6.349)$$

where

$$\log \gamma_{FeO} = A + B \cdot x_{O^{2-}} + C \cdot x_{O^{2-}}^2 \tag{6.350}$$

where the values of A, B, and C are calculated using the following equations:

$$A = 5.50 - 1.12 \cdot S - \frac{5600}{T} \tag{6.351}$$

$$B = -16.07 + 2.85 \cdot S + \frac{20,800}{T} \tag{6.352}$$

$$C = 10.97 - 1.62 \cdot S - \frac{16,100}{T} \tag{6.353}$$

The reaction (6.332) can move to the phosphorus slagging if high oxygen concentrations in the melt are used. However, it is possible that this source of oxygen, which is habitually provided by the slag, could not be exclusively used by the phosphorus but also by the oxidation–slagging of other metals available in the melt: carbon, silicon, and manganese, for instance. The yield of the pig iron dephosphorizing, consequently, is affected by the existence of parallel mechanisms of oxidation of other elements, different from that of the phosphorus, dissolved in the pig iron.

A model of prediction of the partition coefficient supported in an empiric equation of three parameters: temperature, CaO percentage in the slag and total iron in the slag, is the following:

$$\log \frac{C_{P,slag}}{C_{P,pig\ iron}} = \frac{22,350}{T} + 7.0 \cdot \log C_{CaO} + 2.5 \cdot \log C_{Fe} - 24.0 \tag{6.354}$$

Equation (6.354) can be used to estimate the distribution of the phosphorus in the metal and the slag when the CaO content in the slag is greater than 24% CaO. According to Eq. (6.354) when the quantity of phosphorus in the slag is 1.30% of P_2O_5 (0.57% P), at a temperature of 1923 K (1700 °C), the phosphorus percentage in the metal when the slag has the following composition: 50% CaO and 16.79% total iron (18% FeO + 4% Fe_2O_3) is:

$$\log \frac{0.57}{C_{P,pig\ iron}} = \frac{22,350}{1973} + 7.0 \cdot \log 50 + 2.5 \cdot \log 16.79 - 24.0 \rightarrow C_{P,pig\ iron}$$

$$= 0.00297\%(2.97\ \text{thousands or} \simeq 30\ \text{ppm})$$

$$\tag{6.355}$$

As it is indicated in Exercises 6.18 and 6.19, the application of the model developed by the IRSID to a slag of eight oxidized compounds but with 50% CaO,

16.79% total Fe (18% FeO + 4% Fe$_2$O$_3$), and 1.30% P$_2$O$_5$ (0.57% P) predicts a minimum value of phosphorus in the metal of 24.6 ppm. Consequently, the predictions reached with the empiric expression of three parameters, Eq. (6.354), does not differ significantly from the values obtained by the IRSID model, Exercise 6.28.

On the other hand, correlations between the ability to pick up phosphorus of the slags, $C_{PO_4^{3-},g/s}$, according to the gas–slag equilibrium and their optical basicity, Λ_{slag}, have been developed (Ballester et al. 2000, p. 272):

$$\log C_{PO_4^{3-},g/s} = -18.184 + 35.84 \cdot \Lambda_{slag} - 23.35 \cdot \Lambda_{slag}^2 + A \tag{6.356}$$

where

$$A = \frac{22,930}{T} - 0.0625 \cdot C_{FeO} - 0.04256 \cdot C_{MnO} + 0.359 \cdot C_{P_2O_5} \tag{6.357}$$

where C_i is the percentage of the different oxides in the slag. Definitely, the objective that it is pursued is to predict the partition coefficients of the phosphorus between the metal and the slag.

The ability to catch phosphorus of a slag can be defined by the following gas–slag equilibrium (we should think that the elements in the gas always reach an ideal behavior while in the metallic melts or in the slag melts, the elements habitually exhibit strong deviations from the ideality):

$$\frac{1}{2}P_2(g) + \frac{5}{4}O_2(g) + \frac{3}{2}O^{2-}(dis; slag) \leftrightarrow PO_4^{3-}(dis; slag) \tag{6.358}$$

and, thus, the ability of removing phosphorus is given by the expression:

$$C_{PO_4^{3-},g/s} = \frac{C_{slag}}{P_{P_2}^{0.5} \cdot P_{O_2}^{1.25}} \tag{6.359}$$

where the phosphorus concentration in the slag can be expressed of different forms depending on the model considered, ionic or molecular. For instance, for the ionic models, it is possible to use as expression of the concentration the molar fraction of the phosphate anion, $x_{PO_4^{3-}}$, or the equivalent fraction of the anion in the mixture of anions, $N_{e,PO_4^{3-}}$. On the other hand, the utilization of molecular models allows the expression of the concentrations of the slag in the form of molecular fraction of the P$_2$O$_5$, $x_{P_2O_5}$, or molecular fraction of the tricalcic phosphate, $x_{P_2O_5Ca_3}$. The constant of equilibrium of the reaction (6.358) is:

$$K_{(6.358)} = \frac{C_{slag} \cdot \gamma_{slag}}{P_{P_2}^{0.5} \cdot P_{O_2}^{1.25} \cdot a_{O^{2-}}^{1.5}} \tag{6.360}$$

where γ_{slag} is the coefficient of activity of the phosphorus compounds or ions in the slag; in this way, the ability of picking phosphorus by a certain slag will be:

$$C_{PO_4^{3-},g/s} = \frac{K_{(6.358)} \cdot a_{O^2}^{1.5}}{\gamma_{slag}} \qquad (6.361)$$

There is other model connected to the evaluation of the optical basicity of a certain slag, Λ_{slag}. The concept of optical basicity was firstly introduced in the early 1970s, and it was defined by a measurement of the spectral shift of a probe-ion absorption caused by the donation of electrons by the oxygen (or other ligand) in the glass/slag/other to the probe ion (McCloy et al. 2011, page 203). The optical basicity of a pure oxide, MeO (or its capacity to cede electrons), $\Lambda_{M,i}$ can be spectroscopically measured using the (s-p) electronic transitions. The optical basicity, $\Lambda_{M,i}$, can be related with Pauling's electronegativity (see Pero-Sanz et al. 2017, Chap. 1), χ, of the atom Me as follows: $\Lambda_{M,i} = 0.74 \cdot (\chi - 0.26)$. Recently, correlations between the ability of removing phosphorus, sulfur, carbon, or nitrogen by the slags with their basicity have been successfully proposed.

The optical basicity is calculated using the data of Table 6.10 and the following equation:

$$\Lambda_{slag} = \Lambda_{M,1} \cdot X(M_1) + \Lambda_{M,2} \cdot X(M_2) + \cdots + \Lambda_{M,i} \cdot X(M_i) \qquad (6.362)$$

where $\Lambda(M_i)$ is the optical basicity of the oxides that constitute the slag, Table 6.10, and $X(M_i)$ is calculated as follows:

$$X(M_i) = \frac{x_i \cdot n_O}{\sum_1^i x_i \cdot n_O} \qquad (6.363)$$

where x_i is the oxide molar fraction, M_i, in the slag, and n_O is the number of oxygens in the oxide molecule, M_i.

The kinetics of the dephosphorization process can be controlled by the diffusion in the slag or in the metallic phase. It is few probable that the chemical reaction and the interface resistance are the slow steps. The temperature, higher than 1300 °C, makes the chemical reactions, in the metal–slag interface, being very fast. On the other hand, the phosphorus is not an element superficially active, although the P_2O_5 and the FeO of the slag are superficially actives (decreasing of the interface tension metal–slag, $\gamma_{m/s}$, due to the presence of FeO and P_2O_5 in the formulation of the slag; FeO and P_2O_5 tend to accumulate in the metal–slag interface and impede that other reactions could take place). Resistances to the formation of calcium phosphates in the metal–slag interface are not foreseeable as a consequence of the low temperature or the limited concentration of reagents involved in the reaction (6.332).

6.5.3 Solved Exercises: Metal–Slag Equilibrium

Exercise 6.18 A slag has the following chemical composition: 50.0% CaO, 6.0% MgO, 18.0% FeO, 5.0% MnO, 0.7% Al_2O_3, 4.0% Fe_2O_3, 15.0% SiO_2, and 1.3% P_2O_5. We request to calculate the optical basicity, Λ_{slag}, of the slag.

Data: The optical basicity of each one of the oxides and the corresponding molecular weights are collected in Table 6.11 considering the data collected in Table 6.10 (the optical basicity obtained for each one of the oxides in the case of those with different values in Table 6.10 is the mean value).

The process to calculate the optical basicity of the slag has the following steps:

(a) The percentage mass composition of each one of the oxides is transformed in their corresponding molar fraction. Considering as reference 100 g of slag, the number of moles of each one of the oxides will be:

$$\text{Moles CaO } (n_{CaO}) : \frac{50\text{g CaO}}{100 \text{ g slag}} \cdot \frac{1 \text{ mol CaO}}{56.08 \text{ g CaO}} = 0.8916 \frac{\text{mol CaO}}{100 \text{ g slag}} \qquad (6.364)$$

$$\text{Moles MgO } (n_{MgO}) : \frac{6\text{g MgO}}{100 \text{ g slag}} \cdot \frac{1 \text{ mol MgO}}{40.31 \text{ g MgO}} = 0.1488 \frac{\text{mol MgO}}{100 \text{ g slag}} \qquad (6.365)$$

$$\text{Moles FeO } (n_{FeO}) : \frac{18\text{g CaO}}{100 \text{ g slag}} \cdot \frac{1 \text{ mol FeO}}{71.85 \text{ g FeO}} = 0.2505 \frac{\text{mol FeO}}{100 \text{ g slag}} \qquad (6.366)$$

Table 6.10 Data of the optical basicity, $\Lambda_{M,i}$, of the metallic oxides of the slags according to (a) Springorum 1995 and (b) McCloy et al. 2011

Oxide	$\Lambda_{M,i}$(a)	$\Lambda_{M,i}$(b)	Oxide	$\Lambda_{M,i}$(a)	$\Lambda_{M,i}$(b)
Li_2O	1.00	0.84	CoO	0.95	0.98
Na_2O	1.15	1.11	NiO	0.95	0.92
K_2O	1.40	1.32	CuO	0.90	1.10
Cs_2O	1.70	1.52	ZnO	0.90	0.80
MgO	0.78	0.95	PbO	1.15	1.18
CaO	1.00	1.00	Cr_2O_3	0.80	0.80
SrO	1.10	1.08	Fe_2O_3	0.80	0.80
BaO	1.15	1.33	TiO_2		0.91
B_2O_3	0.42	0.40	V_2O_3		1.04
Al_2O_3	0.605	0.61	ZrO_2		0.85
SiO_2	0.48	0.48	SnO_2		0.85
P_2O_5	0.40	0.40	SO_3		0.33
MnO	1.00	0.95	SrO		1.08
FeO	1.00		MoO_3		1.07

Table 6.11 Optical basicity and molecular weight of the oxides of the slag

CaO	MgO	FeO	MnO	Al$_2$O$_3$	Fe$_2$O$_3$	SiO$_2$	P$_2$O$_5$	
56.08	40.31	71.85	70.94	101.96	159.70	60.09	141.94	Molecular weight (g/mol)
1.00	0.865	1.00	0.975	0.608	0.80	0.48	0.40	Λ_i

$$\text{Moles MnO }(n_{\text{MnO}}): \frac{5\text{g MnO}}{100\text{ g slag}} \cdot \frac{1\text{ mol MnO}}{70.94\text{ g MnO}} = 0.0705\frac{\text{mol MnO}}{100\text{ g slag}} \quad (6.367)$$

$$\text{Moles Al}_2\text{O}_3(n_{\text{Al}_2\text{O}_3}): \frac{0.7\text{g Al}_2\text{O}_3}{100\text{ g slag}} \cdot \frac{1\text{ mol Al}_2\text{O}_3}{101.96\text{ g Al}_2\text{O}_3} = 0.0069\frac{\text{mol Al}_2\text{O}_3}{100\text{ g slag}}$$
$$(6.368)$$

$$\text{Moles Fe}_2\text{O}_3(n_{\text{Fe}_2\text{O}_3}): \frac{4\text{g Fe}_2\text{O}_3}{100\text{ g slag}} \cdot \frac{1\text{ mol Fe}_2\text{O}_3}{159.70\text{ g Fe}_2\text{O}_3} = 0.0250\frac{\text{mol Fe}_2\text{O}_3}{100\text{ g slag}}$$
$$(6.369)$$

$$\text{Moles SiO}_2(n_{\text{SiO}_2}): \frac{15\text{g SiO}_2}{100\text{ g slag}} \cdot \frac{1\text{ mol SiO}_2}{60.09\text{ g SiO}_2} = 0.2496\frac{\text{mol SiO}_2}{100\text{ g slag}} \quad (6.370)$$

$$\text{Moles P}_2\text{O}_5(n_{\text{P}_2\text{O}_5}): \frac{1.3\text{g P}_2\text{O}_5}{100\text{ g slag}} \cdot \frac{1\text{ mol P}_2\text{O}_5}{141.94\text{ g P}_2\text{O}_5} = 0.0092\frac{\text{mol P}_2\text{O}_5}{100\text{ g slag}} \quad (6.371)$$

Consequently, the number of moles in 100 g of slag is 1.6521 mol. Thus, the composition expressed as molar fraction of each one of the oxides in the slag will be:

$$\text{Molar fraction CaO }(x_{\text{CaO}}): \frac{0.8916\frac{\text{mol CaO}}{100\text{ g slag}}}{1.6521\frac{\text{mol slag}}{100\text{ g slag}}} = 0.5397\frac{\text{mol CaO}}{\text{mol slag}} \quad (6.372)$$

$$\text{Molar fraction MgO }(x_{\text{MgO}}): \frac{0.1488\frac{\text{mol MgO}}{100\text{ g slag}}}{1.6521\frac{\text{mol slag}}{100\text{ g slag}}} = 0.0901\frac{\text{mol MgO}}{\text{mol slag}} \quad (6.373)$$

$$\text{Molar fraction FeO }(x_{\text{FeO}}): \frac{0.2505\frac{\text{mol FeO}}{100\text{ g slag}}}{1.6521\frac{\text{mol slag}}{100\text{ g slag}}} = 0.1516\frac{\text{mol FeO}}{\text{mol slag}} \quad (6.374)$$

$$\text{Molar fraction MnO } (x_{MnO}) : \frac{0.0705\,\frac{\text{mol MnO}}{100\text{ g slag}}}{1.6521\,\frac{\text{mol slag}}{100\text{ g slag}}} = 0.0427\,\frac{\text{mol MnO}}{\text{mol slag}} \qquad (6.375)$$

$$\text{Molar fraction Al}_2O_3\,(x_{Al_2O_3}) : \frac{0.0069\,\frac{\text{mol Al}_2O_3}{100\text{ g slag}}}{1.6521\,\frac{\text{mol slag}}{100\text{ g slag}}} = 0.0042\,\frac{\text{mol Al}_2O_3}{\text{mol slag}} \qquad (6.376)$$

$$\text{Molar fraction Fe}_2O_3\,(x_{Fe_2O_3}) : \frac{0.0250\,\frac{\text{mol Fe}_2O_3}{100\text{ g slag}}}{1.6521\,\frac{\text{mol slag}}{100\text{ g slag}}} = 0.0152\,\frac{\text{mol Fe}_2O_3}{\text{mol slag}} \qquad (6.377)$$

$$\text{Molar fraction SiO}_2(x_{SiO_2}) : \frac{0.2496\,\frac{\text{mol SiO}_2}{100\text{ g slag}}}{1.6521\,\frac{\text{mol slag}}{100\text{ g slag}}} = 0.1511\,\frac{\text{mol SiO}_2}{\text{mol slag}} \qquad (6.378)$$

$$\text{Molar fraction P}_2O_5(x_{P_2O_5}) : \frac{0.0092\,\frac{\text{mol P}_2O_5}{100\text{ g slag}}}{1.6521\,\frac{\text{mol slag}}{100\text{ g slag}}} = 0.0055\,\frac{\text{mol P}_2O_5}{\text{mol slag}} \qquad (6.379)$$

(b) We calculate now the summation resulting from multiplication of the molar fraction of each oxide by the number of atoms of oxygen in the molecules. The result of the summation is (see equation (6.363)):

$$\sum xO = x_{CaO} \cdot 1 + x_{MgO} \cdot 1 + x_{FeO} \cdot 1 + x_{MnO} \cdot 1 + x_{Al_2O_3} \cdot 3 + x_{Fe_2O_3} \cdot 3$$
$$+ x_{SiO_2} \cdot 2 + x_{P_2O_5} \cdot 5$$
$$= 0.5397 \cdot 1 + 0.0901 \cdot 1 + 0.1516 \cdot 1 + 0.0427 \cdot 1 + 0.0042$$
$$\cdot 3 + 0.0152 \cdot 3 + 0.1511 \cdot 2 + 0.0055 \cdot 5 = 1.2119$$

$$(6.380)$$

(c) We obtain, for each one of the oxides of the slag, the value of $X(M_i)$, consequence of the multiplication of the molar fraction of the oxide by the number of oxygen atoms and dividing the result by the value of the summation obtained in the step (b). The results are:

$$X(M_{CaO}) = \frac{x_{CaO} \cdot 1}{\sum xO} = \frac{0.5397 \cdot 1}{1.2119} = 0.4453 \tag{6.381}$$

$$X(M_{MgO}) = \frac{x_{MgO} \cdot 1}{\sum xO} = \frac{0.0901 \cdot 1}{1.2119} = 0.0743 \tag{6.382}$$

$$X(M_{FeO}) = \frac{x_{FeO} \cdot 1}{\sum xO} = \frac{0.1516 \cdot 1}{1.2119} = 0.1251 \tag{6.383}$$

$$X(M_{MnO}) = \frac{x_{MnO} \cdot 1}{\sum xO} = \frac{0.0427 \cdot 1}{1.2119} = 0.0352 \tag{6.384}$$

$$X(M_{Al_2O_3}) = \frac{x_{Al_2O_3} \cdot 3}{\sum xO} = \frac{0.0042 \cdot 3}{1.2119} = 0.0104 \tag{6.385}$$

$$X(M_{Fe_2O_3}) = \frac{x_{Fe_2O_3} \cdot 3}{\sum xO} = \frac{0.0152 \cdot 3}{1.2119} = 0.0375 \tag{6.386}$$

$$X(M_{SiO_2}) = \frac{x_{SiO_2} \cdot 2}{\sum xO} = \frac{0.1511 \cdot 2}{1.2119} = 0.2493 \tag{6.387}$$

$$X(M_{P_2O_5}) = \frac{x_{P_2O_5} \cdot 5}{\sum xO} = \frac{0.0055 \cdot 5}{1.2119} = 0.0227 \tag{6.388}$$

(d) Finally, we multiply each one of the values of $X(M_i)$ previously calculated and their corresponding value of the optical basicity:

$$\begin{aligned}
\Lambda_{slag} &= \Lambda_{CaO} \cdot X(M_{CaO}) + \Lambda_{MgO} \cdot X(M_{MgO}) + \Lambda_{FeO} \cdot X(M_{FeO}) + \Lambda_{MnO} \\
&\quad \cdot X(M_{MnO}) + \Lambda_{Al_2O_3} \cdot X(M_{Al_2O_3}) + \Lambda_{Fe_2O_3} \cdot X(M_{Fe_2O_3}) \\
&\quad + \Lambda_{SiO_2} \cdot X(M_{SiO_2}) + \Lambda_{P_2O_5} \cdot X(M_{P_2O_5}) \\
&= 1 \cdot 0.4453 + 0.865 \cdot 0.0743 + 1 \cdot 0.1251 + 0.975 \cdot 0.0352 \\
&\quad + 0.608 \cdot 0.0104 + 0.80 \cdot 0.0375 + 0.48 \cdot 0.2493 + 0.40 \\
&\quad \cdot 0.0227 = 0.8341
\end{aligned} \tag{6.389}$$

Commentaries: We compare the value obtained for the optical basicity with the basicity of the slag, habitually expressed by the ratio calcium oxide to silicon oxide:

$$I_B = \frac{\%CaO}{\%SiO_2} = \frac{50}{15} = 3.33 \tag{6.390}$$

We could also suggest a basicity index characteristic of a slag, grouping in the numerator all the basic oxides multiplied each one of them by their optical basicity:

$$I_{B,opt} = \frac{\sum(\%MO)_{i-basic}\cdot\Lambda_i}{\sum(\%MO)_{i-acid}\cdot\Lambda_i} \tag{6.391}$$

In the case of the slag proposed in the enunciation:

$$I_{B,opt} = \frac{50\cdot1.00 + 6.0\cdot0.865 + 18.0\cdot1.00 + 5.0\cdot0.975}{0.70\cdot0.608 + 4.0\cdot0.80 + 15.0\cdot0.48 + 1.30\cdot0.40} = 6.88 \tag{6.392}$$

In the metal–slag reaction in the blast furnace, the basicity of the slag is a fundamental thermodynamic parameter to study the distribution of the elements between the metal and the slag.

Exercise 6.19 For the slag of Exercise 6.18, assuming applicable the ionic model, calculate the ionic fraction and the equivalent fractions of anions and cations.

Data: Apart from using the molar and percentage compositions of Exercise 6.18, assume that the mixture of cations includes: Ca^{2+}, Mg^{2+}, Fe^{2+}, and Mn^{2+} while the mixture of anions includes: O^{2-}, PO_4^{3-}, SiO_4^{4-}, and FeO_3^{3-}.

The process of cations formation from the basic oxides in the melt will be the next:

$$CaO \rightarrow Ca^{2+} + O^{2-} \tag{6.393}$$

$$MgO \rightarrow Mg^{2+} + O^{2-} \tag{6.394}$$

$$FeO \rightarrow Fe^{2+} + O^{2-} \tag{6.395}$$

$$MnO \rightarrow Mn^{2+} + O^{2-} \tag{6.396}$$

In this process of total ionization, oxygen anions are formed, and these are captured by the acid oxides to form the corresponding anions:

$$P_2O_5 + 3O^{2-} \rightarrow 2PO_4^{3-} \tag{6.397}$$

$$SiO_2 + 2O^{2-} \rightarrow SiO_4^{4-} \tag{6.398}$$

$$Fe_2O_3 + 3O^{2-} \rightarrow 2FeO_3^{3-} \tag{6.399}$$

$$Al_2O_3 + 3O^{2-} \rightarrow 2AlO_3^{3-} \tag{6.400}$$

The problem is focused on calculating the corresponding molar fractions or equivalent fractions of both each one of the cations that form the mixture of cations and each one of the anions that form the mixture of anions.

Using the results obtained in Exercise 6.18, in 100 g of slag, the number of moles of cations will be: 0.8916 mol Ca^{2+} in 100 g of slag, 0.1488 mol Mg^{2+} in 100 g of slag, 0.2505 mol Fe^{2+} in 100 g of slag, and 0.0705 mol Mn^{2+} in 100 g of slag.

Consequently, the total number of moles of cations is 1.3614. Thus, the molar fraction of calcium, magnesium, iron, and manganese will be:

$$x_{Ca^{2+}} = \frac{0.8916}{1.3614} = 0.6549 \tag{6.401}$$

$$x_{Mg^{2+}} = \frac{0.1488}{1.3614} = 0.1093 \tag{6.402}$$

$$x_{Fe^{2+}} = \frac{0.2505}{1.3614} = 0.1840 \tag{6.403}$$

$$x_{Mn^{2+}} = \frac{0.0705}{1.3614} = 0.0518 \tag{6.404}$$

The equivalent of the cations Ca^{2+}, Mg^{2+}, Fe^{2+}, and Mn^{2+} in the mixture of cations is:

$$e_{Ca^{2+}} = 0.6549 \cdot 2 = 1.7832 \tag{6.405}$$

$$e_{Mg^{2+}} = 0.1093 \cdot 2 = 0.2977 \tag{6.406}$$

$$e_{Fe^{2+}} = 0.1840 \cdot 2 = 0.5010 \tag{6.407}$$

$$e_{Mn^{2+}} = 0.0518 \cdot 2 = 0.1410 \tag{6.408}$$

and the molar fraction of cations, when the addition of the equivalents of the cations is 2.7229:

$$x_{Ca^{2+}} = \frac{1.7832}{2.7229} = 0.6549 \tag{6.409}$$

$$x_{Mg^{2+}} = \frac{0.2977}{2.7229} = 0.1093 \tag{6.410}$$

$$x_{Fe^{2+}} = \frac{0.5010}{2.7229} = 0.1840 \tag{6.411}$$

$$x_{Mn^{2+}} = \frac{0.1410}{2.7229} = 0.0518 \tag{6.412}$$

as they have all of them the same electric charge, the equivalent fractions of the cations will coincide with the molar fraction of the cations in the slag: $x_{Ca^{2+}} = N_{e,Ca^{2+}} = 0.6549$; $x_{Mg^{2+}} = N_{e,Mg^{2+}} = 0.1093$; $x_{Fe^{2+}} = N_{e,Fe^{2+}} = 0.1840$; $x_{Mn^{2+}} = N_{e,Mn^{2+}} = 0.0518$.

In the process of cations formation, O^{2-} anions are also formed. The O^{2-} anions produced in the dissociation of basic oxides in 100 g of slag are 1.3614 mol of O^{2-}. Consequently, the number of equivalents O^{2-} anions produced is 2.7229 equivalents.

In the process of anions formation, the quantity of oxygen anions consumed by reaction with the acid oxides is of 0.6225 (see below how it is calculated). Consequently, the quantity of oxygen moles to be considered in the mixture of anions will be:

$$[O^{2-}]_{free} = 1.3614 - 0.6225 = 0.7390 \tag{6.413}$$

In 100 g of slag, the quantity of anions moles produced by reaction with the anions of O^{2-} will be (0.6225 mol of anions of O^{2-} react):

$$n_{AlO_3^{3-}} = n_{Al_2O_3} \cdot 2 = 0.0069 \cdot 2 = 0.0137 \tag{6.414}$$

$$n_{SiO_4^{4-}} = n_{SiO_2} = 0.2496 \tag{6.415}$$

$$n_{FeO_3^{3-}} = n_{Fe_2O_3} \cdot 2 = 0.0250 \cdot 2 = 0.0501 \tag{6.416}$$

$$n_{PO_4^{3-}} = n_{P_2O_5} \cdot 2 = 0.0092 \cdot 2 = 0.0183 \tag{6.417}$$

The total quantity of anion moles in the mixture of anions will be: 1.0707 mol. The molar fraction of the anions in the mixture of anions will be:

$$x_{AlO_3^{3-}} = \frac{0.0137}{1.0707} = 0.0128 \tag{6.418}$$

$$x_{SiO_4^{4-}} = \frac{0.2496}{1.0707} = 0.2331 \tag{6.419}$$

$$x_{FeO_3^{3-}} = \frac{0.0501}{1.0707} = 0.0468 \tag{6.420}$$

$$x_{PO_4^{3-}} = \frac{0.0183}{1.0707} = 0.0171 \tag{6.421}$$

$$x_{O^{2-}} = \frac{0.7390}{1.0707} = 0.6901 \tag{6.422}$$

In this case, the corresponding equivalent fractions of the anions, in the mixture of anions, will not be the same that the molar fraction of the anions. First, we calculate the equivalent of cations:

$$e_{AlO_3^{3-}} = 0.0137 \cdot 3 = 0.0412 \tag{6.423}$$

$$e_{SiO_4^{4-}} = 0.2496 \cdot 4 = 0.9985 \tag{6.424}$$

$$e_{FeO_3^{3-}} = 0.0501 \cdot 3 = 0.1503 \tag{6.425}$$

$$e_{PO_4^{3-}} = 0.0183 \cdot 3 = 0.0550 \tag{6.426}$$

And the equivalent fractions of the anions are:

$$N_{e,AlO_3^{3-}} = \frac{e_{AlO_3^{3-}}}{2.7229} = \frac{0.0412}{2.7229} = 0.0151 \tag{6.427}$$

$$N_{e,SiO_4^{4-}} = \frac{e_{SiO_4^{4-}}}{2.7229} = \frac{0.9985}{2.7229} = 0.3667 \tag{6.428}$$

$$N_{e,FeO_3^{3-}} = \frac{e_{FeO_3^{3-}}}{2.7229} = \frac{0.1503}{2.7229} = 0.0552 \tag{6.429}$$

$$N_{e,PO_4^{3-}} = \frac{e_{PO_4^{3-}}}{2.7229} = \frac{0.0550}{2.7229} = 0.0202 \tag{6.430}$$

Commentaries: In the ionic models, the behavior of the ions (anions or cations) is usually identified with the ideal behavior (activity of the ions equivalent to their concentrations expressed in molar fraction).

On the other hand, in the ionic models, although it is usually employed the molar fraction to express the contribution–concentration of the anion–cation in the melt, the equivalent fraction is also used. The activity of a neutral molecule, according to the ionic model, is usually identified as both the product of the molar fraction of the anion in the mixture of anions and the product of the molar fraction of the cation in the mixture of cations. For instance, the activity of the CaO would be:

$$a(CaO) = x(Ca^{2+}) \cdot x(O^{2-}) \tag{6.431}$$

Caution! The equivalent fractions, in the text, to avoid mistakes with the molar fractions, are expressed as: $N_{e,Ca^{2+}}$ and not as $x_{e,Ca^{2+}}$.

Exercise 6.20 Analyze the maximum concentration (to achieve the thermodynamic equilibrium) that the manganese and the silicon can reach in the pig iron at 1450 and 1500 °C as result of the equilibriums metal–slag in the zone of elaboration of the furnace, assuming:

(a) That the partial pressure of the CO in the gas is 1.0 atm and that the activity of Raoult of the carbon in the pig iron is 0.75 (Exercise 6.7).
(b) That the behavior of the manganese in the metallic melt is ideal, while the silicon shows a negative deviation with a coefficient of activity of Raoult of 0.01. The average molecular weight of the pig iron is 53.73 g/mol (94% Fe, 4.5% C, 1.0% Mn, and 0.5% Si).
(c) In the slag, the manganese oxide (0.80% Mn) has a slight positive deviation with a coefficient of activity of Raoult of 1.60, while for the silicon oxide, it takes values within 0.50 (acid slag with 48% molar SiO_2) and 0.12 (basic slag with 33% molar SiO_2). The average molecular weight of the slag is, in this case, 62.0 g/mol.

The reduction of the manganese oxide, MnO, of the slag by the action of the carbon (coke) in the zone of elaboration of the furnace (hearth and bosh) is:

$$MnO(f) + C(s) \leftrightarrow CO(g) + Mn(f) \tag{6.432}$$

Considering the values of Appendix 1, the standard free energy associated with the carbothermal reduction of MnO is:

$$\Delta_r G^0(T) = 284.0 - 0.170 \cdot T(K) \left[kJ \, (mol \, CO)^{-1} \right] \tag{6.433}$$

At 1450 °C (1723 K), the equilibrium constant of the MnO reduction is:

$$\Delta_r G^0(T) = -R \cdot T(K) \cdot \ln K_{eq} = 284.0 - 0.170 \cdot T(K) \rightarrow K_{eq}$$
$$= \exp\left(-\frac{284.0 - 0.170 \cdot T(K)}{R \cdot T(K)} \right)$$
$$= \exp\left(-\frac{\dfrac{(284.0 - 0.170 \cdot 1723)kJ}{mol} \cdot \dfrac{1000 \, J}{1 \, kJ}}{8.314 \dfrac{J}{mol \, K} \cdot 1723 \, K} \right) = 1.86 \tag{6.434}$$

and at 1500 °C (1773 K):

$$\Delta_r G^0(T) = -R \cdot T(\text{K}) \cdot \ln K_{eq} = 284.0 - 0.170 \cdot T(\text{K}) \rightarrow K_{eq}$$
$$= \exp\left(-\frac{284.0 - 0.170 \cdot T(\text{K})}{R \cdot T(\text{K})}\right)$$
$$= \exp\left(-\frac{\dfrac{(284.0 - 0.170 \cdot 1773)\text{kJ}}{\text{mol}} \cdot \dfrac{1000 \text{ J}}{1 \text{ kJ}}}{8.314 \dfrac{\text{J}}{\text{mol K}} \cdot 1773 \text{ K}}\right) = 3.26 \tag{6.435}$$

and the equilibrium constant is:

$$K_{\text{MnO}-\text{Mn}} = \frac{P_{\text{CO}} \cdot \gamma_{\text{Mn}} \cdot x_{\text{Mn}}}{a_\text{C} \cdot \gamma_{\text{MnO}} \cdot x_{\text{MnO}}} \tag{6.436}$$

When the pressure of CO is 1.0 atm, γ_{MnO} is 1.60, γ_{Mn} is 1.0, and the carbon activity in the pig iron, a_C is 0.75, the partition coefficient of the manganese between the metal and the slag is:

- at 1450 °C:

$$1.86 = \frac{1 \text{ atm} \cdot 1.0 \cdot x_{\text{Mn}}}{0.75 \cdot 1.60 \cdot x_{\text{MnO}}} \rightarrow \frac{x_{\text{Mn}}}{x_{\text{MnO}}} = 2.2351 \tag{6.437}$$

- at 1500 °C:

$$3.26 = \frac{1 \text{ atm} \cdot 1.0 \cdot x_{\text{Mn}}}{0.75 \cdot 1.60 \cdot x_{\text{MnO}}} \rightarrow \frac{x_{\text{Mn}}}{x_{\text{MnO}}} = 3.9093 \tag{6.438}$$

where x_{Mn} and x_{MnO} are, respectively, the molar fraction of the manganese in the pig iron and the molar fraction in the slag. If we analyze the values of the average molecular weight of the slag and of the pig iron (62 g/mol and 53.73 g/mol, respectively), the concentration of MnO in the slag of 0.80% in weight and the maximum percentage of manganese in the pig iron (atomic weights of the manganese and the oxygen are 54.94 g/mol and 16.00 g/mol, respectively), are:

- at 1450 °C:

$$\frac{x_{Mn}}{x_{MnO}} = 2.2351 \frac{\text{mol Mn/mol pig iron}}{\text{mol MnO/mol slag}} \rightarrow x_{Mn}$$

$$= 2.2351 \frac{\text{mol Mn/mol pig iron}}{\text{mol MnO/mol slag}} \cdot \frac{0.8 \text{ g MnO} \cdot \dfrac{1 \text{ mol MnO}}{70.94 \text{ g MnO}}}{100 \text{ g slag} \cdot \dfrac{\text{mol slag}}{62 \text{ g slag}}} \qquad (6.439)$$

$$= 1.562 \times 10^{-2} \frac{\text{mol Mn}}{\text{mol pig iron}}$$

$$\%Mn = 1.562 \times 10^{-2} \frac{\text{mol Mn}}{\text{mol pig iron}} \cdot \frac{54.94 \text{ g Mn}}{1 \text{ mol Mn}} \cdot \frac{1 \text{ mol pig iron}}{53.73 \text{ g pig iron}} \cdot 100 = 1.60\%$$

$$(6.440)$$

- at 1500 °C:

$$\frac{x_{Mn}}{x_{MnO}} = 3.9093 \frac{\text{mol Mn/mol pig iron}}{\text{mol MnO/mol slag}} \rightarrow x_{Mn}$$

$$= 3.9093 \frac{\text{mol Mn/mol pig iron}}{\text{mol MnO/mol slag}}$$

$$\cdot \frac{0.8 \text{ g MnO} \cdot \dfrac{1 \text{ mol MnO}}{70.94 \text{ g MnO}}}{100 \text{ g slag} \cdot \dfrac{\text{mol slag}}{62 \text{ g slag}}} = 2.733 \times 10^{-2} \frac{\text{mol Mn}}{\text{mol pig iron}} \qquad (6.441)$$

$$\%Mn = 2.733 \times 10^{-2} \frac{\text{mol Mn}}{\text{mol pig iron}} \cdot \frac{54.94 \text{ g Mn}}{1 \text{ mol Mn}} \cdot \frac{1 \text{ mol pig iron}}{53.73 \text{ g pig iron}} \cdot 100 = 2.79\%$$

$$(6.442)$$

The reduction of the silicon oxide of the slag in a blast furnace takes place through the following reaction:

$$SiO_2(f) + 2C(s) \leftrightarrow 2CO(g) + Si(f) \qquad (6.443)$$

The standard free energy associated with the reaction (Appendix 1) is:

$$\Delta_r G^0(T) = 709.0 - 0.366 \cdot T(K) \left[kJ \left(mol \ SiO_2 \right)^{-1} \right] \qquad (6.444)$$

At 1450 °C (1723 K), the equilibrium constant of the reduction of the MnO is:

$$\Delta_r G^0(T) = -R \cdot T(\mathrm{K}) \cdot \ln K_{eq} = 709.0 - 0.366 \cdot T(\mathrm{K}) \rightarrow K_{eq}$$
$$= \exp\left(-\frac{709.0 - 0.366 \cdot T(\mathrm{K})}{R \cdot T(\mathrm{K})}\right)$$
$$= \exp\left(-\frac{\dfrac{(709.0 - 0.366 \cdot 1723)\mathrm{kJ}}{\mathrm{mol}} \cdot \dfrac{1000\,\mathrm{J}}{1\,\mathrm{kJ}}}{8.314\dfrac{\mathrm{J}}{\mathrm{mol\,K}} \cdot 1723\,\mathrm{K}}\right) = 4.20 \times 10^{-3} \tag{6.445}$$

and at 1500 °C (1773 K):

$$\Delta_r G^0(T) = -R \cdot T(\mathrm{K}) \cdot \ln K_{eq} = 709.0 - 0.366 \cdot T(\mathrm{K}) \rightarrow K_{eq}$$
$$= \exp\left(-\frac{709.0 - 0.366 \cdot T(\mathrm{K})}{R \cdot T(\mathrm{K})}\right)$$
$$= \exp\left(-\frac{\dfrac{(709.0 - 0.366 \cdot 1723)\mathrm{kJ}}{\mathrm{mol}} \cdot \dfrac{1000\,\mathrm{J}}{1\,\mathrm{kJ}}}{8.314\dfrac{J}{\mathrm{mol\cdot K}} \cdot 1723\,\mathrm{K}}\right) = 4.20 \times 10^{-3} \tag{6.446}$$

and the equilibrium constant is:

$$K_{\mathrm{SiO_2-Si}} = \frac{P_{\mathrm{CO}}^2 \cdot \gamma_{\mathrm{Si}} \cdot x_{\mathrm{Si}}}{a_C^2 \cdot \gamma_{\mathrm{SiO_2}} \cdot x_{\mathrm{SiO_2}}} \tag{6.447}$$

When the pressure of CO is 1.0 atm and the activity of the carbon in the pig iron is 0.75, the partition coefficients between the metallic melt and the acid slag (molar fraction of the SiO_2 in the slag of 0.48 and activity coefficient of 0.50) are:

- at 1450 °C:

$$4.20 \times 10^{-3} = \frac{(1\,\mathrm{atm})^2 \cdot 0.01 \cdot x_{\mathrm{Si}}}{(0.75)^2 \cdot 0.50 \cdot x_{\mathrm{SiO_2}}} \rightarrow \frac{x_{\mathrm{Si}}}{x_{\mathrm{SiO_2}}} = 0.1181 \tag{6.448}$$

- at 1500 °C:

$$1.70 \times 10^{-2} = \frac{(1\,\mathrm{atm})^2 \cdot 0.01 \cdot x_{\mathrm{Si}}}{(0.75)^2 \cdot 0.50 \cdot x_{\mathrm{SiO_2}}} \rightarrow \frac{x_{\mathrm{Si}}}{x_{\mathrm{SiO_2}}} = 0.4781 \tag{6.449}$$

while for a basic slag (molar fraction of SiO_2 of 0.33 and coefficient of activity 0.12), the partition coefficients are:

- at 1450 °C:

$$4.20 \times 10^{-3} = \frac{(1 \text{ atm})^2 \cdot 0.01 \cdot x_{Si}}{(0.75)^2 \cdot 0.12 \cdot x_{SiO_2}} \rightarrow \frac{x_{Si}}{x_{SiO_2}} = 0.0284 \qquad (6.450)$$

- at 1500 °C:

$$1.70 \times 10^{-2} = \frac{(1 \text{ atm})^2 \cdot 0.01 \cdot x_{Si}}{(0.75)^2 \cdot 0.12 \cdot x_{SiO_2}} \rightarrow \frac{x_{Si}}{x_{SiO_2}} = 0.1148 \qquad (6.451)$$

As the average atomic weight of the pig iron is 53.73 g/mol and the molar fraction of the silicon oxide is 0.48 (acid slag) or 0.33 (basic slag), the maximum percentage that the silicon can dissolve in the pig iron will be (atomic weight of the silicon is 28.09 g/mol)

- At 1450 °C (acid slag):

$$\frac{x_{Si}}{x_{SiO_2}} = 0.1181 \frac{\text{mol Si/mol pig iron}}{\text{mol SiO}_2/\text{mol slag}} \rightarrow x_{Si}$$

$$= 0.1181 \frac{\text{mol Si/mol pig iron}}{\text{mol SiO}_2/\text{mol slag}} \cdot 0.48 \qquad (6.452)$$

$$= 0.0567 \frac{\text{mol Si}}{\text{mol pig iron}}$$

$$\%Si = 0.0567 \frac{\text{mol Si}}{\text{mol pig iron}} \cdot \frac{28.09 \text{ g Si}}{1 \text{ mol Si}} \cdot \frac{1 \text{ mol pig iron}}{53.73 \text{ g pig iron}} \cdot 100 = 2.96\% \quad (6.453)$$

- At 1500 °C (acid slag):

$$\frac{x_{Si}}{x_{SiO_2}} = 0.4781 \frac{\text{mol Si/mol pig iron}}{\text{mol SiO}_2/\text{mol slag}} \rightarrow x_{Si}$$

$$= 0.4781 \frac{\text{mol Si/mol pig iron}}{\text{mol SiO}_2/\text{mol slag}} \cdot 0.48 \qquad (6.454)$$

$$= 0.2295 \frac{\text{mol Si}}{\text{mol pig iron}}$$

$$\%Si = 0.2295 \frac{\text{mol Si}}{\text{mol pig iron}} \cdot \frac{28.09 \text{ g Si}}{1 \text{ mol Si}} \cdot \frac{1 \text{ mol pig iron}}{53.73 \text{ g pig iron}} \cdot 100 = 12\% \quad (6.455)$$

- At 1450 °C (basic slag):

$$\frac{x_{Si}}{x_{SiO_2}} = 0.0284 \frac{\text{mol Si/mol pig iron}}{\text{mol SiO}_2/\text{mol slag}} \rightarrow x_{Si}$$

$$= 0.0284 \frac{\text{mol Si/mol pig iron}}{\text{mol SiO}_2/\text{mol slag}} \cdot 0.33 \qquad (6.456)$$

$$= 9.372 \times 10^{-3} \frac{\text{mol Si}}{\text{mol pig iron}}$$

$$\%Si = 9.372 \times 10^{-3} \frac{\text{mol Si}}{\text{mol pig iron}} \cdot \frac{28.09 \text{ g Si}}{1 \text{ mol Si}} \cdot \frac{1 \text{ mol pig iron}}{53.73 \text{ g pig iron}} \cdot 100 = 0.49\%$$

$$(6.457)$$

- At 1500 °C (basic slag):

$$\frac{x_{Si}}{x_{SiO_2}} = 0.1148 \frac{\text{mol Si/mol pig iron}}{\text{mol SiO}_2/\text{mol slag}} \rightarrow x_{Si}$$

$$= 0.1148 \frac{\text{mol Si/mol pig iron}}{\text{mol SiO}_2/\text{mol slag}} \cdot 0.33 \qquad (6.458)$$

$$= 0.0379 \frac{\text{mol Si}}{\text{mol pig iron}}$$

$$\%Si = 0.0379 \frac{\text{mol Si}}{\text{mol pig iron}} \cdot \frac{28.09 \text{ g Si}}{1 \text{ mol Si}} \cdot \frac{1 \text{ mol pig iron}}{53.73 \text{ g pig iron}} \cdot 100 = 1.98\%$$

$$(6.459)$$

Commentaries: The higher the temperature in the hearth, the greater the quantity of silicon and manganese in the pig iron could be dissolved (into solution). Equally, we can appreciate in the calculations a greater sensitivity of the silicon with respect to the manganese against the temperature changes and the composition of the slag. This can be one of the reasons to use the evolution of the silicon (it takes the mathematical form of a periodic function in the time and in the temperature) to predict or forecast the heating and cooling that the hearth will suffer in the tapping to be performed in the future.

Exercise 6.21 Estimate the percentage of silicon and manganese that can be incorporated from the slag to the pig iron in the hearth of a blast furnace that operates in the range of temperature 1450–1550 °C, assuming that:

(a) The controlling mechanism of the reaction is the diffusion of the oxides in the slag (Eq. (6.29) and Fig. 6.10).
(b) The geometrical and operational conditions of the furnace are: mean time/period of tapping equal to 90 min, average height of the slag layer over the pig iron of 40 cm, height of the hearth equal to 6.0 m, diameter of 10.0 m, and

relative displacement rate of the slag in contact with the pig iron equals to 0.035 m s^{-1}.

(c) The physical–chemical properties of the slag are: density, 2650 kg m^{-3}; viscosity, 5.0 Pa s; average molecular weight of the slag, 62.0 g/mol; and, according to Kawai and Shiraishi (1988) (pages 195–197) and Ballester et al. 2000 (page 277), the coefficient of diffusion of the silicon oxide in the slag has a value of 1.724×10^{-11} m^2 s^{-1} while that of the manganese oxide in the slag is 1.767×10^{-10} m^2 s^{-1}.

The incorporation rate of the silicon and the manganese by the carbothermal reduction (Exercise 6.20) to the molten iron can be analyzed by integration of a differential equation of first order similar to Eq. (6.30), which predicts the evolution of the metals percentage in the pig iron with the time:

$$\%C_{i/m} = k_d \cdot \left(C^0_{i/m} - C^=_{i/m}\right) \cdot \left(\frac{A}{V}\right) \cdot \left(\frac{PA_i}{PM_e}\right) \cdot 100 \cdot t \qquad (6.460)$$

where PA_i is the atomic weight of the considered metal (Si or Mn), PM_e is the average molecular weight of the oxides that form the slag, $\%C_{i/m}$ is the concentration of the metal in the melt as a function of the time t and $C^0_{i/m} - C^=_{i/m}$ is the driving force for the reaction, which indicates the difference between the molar fractions or the activities of the reactive oxides of the slag with respect to the activities that correspond to the thermodynamic equilibrium metal–slag. We assume that this value in the case of the silicon oxide is 0.13, while in the case of the manganese oxide is 0.045 (Exercises 6.29 to 6.32).

Using the previous equation to calculate the percentage of metal that can incorporate to the melt after 90 min (duration of the period of tapping), the ratio area/volume of the hearth is also considered, which is inversely proportional to the height of the hearth:

$$\frac{A}{V} = \frac{\pi \cdot \left(\frac{D}{2}\right)^2}{\pi \cdot \left(\frac{D}{2}\right)^2 \cdot h} = \frac{1}{h} = \frac{1}{6\,\mathrm{m}} = 0.167\ \mathrm{m}^{-1} \qquad (6.461)$$

Thus, the only variable that was still not evaluated is the diffusional mass transfer coefficient in the slag, k_d. Assuming the model of boundary layer and the theory of tight boundary layer, the steps to determine k_d are:

(a) Determine the thickness of the boundary layer of concentrations, δ, for the transportation of the manganese and silicon into the slag (Eq. 6.35):

$$\delta_{Si} = 1.5 \cdot \left(1.724 \times 10^{-11} \frac{m^2}{s}\right)^{1/3} \cdot \left(\frac{5.0 \text{ Pa s}}{2650 \text{ kg m}^3}\right)^{1/6} \cdot \left(\frac{0.4 \text{ m}}{0.035 \text{ m s}^{-1}}\right)^{1/2}$$

$$= 4.6047 \times 10^{-4} \text{ m}$$

$$(6.462)$$

$$\delta_{Mn} = 1.5 \cdot \left(1.767 \times 10^{-10} \frac{m^2}{s}\right)^{1/3} \cdot \left(\frac{5.0 \text{ Pa s}}{2650 \text{ kg m}^3}\right)^{1/6} \cdot \left(\frac{0.4 \text{ m}}{0.035 \text{ m s}^{-1}}\right)^{1/2}$$

$$= 1.0002 \times 10^{-3} \text{ m}$$

$$(6.463)$$

(b) Knowing the values of δ_{Si} and δ_{Mn}, we calculate the transfer coefficients of the metals in the slag, using the correlation between the dimensionless numbers of Nusselt, Nu, Reynolds, Re and Schmidt, Sc (Eq. 6.32):

$$\text{Nu} = \frac{k_i \cdot \delta_i}{D_i} = 0.332 \cdot \text{Re}^{1/2} \cdot \text{Sc}^{1/3} \rightarrow k_i = 0.332 \cdot \frac{D_i}{\delta_i} \cdot \text{Re}^{1/2} \cdot \text{Sc}^{1/3} \rightarrow k_i$$

$$= 0.332 \cdot \frac{D_i}{\delta_i} \cdot \left(\frac{L \cdot \rho \cdot v}{\eta}\right)^{1/2} \cdot \left(\frac{\eta}{\rho \cdot D_i}\right)^{1/3}$$

$$(6.464)$$

$$k_{Si} = 0.332 \cdot \left(\frac{1.724 \times 10^{-11} \frac{m^2}{s}}{4.6047 \times 10^{-4} m}\right) \cdot \left(\frac{10 \text{ m} \cdot 2650 \text{ kg m}^3 \cdot 0.035 \text{ m s}^{-1}}{5.0 \text{ Pa s}}\right)^{1/2}$$

$$\cdot \left(\frac{5.0 \text{ Pa s}}{2650 \text{ kg m}^3 \cdot 1.724 \times 10^{-11} \frac{m^2}{s}}\right)^{1/3} = 8.1 \times 10^{-5} \frac{m}{s}$$

$$(6.465)$$

$$k_{Mn} = 0.332 \cdot \left(\frac{1.767 \times 10^{-10} \frac{m^2}{s}}{1.0002 \times 10^{-3} m}\right) \cdot \left(\frac{10 \text{ m} \cdot 2650 \text{ kg m}^3 \cdot 0.035 \text{ m s}^{-1}}{5.0 \text{ Pa s}}\right)^{1/2}$$

$$\cdot \left(\frac{5.0 \text{ Pa s}}{2650 \text{ kg m}^3 \cdot 1.767 \times 10^{-10} \frac{m^2}{s}}\right)^{1/3} = 1.76 \times 10^{-4} \frac{m}{s}$$

$$(6.466)$$

(c) Finally, we estimate the quantity of silicon and manganese that can incorporate to the metallic melt after 90 min (5400 s):

$$\%\text{Si} = 8.1 \times 10^{-5}\frac{\text{m}}{\text{s}} \cdot 0.13 \cdot \frac{0.167}{\text{m}} \cdot \left(\frac{28.09\ \text{g/mol}}{62.00\ \text{g/mol}}\right) \cdot 100 \cdot 5400\ \text{s} = 0.43\%$$

$$(6.467)$$

$$\%\text{Mn} = 1.76 \times 10^{-4}\frac{\text{m}}{\text{s}} \cdot 0.045 \cdot \frac{0.167}{\text{m}} \cdot \left(\frac{54.94\ \text{g/mol}}{62.00\ \text{g/mol}}\right) \cdot 100 \cdot 5400\ \text{s} = 0.63\%$$

$$(6.468)$$

Commentaries: The diffusion rate of the silicon in the slag, as it is associated with the movement, not only of the atom of silicon but also of the anionic complex structures of silicates, SiO_4^{4-}, is slower than that of the simple anions (as the O^{2-}) or the cations as the Ca^{2+} or the Mg^{2+}. The weight percentage of Mn and Si in the typical pig iron is 0.50% Si and 0.70% Mn, near to those previously calculated.

In the area/volume ratio that participates in the kinetical equation, apart from the indicated geometrical considerations, also participate the surface energies of the melts (Exercise 6.34). A deficient contact (bad wetting between the slag and the metal) will reduce the value attributed to this relation in the exercise ($0.167\ \text{m}^{-1}$).

Exercise 6.22 Analyze what is the most stable combination between the sulfur and the calcium under the thermodynamic conditions that can be found in a blast furnace at 1500 °C and partial pressure of oxygen in equilibrium with the carbon (coke) of 1.8704×10^{-16} atm. Determine the partition coefficient between the sulfur of the slag and the sulfur of the pig iron (expressed in percentage), considering the following conditions:

(a) The activity of the CaO is 5.0×10^{-3} in acid slags and 4.56×10^{-2} in basic slags (Exercise 6.30).
(b) The potential of the oxygen in the molten pig iron is 80 ppm.
(c) The coefficient of Raoult's activity of the calcium sulfide, CaS, in the slag is 18 (Coudurier et al. 1985, page 368).
(d) The atomic weight of the sulfur is 32.06 and the average molecular weight of the slag is 62.00 g/mol (Exercise 6.33).

According to Ballester et al. 2000 (page 241), the standard free energy of the reaction of calcium sulfate formation from the calcium sulfide is:

$$\text{CaS (s)} + 2O_2(g) \leftrightarrow CaSO_4(s) \qquad (6.469)$$

$$\Delta_r G^0(T) = -923.4 + 0.324 \cdot T(\text{K})\left[\text{kJ (mol CaS)}^{-1}\right] \qquad (6.470)$$

The standard free energy and the partial pressure of oxygen in equilibrium with the CaS and the $CaSO_4$, at 1500 °C, are:

$$\Delta_r G^0 (1773 \text{ K}) = -923.4 + 0.324 \cdot 1773 = -348.95 \text{ kJ} (\text{mol CaS})^{-1} \quad (6.471)$$

$$\Delta_r G^0 (1773 \text{ K}) = -R \cdot T \cdot \ln K_{eq} \rightarrow K_{eq} = \exp\left(-\frac{-348{,}950 \dfrac{\text{J}}{\text{mol}}}{8.314 \dfrac{\text{J}}{\text{mol K}} \cdot 1773 \text{ K}}\right)$$
$$= 1.909 \times 10^{10}$$

$$(6.472)$$

$$K_{eq} = \frac{a_{CaSO_4}}{a_{CaS} \cdot P_{O_2}^2} \rightarrow P_{O_2} = \sqrt{\frac{1}{1 \cdot K_{eq}}} = 7.237 \times 10^{-6} \text{ atm} \quad (6.473)$$

If we compare the pressure of oxygen, previously calculated, with the condensed phases of sulfate–sulfide, with that registered in the elaboration zone of the blast furnace, the stable combination of the sulfur with the calcium is the calcium sulfide. The partition coefficient of one harmful element for the finished product is usually evaluated through the relation in weight of it that can be retained by the slag with respect to its proportion in the metal. In the hearth of the blast furnace, it is possible to set out the distribution of the sulfur between the slag and the pig iron, and, we show in Eqs. (6.21) and (6.22) the reaction of sulfur exchange between the slag and the metal, in addition to the variables that participate in the thermodynamic equilibrium. In this exercise, for the elements dissolved in the metal (sulfur and oxygen) we make the change of Raoult's activity to Henry's activity. When calculating the equilibrium constant, the values that we should take for the activities of the sulfur and the oxygen will not be those of Raoult but those of Henry. That is to say, the exchange reaction:

$$\text{CaO}(\text{dis; slag}) + \frac{1}{2}\text{O}_2(\text{g}) \leftrightarrow \text{CaS}(\text{dis; slag}) + \frac{1}{2}\text{O}_2(\text{g}) \quad (6.474)$$

is transformed in:

$$\text{CaO}(\text{dis; slag}) + \text{S}\left(\text{dis; } a_{H,S} = 1\right) \leftrightarrow \text{CaS}(\text{dis; slag}) + \text{O}\left(\text{dis; } a_{H,O} = 1\right) \quad (6.475)$$

And the equilibrium constant (Eq. 6.22) is:

$$K = \frac{a_{H,O} \cdot a_{CaS}}{a_{H,S} \cdot a_{CaO}} \quad (6.476)$$

To estimate the value of the standard free energy and the equilibrium constant, we work with the data of Appendixes 1 and 3, setting the following equilibriums:

$$\text{Ca}(\text{g}) + \frac{1}{2}\text{O}_2(\text{g}) \leftrightarrow \text{CaO}(\text{s}) \quad (6.477)$$

$$\Delta_r G^0(\text{reaction 6.477}) = -788.0 + 0.193 \cdot T(\text{K}) \left[\text{kJ (mol CaO)}^{-1}\right] \qquad (6.478)$$

$$\text{Ca(g)} + \frac{1}{2}\text{S}_2(\text{g}) \leftrightarrow \text{CaS(s)} \qquad (6.479)$$

$$\Delta_r G^0(\text{reaction 6.479}) = -696.0 + 0.189 \cdot T(\text{K}) \left[\text{kJ (mol CaS)}^{-1}\right] \qquad (6.480)$$

$$\frac{1}{2}\text{O}_2(\text{g}) \leftrightarrow \text{O}\left(\text{dis}; a_{\text{H,O}} = 1\right) \qquad (6.481)$$

$$\Delta_r G^0(\text{reaction 6.481}) = -117.0 - 3.14 \times 10^{-3} \cdot T(\text{K}) \left[\text{kJ (mol O)}^{-1}\right] \qquad (6.482)$$

$$\frac{1}{2}\text{S}_2(\text{g}) \leftrightarrow \text{S}\left(\text{dis}; a_{\text{H,S}} = 1\right) \qquad (6.483)$$

$$\Delta_r G^0(\text{reaction 6.483}) = -130.0 + 2.10 \times 10^{-2} \cdot T(\text{K}) \left[\text{kJ (mol S)}^{-1}\right] \qquad (6.484)$$

Thus, to estimate the standard free energy of the reaction (6.475) we must add:

$$\Delta_r G^0(\text{reaction 6.475})$$
$$= -\Delta_r G^0(\text{reaction 6.477}) + \Delta_r G^0(\text{reaction 6.479}) \qquad (6.485)$$
$$+ \Delta_r G^0(\text{reaction 6.481}) - \Delta_r G^0(\text{reaction 6.483})$$

The standard free energy as a function of the temperature for reactions in the hearth of the blast furnace, electric furnace, LD converter or secondary metallurgy, in an interval between 1487 and 1700 °C, is:

$$\Delta_r G^0(\text{reaction 6.475})$$
$$= -(-788.0 + 0.193 \cdot T(\text{K})) + (-696.0 + 0.189 \cdot T(\text{K}))$$
$$+ \left(-117.0 - 3.14 \times 10^{-3} \cdot T(\text{K})\right) \qquad (6.486)$$
$$- \left(-130.0 + 2.10 \times 10^{-2} \cdot T(\text{K})\right)$$
$$= 105.0 - 2.814 \times 10^{-2} \cdot T(\text{K}) \left[\text{kJ (mol CaO)}^{-1}\right]$$

The variation of the standard free energy for the exchange of the sulfur and oxygen between the pig iron and the slag would be slightly different for an interval of temperatures between 850 and 1487 °C (pretreatment of the pig iron, Sect. 6.5.1):

$$\Delta_r G^0(\text{reaction } 6.475) = 104.0 - 2.714 \times 10^{-2} \cdot T(K) \left[kJ \, (\text{mol CaO})^{-1} \right]$$

$$(6.487)$$

To calculate the partition coefficient, we require the value of the equilibrium constant and the variables involved in it. At 1500 °C, the value of the standard free energy of the reaction (6.475), where the sulfur and the oxygen dissolved in the metal are considered through Henry's activity, is:

$$\Delta_r G^0(\text{reaction } 6.475) = 104.0 - 2.714 \times 10^{-2} \cdot 1773 = 55.88 \frac{kJ}{\text{mol CaO}} \quad (6.488)$$

and the value of the equilibrium constant is:

$$\Delta_r G^0(\text{reaction } 6.475) = -R \cdot T \cdot \ln K_{eq} \rightarrow K_{eq}$$
$$= \exp\left(-\frac{\Delta_r G^0(\text{reaction } 6.475)}{R \cdot T} \right)$$
$$= \exp\left(-\frac{55{,}880 \frac{J}{\text{mol CaO}}}{8.314 \frac{J}{\text{mol K}} \cdot 1773 \, K} \right) = 2.258 \times 10^{-2}$$

$$(6.489)$$

When Henry's coefficients of activity of the oxygen and the sulfur are equal to one:

$$K_{eq} = \frac{a_O \cdot a_{CaS}}{a_S \cdot a_{CaO}} = \frac{(\%O) \cdot \gamma_{CaS} \cdot x_{CaS}}{(\%S)_{metal} \cdot a_{CaO}} \quad (6.490)$$

$$\frac{x_{CaS}}{(\%S)_{metal}} = 2.258 \times 10^{-2} \cdot \left(\frac{a_{CaO}}{(\%O) \cdot \gamma_{CaS}} \right) \quad (6.491)$$

$$\frac{X \text{ mol CaS}}{1 \text{ mol slag}} \cdot \frac{1 \text{ mol slag}}{62 \text{ g slag}} \cdot \frac{1 \text{ mol S}}{1 \text{ mol CaS}} \cdot \frac{32 \text{ g S}}{1 \text{ mol S}} \cdot 100 = (\%S)_{slag} \quad (6.492)$$

$$x_{CaS} = \left(\frac{62.00}{100 \cdot 32.06} \right) \cdot (\%S)_{slag} \quad (6.493)$$

$$\eta = \frac{(\%S)_{slag}}{(\%S)_{metal}} = 2.258 \times 10^{-2} \cdot \left(\frac{a_{CaO}}{(\%O) \cdot \gamma_{CaS}} \right) \cdot \left(\frac{100 \cdot 32.06}{62.00} \right) \quad (6.494)$$

Entering in the previous equation, the values of the activity of the calcium oxide, the coefficient of activity of the calcium sulfide in the slag and the concentration of oxygen in the melt, we obtain:

$$\eta = \frac{(\%S)_{\text{slag}}}{(\%S)_{\text{metal}}} = 2.258 \times 10^{-2} \cdot \left(\frac{4.56 \times 10^{-2}}{80 \times 10^{-4} \cdot 18}\right) \cdot \left(\frac{100 \cdot 32.06}{62.00}\right)$$

$$= 3.697 \times 10^{-1} \text{ for basic slag with } a_{\text{CaO}} = 4.56 \times 10^{-2}$$

$$(6.495)$$

$$\eta = \frac{(\%S)_{\text{slag}}}{(\%S)_{\text{metal}}} = 2.258 \times 10^{-2} \cdot \left(\frac{5.0 \times 10^{-3}}{80 \times 10^{-4} \cdot 18}\right) \cdot \left(\frac{100 \cdot 32.06}{62.00}\right) \quad (6.496)$$

$$= 4.054 \times 10^{-2} \text{ for acid slag with } a_{\text{CaO}} = 5.0 \times 10^{-3}$$

Commentaries: The values calculated for the partition coefficients between the melts in the hearth of the blast furnace show that only the basic slags are able of retaining part of the total sulfur that is charged into the blast furnace with the raw materials (iron ore and ironmaking coke). It is estimated that 70–80% of the sulfur that enters in the furnace returns to the pig iron.

Exercise 6.23 Determine the value of Henry's activity of the sulfur in the following melts:

- Melt A: 0.40% C, 0.08% Si, 0.040% S, and 99.48% Fe.
- Melt B: 4.50% C, 0.60% Si, 0.040% S, and 94.86% Fe.

Determine the best thermodynamic conditions for the desulfurizing (removal or reduction) of the 40 thousandths of sulfur dissolved in both melts. Assume in both cases the same slag (habitually, synthetic commercial products manufactured by specialized companies) to remove sulfur and the same concentration of oxygen (potential of oxygen of the system, degree of oxidation of the metal).

For the melt A, which has similar conditions to those of a metallic melt that, by solidification, will give a mild steel of ferritic–pearlitic structure, Henry's activity of the sulfur, $a_{\text{H,S}}$, will be:

$$a_{\text{H,S}} = (\%S) \cdot f_S \quad (6.497)$$

$$\log f_S = e_S^S \cdot (\%S) + e_S^C \cdot (\%C) + e_S^{\text{Si}} \cdot (\%Si) \quad (6.498)$$

For the value of each one of the interaction parameters, e_i^j, we use the mean data of Table 6.2 at 1600 °C, that are:

$$e_S^S = \frac{-0.028 - 0.046}{2} = -0.037; e_S^C = \frac{0.11 + 0.12}{2} = 0.115; e_S^{\text{Si}}$$

$$= \frac{0.063 + 0.075}{2} = 0.069 \quad (6.499)$$

Thus, the value of Henry's coefficient of activity, f_S, and its activity are:

$$\log f_S = -0.037 \cdot 0.04 + 0.115 \cdot 0.40 + 0.069 \cdot 0.08 = 0.050 \rightarrow f_S = 1.122$$
(6.500)

$$a_{H,S} = (\%S) \cdot f_S = 0.04 \cdot 1.122 = 0.045 \tag{6.501}$$

For the melt B, with the characteristics of a pig iron tapped from a blast furnace and stored in some transport equipment (ladle or torpedo) to be brought to the pretreatment station or steelworks, it is necessary to correct with the temperature the values of the interaction parameters of Table 6.2. Using the same criteria that in Exercise 6.6, we obtain at 1400 °C:

$$e_S^S = -0.037 \cdot \frac{1673}{1873} = -0.033 \tag{6.502}$$

$$e_S^C = 0.115 \cdot \frac{1873}{1673} = 0.129 \tag{6.503}$$

$$e_S^{Si} = 0.069 \cdot \frac{1873}{1673} = 0.077 \tag{6.504}$$

Consequently, the value of Henry's coefficient of activity and its activity are:

$$\log f_S = -0.033 \cdot 0.04 + 0.129 \cdot 0.40 + 0.077 \cdot 0.60 - 0.6244 \rightarrow f_S = 4.211$$
(6.505)

$$a_{H,S} = (\%S) \cdot f_S = 0.04 \cdot 4.211 = 0.168 \tag{6.506}$$

As it was indicated in Exercise 6.22, Raoult's activity of the calcium sulfide, CaS, in the slag and, for that reason, the concentration of sulfur in the slag is increased with Henry's activity of the sulfur in the melt. If the activity of the calcium oxide and the concentration of oxygen in the metal are the same in both melts, apparently, the best thermodynamic conditions are verified in the melt B with activity of the sulfur 3.73 times bigger than in the melt A.

However, considering the effect of the temperature on the constant of the oxygen–sulfur thermodynamic equilibrium in the hearth of the blast furnace (Eq. (6.22), Exercise 6.22), its value is 3.480×10^{-2} at 1600 °C while the value is 1.481×10^{-2} at 1400 °C, that is to say, when the temperature is decreased, the equilibrium constant is 2.34 times smaller. Definitely, the thermodynamic conditions to desulfurize (slagging the sulfur) in the melt B are slightly better (1.59 times greater) than in the melt A.

Commentaries: To remove the sulfur dissolved in the metallic melt, two mechanisms are activated at the same time:

(a) Finding a good basic slag, to fix the calcium sulfide or to dissolve other metallic sulfides.

(b) Facilitating the participation of active metals with high affinity by the sulfur such as manganese, magnesium, calcium, or rare earth metals.

Exercise 6.24 It is possible to perform the desulfurizing of the pig iron, as a previous refinement process of the carbon in the converter, by injecting calcium carbide, CaC_2, in the torpedo or in the ladle. To estimate the value, according to the thermodynamic equilibrium, of the partition coefficient of the sulfur at 1400 °C in a slag and a pig iron of the following characteristics:

(a) Slag with an average molecular weight of 62.00 g/mol, where Raoult's coefficient of activity of the CaS is 18.0 (Exercise 6.22).
(b) A pig iron with the following composition: 4.50% C, 1.00% Mn, 0.50% Si, 0.04% P, 0.04% S, and 93.92% Fe.

According to Eq. (6.313), the slagging of the sulfur dissolved in the pig iron takes place by the following reaction:

$$CaC_2(s) + S(dis; a_{H,S} = 1) \leftrightarrow 2C(dis; a_{H,C} = 1) + CaS(dis; slag) \quad (6.507)$$

We must consider the values of $\Delta_r G^0(T)$ for the following reactions (Appendixes 1 and 3) to estimate the standard free energy of this reaction:

$$Ca(l) + 2C(s) \leftrightarrow CaC_2(s) \quad (6.508)$$

$$C(l) + \frac{1}{2}S_2(g) \leftrightarrow CaS(s) \quad (6.509)$$

$$\frac{1}{2}S_2(g) \leftrightarrow S(dis; a_{H,S} = 1) \quad (6.510)$$

$$C(s) \leftrightarrow C(dis; a_{H,C} = 1) \quad (6.511)$$

Consequently,

$$\Delta_r G^0(T)(\text{reaction } 6.313) = -318.6 + 2.78 \times 10^{-2} \cdot T(K) \left[kJ \, (\text{mol CaS})^{-1} \right] \quad (6.512)$$

At 1400 °C, the value of the $\Delta_r G^0(T)$(reaction 6.313) is:

$$\Delta_r G^0(\text{reaction } 6.313) = -318.6 + 2.78 \times 10^{-2} \cdot 1673 = -272.09 \frac{kJ}{\text{mol CaS}} \quad (6.513)$$

and the value of the equilibrium constant is:

$$\Delta_r G^0 (\text{eq.6.313}) = -R \cdot T \cdot \ln K_{eq} \rightarrow K_{eq} = \exp\left(-\frac{\Delta_r G^0 (\text{reaction } 6.313)}{R \cdot T}\right)$$

$$= \exp\left(-\frac{-272{,}090 \frac{J}{\text{mol CaO}}}{8.314 \frac{J}{\text{mol K}} \cdot 1673 \text{ K}}\right) = 3.13 \times 10^8$$

$$(6.514)$$

The corresponding partition coefficient is:

$$K_{eq} = \frac{a_C^2 \cdot a_{CaS}}{a_S \cdot a_{CaC_2}} = \frac{a_C^2 \cdot \gamma_{CaS} \cdot x_{CaS}}{f_S \cdot (\%S)_{metal} \cdot a_{CaC_2}} \qquad (6.515)$$

$$\frac{x_{CaS}}{(\%S)_{metal}} = 3.13 \times 10^8 \cdot \left(\frac{a_{CaC_2}}{\gamma_{CaS}}\right) \cdot \left(\frac{f_S}{a_{H,C}^2}\right) \qquad (6.516)$$

$$\frac{X \text{ mol CaS}}{1 \text{ mol slag}} \cdot \frac{1 \text{ mol slag}}{62 \text{ g slag}} \cdot \frac{1 \text{ mol S}}{1 \text{ mol CaS}} \cdot \frac{32 \text{ g S}}{1 \text{ mol S}} \cdot 100 = (\%S)_{slag} \qquad (6.517)$$

$$x_{CaS} = \left(\frac{62.00}{100 \cdot 32.06}\right) \cdot (\%S)_{slag} \qquad (6.518)$$

$$\eta = \frac{(\%S)_{slag}}{(\%S)_{metal}} = 3.13 \times 10^8 \cdot \left(\frac{a_{CaC_2}}{\gamma_{CaS}}\right) \cdot \left(\frac{f_S}{a_{H,C}^2}\right) \cdot \left(\frac{100 \cdot 32.06}{62.00}\right) \qquad (6.519)$$

$$\eta = \frac{(\%S)_{slag}}{(\%S)_{metal}} = 3.13 \times 10^8 \cdot \left(\frac{100 \cdot 32.06}{62.00 \cdot 18}\right) \cdot \left(\frac{f_S}{a_{H,C}^2}\right) \qquad (6.520)$$

To calculate f_S, the values of the interaction parameters of the carbon, manganese, silicon, phosphorus, and sulfur on the dissolved sulfur at 1600 °C (Table 6.2) are considered, corrected at 1400 °C, using the same criterion that in previous exercises:

$$e_S^S = \left(\frac{-0.046 - 0.028}{2}\right) \frac{1673}{1873} = -0.033 \qquad (6.521)$$

$$e_S^P = \left(\frac{0.24 + 0.035}{2}\right) \cdot \frac{1873}{1673} = 0.154 \qquad (6.522)$$

$$e_S^{Si} = \left(\frac{0.063 + 0.075}{2}\right) \cdot \frac{1873}{1673} = 0.077 \qquad (6.523)$$

$$e_S^{Mn} = \left(\frac{-0.026 - 0.027}{2}\right) \cdot \frac{1673}{1873} = -0.024 \qquad (6.524)$$

$$e_S^{C} = \left(\frac{0.11 + 0.12}{2}\right) \cdot \frac{1873}{1673} = 0.129 \qquad (6.525)$$

Consequently, the value of Henry's coefficient of activity and its activity are:

$$\log f_S = -0.033 \cdot 0.04 + 0.154 \cdot 0.04 + 0.077 \cdot 0.50 - 0.024 \cdot 1.00 + 0.129$$
$$\cdot 4.5 = 0.5998 \rightarrow f_S = 3.9796$$
$$(6.526)$$

Similarly, the value of Henry's activity of the carbon in the pig iron, $a_{H,C}$, at 1673 K, will be:

$$e_C^S = \left(\frac{0.046 + 0.044}{2}\right) \cdot \frac{1873}{1673} = 0.050 \qquad (6.527)$$

$$e_C^P = \left(\frac{0.051 + 0.051}{2}\right) \cdot \frac{1873}{1673} = 0.057 \qquad (6.528)$$

$$e_C^{Si} = \left(\frac{0.08 + 0.08}{2}\right) \cdot \frac{1873}{1673} = 0.0896 \qquad (6.529)$$

$$e_C^{Mn} = \left(\frac{-0.012 - 0.008}{2}\right) \cdot \frac{1673}{1873} = -0.0089 \qquad (6.530)$$

$$e_C^C = \left(\frac{0.14 + 0.243}{2}\right) \cdot \frac{1873}{1673} = 0.214 \qquad (6.531)$$

Consequently, the value of Henry's coefficient of activity and its activity are:

$$\log f_C = 0.050 \cdot 0.04 + 0.057 \cdot 0.04 + 0.0896 \cdot 0.50 - 0.0089 \cdot 1.00 + 0.214$$
$$\cdot 4.5 = 1.0032 \rightarrow f_C = 10.07 \rightarrow a_{H,C} = 4.5 \cdot 10.07 = 45.33$$
$$(6.532)$$

Thus, the partition coefficient is:

$$
\eta = \frac{(\%S)_{slag}}{(\%S)_{metal}} = 3.13 \times 10^8 \cdot \left(\frac{100 \cdot 32.06}{62.00 \cdot 18}\right) \cdot \left(\frac{f_S}{a_{H,C}^2}\right)
$$

$$
= 3.13 \times 10^8 \cdot \left(\frac{100 \cdot 32.06}{62.00 \cdot 18}\right) \cdot \left(\frac{3.9796}{45.33^2}\right) = 1.74 \times 10^6
$$

$$(6.533)$$

Commentaries: As it is possible to check, the values of the partition coefficient cause thermodynamic conditions very favorable for the slagging of the sulfur in the pig iron. However, the possible parallel reactions of the calcium carbide powders with the oxidizing gases (oxygen and water vapor) should be also considered; as apart from reducing the performance of the treatment, they can form explosive mixtures (methane–hydrogen–acetylene). Consequently, the storage, transport, and injection of the CaC_2 must be carried out under the best conditions of tightness, to avoid filtrations or contact with the environmental humid air.

Exercise 6.25 One of the economic and safe routes for the desulfurizing of the pig iron is the treatment with ferromanganese powders and a basic slag that could pick up the MnS (Tables 6.7 and 6.8). Calculate the partition coefficient of the sulfur between the slag and the metal, at 1400 °C, considering the following reaction:

$$
Mn(dis;a_{H,Mn}= 1) + S(dis;a_{H,S}= 1) \leftrightarrow MnS(dis;slag) \tag{6.534}
$$

The characteristics of the slag include an average molecular weight of 62.00 g/mol, Raoult's coefficient of activity of the MnS of 1.6, and composition of the pig iron of 4.50% C, 1.00% Mn, 0.50% Si, 0.04% S, and 93.96% Fe.

To estimate the standard free energy of the chemical reaction indicated in the enunciation, $\Delta_{MnS}G^0(T)$, we take into account the following reactions (Appendixes 1 and 2):

$$
Mn(l) + \frac{1}{2}S_2(g) \leftrightarrow MnS(s) \tag{6.535}
$$

$$
Mn(l) \leftrightarrow Mn(dis; a_{H,Mn} = 1) \tag{6.536}
$$

$$
\frac{1}{2}S_2(g) \leftrightarrow S(dis; a_{H,S} = 1) \tag{6.537}
$$

where

$$
\begin{aligned}
\Delta_{\text{MnS}} G^0(T) &= \Delta_r G^0(\text{reaction } 6.535) - \Delta_r G^0(\text{reaction } 6.536) \\
&\quad - \Delta_r G^0(\text{reaction } 6.537) \\
&= (-291{,}000 + 78 \cdot T) - (4084 - 38.16 \cdot T) \\
&\quad - (-130{,}000 + 21 \cdot T) = -165{,}084 + 95.16 \cdot T
\end{aligned}
\tag{6.538}
$$

$$
\Delta_{\text{MnS}} G^0(1673\ \text{K}) = -165{,}084 + 95.16 \cdot 1673 = -5881.32\,\frac{\text{J}}{\text{mol MnS}}
\tag{6.539}
$$

and the value of the equilibrium constant is:

$$
\begin{aligned}
\Delta_r G^0(\text{reaction } 6.534) &= -R \cdot T \cdot \ln K_{\text{eq}} \rightarrow K_{\text{eq}} \\
&= \exp\left(-\frac{\Delta_r G^0(\text{reaction } 6.534)}{R \cdot T} \right) \\
&= \exp\left(-\frac{-5881.32\,\dfrac{\text{J}}{\text{mol Mns}}}{8.314\,\dfrac{\text{J}}{\text{mol K}} \cdot 1673\ \text{K}} \right) = 1.526
\end{aligned}
\tag{6.540}
$$

The corresponding partition coefficient is:

$$
K_{\text{eq}} = \frac{a_{\text{MnS}}}{a_{\text{S}} \cdot a_{\text{Mn}}} = \frac{a_{\text{MnS}}}{(\%\text{S})_{\text{metal}} \cdot (\%\text{Mn}) \cdot f_{\text{S}} \cdot f_{\text{Mn}}}
\tag{6.541}
$$

$$
\frac{x_{\text{MnS}}}{(\%\text{S})_{\text{metal}}} = 1.526 \cdot \left[\frac{(\%\text{Mn})}{\gamma_{\text{MnS}}} \right] \cdot (f_{\text{S}} \cdot f_{\text{Mn}})
\tag{6.542}
$$

$$
\frac{X\ \text{mol MnS}}{1\ \text{mol slag}} \cdot \frac{1\ \text{mol slag}}{62\ \text{g slag}} \cdot \frac{1\ \text{mol S}}{1\ \text{mol MnS}} \cdot \frac{32.06\ \text{g S}}{1\ \text{mol S}} \cdot 100 = (\%\text{S})_{\text{slag}}
\tag{6.543}
$$

$$
x_{\text{MnS}} = \left(\frac{62.00}{100 \cdot 32.06} \right) \cdot (\%\text{S})_{\text{slag}}
\tag{6.544}
$$

$$
\eta = \frac{(\%\text{S})_{\text{slag}}}{(\%\text{S})_{\text{metal}}} = 1.526 \cdot \left[\frac{(\%\text{Mn})}{\gamma_{\text{MnS}}} \right] \cdot (f_{\text{S}} \cdot f_{\text{Mn}}) \cdot \left(\frac{100 \cdot 32.06}{62.00} \right)
\tag{6.545}
$$

To calculate f_{S}, the values of the interaction parameters of the carbon, manganese, silicon, and sulfur on the dissolved sulfur at 1600 °C (Table 6.2) are considered, corrected at 1400 °C, using the same criterion that in previous exercises:

$$e_S^S = \left(\frac{-0.046 - 0.028}{2}\right) \cdot \frac{1673}{1873} = -0.033 \tag{6.546}$$

$$e_S^{Si} = \left(\frac{0.063 + 0.075}{2}\right) \cdot \frac{1873}{1673} = 0.077 \tag{6.547}$$

$$e_S^{Mn} = \left(\frac{-0.026 - 0.027}{2}\right) \cdot \frac{1673}{1873} = -0.024 \tag{6.548}$$

$$e_S^C = \left(\frac{0.11 + 0.12}{2}\right) \cdot \frac{1873}{1673} = 0.129 \tag{6.549}$$

Consequently, the value of Henry's coefficient of activity and its activity are:

$$\log f_S = -0.033 \cdot 0.04 + 0.077 \cdot 0.50 - 0.024 \cdot 1.00 + 0.129 \cdot 4.5 = 0.5937$$
$$\rightarrow f_S = 3.9236$$

$$\tag{6.550}$$

To calculate f_{Mn}, the values of the interaction parameters of the carbon, manganese, silicon, and sulfur on the dissolved sulfur at 1600 °C (Table 6.2) are considered, corrected at 1400 °C, using the same criterion that in previous exercises:

$$e_{Mn}^S = \left(\frac{-0.048 - 0.047}{2}\right) \cdot \frac{1673}{1873} = -0.042 \tag{6.551}$$

$$e_{Mn}^{Si} = \left(\frac{0 - 0.033}{2}\right) \cdot \frac{1673}{1873} = -0.015 \tag{6.552}$$

$$e_{Mn}^{Mn} = 0.000 \tag{6.553}$$

$$e_{Mn}^C = \left(\frac{-0.07 - 0.054}{2}\right) \cdot \frac{1673}{1873} = -0.055 \tag{6.554}$$

Consequently, the value of Henry's coefficient of activity and its activity are:

$$\log f_{Mn} = -0.042 \cdot 0.04 - 0.015 \cdot 0.5 + 0.000 \cdot 1.00 - 0.055 \cdot 4.5 = -0.2567$$
$$\rightarrow f_{Mn} = 0.5538$$

$$\tag{6.555}$$

Thus, the partition coefficient is:

$$
\eta = \frac{(\%S)_{\text{slag}}}{(\%S)_{\text{metal}}} = 1.526 \cdot \left[\frac{(\%Mn)}{\gamma_{MnS}}\right] \cdot (f_S \cdot f_{Mn}) \cdot \left(\frac{100 \cdot 32.06}{62.00}\right)
$$

$$
= 1.526 \cdot \left[\frac{1.00}{1.6}\right] \cdot (3.9236 \cdot 0.5538) \cdot \left(\frac{100 \cdot 32.06}{62.00}\right) = 107.16
$$

$$(6.556)$$

Commentaries: The partition coefficient that was obtained is favorable if we want to make a pretreatment of the pig iron with ferromanganese powders and a basic slag that could dissolve the MnS. However, the form of achieving the contact or the addition of these reagents is important because secondary reactions can be avoided, as the oxidation of the manganese, which can have influence in the yield of the sulfur slag formation. If it is compared with the desulfurization with calcium carbide, CaC_2, the treatments with ferromanganese are thermodynamically less effective, but they offer safer conditions to avoid risks of explosion/bursting.

Exercise 6.26 The elimination of the phosphorus from the pig iron can be performed via mechanisms that could form calcium phosphates: tricalcium phosphate $(PO_4)_2Ca_3$(3CaP) and tetracalcium phosphate $P_2O_9Ca_4$(4CaP). If we assume that the behavior of the slag is similar to the molecular model (Coudurier et al. 1985, page 373; Ballester et al. 2000, page 283), determine the partition coefficient of the phosphorus between the slag and the pig iron at 1400 °C when the composition of the metal is 4.50% C, 1.00% Mn, 0.50% Si, 0.04% S, 0.04% P, and 93.92% Fe. The average molecular weight of the slag is 62.00 g/mol.

The reaction of the tricalcium phosphate (3CaP) formation that appears in Eq. (6.324) can represent a mechanism to attach the phosphorus of the pig iron in a slag. Moreover, the formation of 4CaP can be also a method to turn into slag this impurity. The partition coefficient is calculated in both situations. The reaction that should be considered for the 3CaP is:

$$
2P(\text{dis}; a_{H,P} = 1) + 5O(\text{dis}; a_{H,S} = 1) + 3CaO(\text{dis}; \text{slag}) \leftrightarrow Ca_3P_2O_8(\text{dis}; \text{slag})
$$

$$(6.557)$$

The standard free energy of this reaction is calculated using the data of Appendixes 1 and 3 changing the activities of the phosphorus and the oxygen in the reaction of the 3CaP formation:

$$
\Delta_{3CaP}G^0 = -1441.0 + 0.630 \cdot T \left[\text{kJ (mol 3CaP)}^{-1}\right]
$$

$$(6.558)$$

and the value of the equilibrium constant is:

$$\Delta_r G^0(\text{reaction } 6.557) = -R \cdot T \cdot \ln K_{eq} \rightarrow K_{eq}$$

$$= \exp\left(-\frac{\Delta_r G^0(\text{reaction } 6.557)}{R \cdot T}\right)$$

$$= \exp\left(-\frac{-387{,}010\,\dfrac{J}{\text{mol 3CaP}}}{8.314\,\dfrac{J}{\text{mol K}} \cdot 1673K}\right) = 1.213 \times 10^{12}$$

$$(6.559)$$

In the case of the 4CaP, the reaction and the values of the equilibrium constant at 1400 °C are:

$$2P(\text{dis}; a_{H,P} = 1) + 5O(\text{dis}; a_{H,S} = 1) + 4CaO(\text{dis}; \text{slag}) \leftrightarrow Ca_4P_2O_9(\text{dis}; \text{slag})$$

$$(6.560)$$

The standard free energy of this reaction is calculated using the data of Appendixes 1 and 3 changing the activities of the phosphorus and the oxygen in the reaction of the 4CaP formation:

$$\Delta_{4CaP}G^0 = -1483.0 + 0.630 \cdot T\left[kJ\,(\text{mol } 4CaP)^{-1}\right] \qquad (6.561)$$

and the value of the equilibrium constant is:

$$\Delta_r G^0(\text{reaction } 6.560) = -R \cdot T \cdot \ln K_{eq} \rightarrow K_{eq}$$

$$= \exp\left(-\frac{\Delta_r G^0(\text{reaction } 6.560)}{R \cdot T}\right)$$

$$= \exp\left(-\frac{-429{,}010\,\dfrac{J}{\text{mol 4CaP}}}{8.314\,\dfrac{J}{\text{mol K}} \cdot 1673\,K}\right) = 2.484 \times 10^{13}$$

$$(6.562)$$

In this case, it is not possible to directly calculate the partition coefficient from the equilibrium constant, but a first calculation of the maximum quantity of phosphorus in the slag (of equilibrium) compatible with defined both pig iron and slag can be obtained. The equation of the maximum percentage of phosphorus in the slag when $Ca_3P_2O_8$ is formed is:

$$K_{eq} = \frac{x_{Ca_3P_2O_8}}{a_{H,P}^2 \cdot a_{CaO}^3 \cdot a_{H,O}^5} \qquad (6.563)$$

$$\frac{X \text{ mol } Ca_3P_2O_8}{1 \text{ mol slag}} \cdot \frac{1 \text{ mol slag}}{62 \text{ g slag}} \cdot \frac{2 \text{ mol P}}{1 \text{ mol } Ca_3P_2O_8} \cdot \frac{30.97 \text{ g P}}{1 \text{ mol P}} \cdot 100 = (\%P)_{slag}$$

(6.564)

$$x_{Ca_3P_2O_8} = \left(\frac{62.00}{2 \cdot 100 \cdot 30.917}\right) \cdot (\%P)_{slag}$$

(6.565)

$$K_{eq} = \frac{\left(\frac{62.00}{2 \cdot 100 \cdot 30.917}\right) \cdot (\%P)_{slag}}{a_{H,P}^2 \cdot a_{CaO}^3 \cdot a_{H,O}^5} \rightarrow (\%P)_{slag}$$

$$= K_{eq} \cdot \frac{1}{\left(\frac{62.00}{2 \cdot 100 \cdot 30.917}\right)} \cdot a_{H,P}^2 \cdot a_{CaO}^3 \cdot a_{H,O}^5$$

(6.566)

In the same way, for the formation of the $Ca_4P_2O_9$:

$$K_{eq} = \frac{x_{Ca_4P_2O_9}}{a_{H,P}^2 \cdot a_{CaO}^4 \cdot a_{H,O}^5}$$

(6.567)

$$\frac{X \text{ mol } Ca_3P_2O_8}{1 \text{ mol slag}} \cdot \frac{1 \text{ mol slag}}{62 \text{ g slag}} \cdot \frac{2 \text{ mol P}}{1 \text{ mol } Ca_3P_2O_8} \cdot \frac{30.97 \text{ g P}}{1 \text{ mol P}} \cdot 100 = (\%P)_{slag}$$

(6.568)

$$x_{Ca_3P_2O_9} = \left(\frac{62.00}{2 \cdot 100 \cdot 30.917}\right) \cdot (\%P)_{slag}$$

(6.569)

$$K_{eq} = \frac{\left(\frac{62.00}{2 \cdot 100 \cdot 30.917}\right) \cdot (\%P)_{slag}}{a_{H,P}^2 \cdot a_{CaO}^4 \cdot a_{H,O}^5} \rightarrow (\%P)_{slag}$$

$$= K_{eq} \cdot \frac{1}{\left(\frac{62.00}{2 \cdot 100 \cdot 30.917}\right)} \cdot a_{H,P}^2 \cdot a_{CaO}^4 \cdot a_{H,O}^5$$

(6.570)

For this, we must calculate Henry's activities of the oxygen and the phosphorus in the pig iron in both cases. The same way, the quantity of phosphorus that can be removed depends on the basicity of the slag (Raoult's activity of the CaO) that is used. The interaction parameters of the elements dissolved in the pig iron over the oxygen and the phosphorus are obtained using the method described in the previous exercises. Now, we show the values of the partition coefficient of the phosphorus, η, between the slag and the pig iron, considering the transfer of the phosphorus to the slag either as $Ca_3P_2O_8$ or as $Ca_4P_2O_9$, as a function of both the oxidizing potential of the system, $(\%O)_{metal}$, and Raoult's activity of the CaO in the slag, a_{CaO}, when the temperature is constant (1400 °C) and Henry's activity of the phosphorus in the pig iron (0.1989), compatible with 0.04%P (Tables 6.12 and 6.13).

Table 6.12 Formation of the $Ca_3P_2O_8$

$(\%O)_{metal}$	a_{CaO}	$(\%P)_{slag}$	η
0.010	5.00×10^{-2}	8.73×10^{-11}	2.18×10^{-9}
0.100	1.00×10^{-1}	6.99×10^{-5}	1.75×10^{-3}
0.150	1.50×10^{-1}	1.79×10^{-3}	4.48×10^{2}
0.200	4.00×10^{-1}	0.143	3.58

Table 6.13 Formation of the $Ca_4P_2O_9$

$(\%O)_{metal}$	a_{CaO}	$(\%P)_{slag}$	η
0.010	5.00×10^{-2}	8.94×10^{-11}	2.24×10^{-9}
0.100	1.00×10^{-1}	1.43×10^{-4}	3.58×10^{-3}
0.150	1.50×10^{-1}	5.51×10^{-3}	1.37×10^{-1}
0.200	4.00×10^{-1}	1.172	29.30

To calculate f_O (and later the a_{H_2O}), the values of the interaction parameters of the carbon, manganese, silicon, phosphorus, and sulfur on the dissolved sulfur at 1600 °C (Table 6.2) are considered, corrected at 1400 °C, using the same criterion that in the previous exercises:

$$e_O^S = \left(\frac{-0.133 - 0.133}{2}\right) \cdot \frac{1673}{1873} = -0.119 \tag{6.571}$$

$$e_O^P = \left(\frac{0.07 + 0.07}{2}\right) \cdot \frac{1873}{1673} = 0.078 \tag{6.572}$$

$$e_O^{Si} = \left(\frac{-0131 - 0.066}{2}\right) \cdot \frac{1673}{1873} = -0.088 \tag{6.573}$$

$$e_O^{Mn} = \left(\frac{-0.021 - 0.021}{2}\right) \cdot \frac{1673}{1873} = -0.019 \tag{6.574}$$

$$e_O^C = \left(\frac{-0.45 - 0.421}{2}\right) \cdot \frac{1673}{1873} = -0.389 \tag{6.575}$$

Consequently, the value of Henry's coefficient of activity and its activity are:

$$\log f_O = -0.119 \cdot 0.04 + 0.078 \cdot 0.04 - 0.088 \cdot 0.50 - 0.019 \cdot 1.00 - 0.389$$
$$\cdot 4.5 = -1.8721 \rightarrow f_O = 0.0134$$

$$\tag{6.576}$$

Similarly, the value of Henry's activity of the phosphorus in the pig iron, $a_{H,C}$, will be:

$$e_P^S = \left(\frac{0.028 + 0.034}{2}\right) \cdot \frac{1873}{1673} = 0.035 \tag{6.577}$$

$$e_P^P = \left(\frac{0.062 + 0.054}{2}\right) \cdot \frac{1873}{1673} = 0.065 \tag{6.578}$$

$$e_P^{Si} = \left(\frac{0.12 + 0.099}{2}\right) \cdot \frac{1873}{1673} = 0.123 \tag{6.579}$$

$$e_P^{Mn} = \left(\frac{0 - 0.032}{2}\right) \cdot \frac{1673}{1873} = -0.014 \tag{6.580}$$

$$e_P^C = \left(\frac{0.13 + 0.126}{2}\right) \cdot \frac{1873}{1673} = 0.143 \tag{6.581}$$

Consequently, the value of Henry's coefficient of activity and its activity are:

$$\log f_P = 0.035 \cdot 0.04 + 0.065 \cdot 0.04 + 0.126 \cdot 0.50 - 0.014 \cdot 1.00 + 0.143 \cdot 4.5$$
$$= 0.6965 \rightarrow f_P = 4.9716 \rightarrow a_{H,P} = 0.04 \cdot 4.9716 = 0.1989 \tag{6.582}$$

Commentaries: The mechanism of the transfer to the slag of the phosphorus by the calcium phosphates is thermodynamically conditioned (if the temperature is kept constant) by two variables:

(a) The potential of oxygen in the melt (concentration of oxygen in the pig iron).
(b) The activity of the calcium in the slag.

Keeping a high potential in the system propitiates other parallel reactions that have influence in the dephosphorization yield (for instance, oxidation of the carbon). Thus, the dephosphorization of the pig iron is not a usual practice, such as the pass of the sulfur to the slag, as well.

Exercise 6.27 For the slag used in previous exercises (Exercises 6.18 and 6.19), calculate the value of the equilibrium constant, according to the IRSID model, for the following reaction of the phosphorus slagging at 1700 °C:

$$2P(\text{dis}; a_H) + 5O(\text{dis}; a_H) + 3O^{2-}(\text{dis}; \text{slag}) \leftrightarrow 2PO_4^{3-}(\text{dis}; \text{slag}) \tag{6.583}$$

In the above-indicated reaction, it is possible to check that some of the elements, compounds, or ions that are involved in it are connected either with Raoult's activities (O^{2-} and $PO_4{}^{3-}$) or with Henry's activities (P and O). Consequently, the value of the equilibrium constant is:

$$K_{eq-PO_4^{3-}} = \frac{\left(x\left(PO_4^{3-}\right)\right)^2}{\left(x\left(O^{2-}\right)\right)^3 \cdot a_{H,O}^5 \cdot a_{H,P}^2} \tag{6.584}$$

The model developed by the IRSID provides a method to calculate the value of the decimal logarithm of the equilibrium constant in the phosphorus slagging. The model is supported in a single ionic conception of the slag and, it is a first approximation to calculate the value of the partition coefficient; all the possible reactions of metallic phosphates formation by combination–participation of the basic oxides–cations of the slag with the P_2O_5 are considered. In this case, we analyzed and proceeded to calculate the contributions of the Ca^{2+}, Mg^{2+}, Mn^{2+}, and Fe^{2+} and their combinations with the $PO_4{}^{3-}$:

- In the case of the calcium, Ca^{2+} (whose molar fraction in the mixture of cations, Exercise 6.19, is 0.6549):

$$\log K_{Ca^{2+}} = x_e\left(Ca^{2+}\right) \cdot \left(\frac{82,200}{T} - 34.2\right) = 0.6549 \cdot \left(\frac{82,200}{1973} - 34.2\right)$$
$$= 0.6549 \cdot 7.4624 = 4.8871$$

$$\tag{6.585}$$

- In the case of the magnesium, Mg^{2+}:

$$\log K_{Mg^{2+}} = x_e\left(Mg^{2+}\right) \cdot \left(\frac{75,200}{T} - 34.2\right) = 0.1093 \cdot \left(\frac{75,200}{1973} - 34.2\right)$$
$$= 0.1093 \cdot 3.9146 = 0.4280$$

$$\tag{6.586}$$

- In the case of the manganese, Mn^{2+}:

$$\log K_{Mn^{2+}} = x_e\left(Mn^{2+}\right) \cdot \left(\frac{64,370}{T} - 31.4\right) = 0.0518 \cdot \left(\frac{64,370}{1973} - 31.4\right)$$
$$= 0.0518 \cdot 1.2254 = 0.0634$$

$$\tag{6.587}$$

- Finally, in the case of the Fe^{2+}:

$$\log K_{Fe^{2+}} = x_e(Fe^{2+}) \cdot \left(\frac{59,300}{T} - 33.9\right) = 0.1840 \cdot \left(\frac{59,300}{1973} - 33.9\right)$$
$$= 0.1840 \cdot (-3.8442) = -0.7074$$

$$(6.588)$$

Consequently, the total contribution of these four elements to the value of the equilibrium constant is of 4.6711.

To correct the deviations of the ideal mixture of ions, the IRSID model adds a correction factor referenced as g(P) :

$$g(P) = -8.18 + 1.41 \cdot S + \frac{13,200}{T} + A \cdot x(O^{2-}) + B \cdot (x(O^{2-}))^2 \quad (6.589)$$

where the values of A and B are:

$$A = 40.55 - 6.95 \cdot S - \frac{64,500}{T} \quad (6.590)$$

$$B = -13.10 + 5.91 \cdot S + \frac{22,100}{T} \quad (6.591)$$

where S is the relation between the anionic molar fractions shown as follows (Exercise 6.19):

$$S = \frac{x(SiO_4^{4-})}{x(SiO_4^{4-}) + x(PO_4^{3-})} = \frac{0.2331}{0.2331 + 0.0171} = 0.9316 \quad (6.592)$$

and $x(O^{2-})$, according to the data of Exercise 6.19, is equal to 0.6901.

The value of the terms included in g(P) is:

$$-8.18 + 1.41 \cdot 0.9316 + \frac{13,200}{1973} = -0.1761 \quad (6.593)$$

$$A \cdot x(O^{2-}) = 0.9550 \quad (6.594)$$

$$B \cdot (x(O^{2-}))^2 = 17,181 \quad (6.595)$$

Consequently, the total addition of the three terms in which g(P) was divided is equal to 2.4971.

Thus, the equilibrium constant is:

$$\log K_{eq-PO_4^{3-}} = 4.6711 + 2.4971 = 7.1682 \rightarrow K_{eq-PO_4^{3-}} = 1.473 \times 10^7 \quad (6.596)$$

Commentaries: Habitually, the deviation from the ideal behavior is matched, both for the ionic and molecular models associated with the structure of the slags, with the activity coefficients. In this case, the deviation from the ideal situation is identified with the behavior in the melt of the PO_4^{3-} and O^{2-} anions. The concentration $x(PO_4^{3-})$ and $x(O^{2-})$ should be corrected, for non-ideal behaviors, with the activity coefficients. The activity coefficients are at the same time function of both the concentration and the temperature. In the IRSID model, this deviation from the ideal situation for the PO_4^{3-} and O^{2-} anions is corrected through the coefficient $g(P)$ where it is collected the correction of the activity coefficient for the PO_4^{3-} that, only depends on the temperature; and that of the O^{2-} that not only depends on the temperature but also on the concentration of oxygen anions in the slag.

Exercise 6.28 Calculate the minimum phosphorus concentration in an iron melt in equilibrium with the slag of Exercise 6.18, at the temperature of 1700 °C.

Data: We apply the model of the phosphorus slagging proposed by the IRSID, Exercises 6.19 and 6.27, for the exchange of the phosphorus between the metal and the slag.

If we know, by the results of Exercise 6.27, that the value of the equilibrium constant of the IRSID ionic model proposed for the phosphorus slagging is of 1.473×10^7:

$$K_{eq-PO_4^{3-}} = \frac{\left(x(PO_4^{3-})\right)^2}{\left(x(O^{2-})\right)^3 \cdot a_{H,O}^5 \cdot a_{H,P}^2} = 1.473 \times 10^7 \quad (6.597)$$

From Exercises 6.18 and 6.19, it is possible to deduce that:

$$x(O^{2-}) = 0.6901; x(PO_4^{3-}) = 0.0171 \quad (6.598)$$

the values of Henry's activities of the oxygen and the phosphorus in the iron are the only two variables that are still unknown. As general criterion, it is assumed that, both the dissolutions of Henry of the oxygen and of the phosphorus in the metal are ideal; in other words, the coefficients of activity of Henry for the oxygen and the phosphorus in the metal are equal to one ($f_O = f_P = 1$), and thus,

$$a_{H,O} = (\%O); a_{H,P} = (\%P) \quad (6.599)$$

On the other hand, the value of $(\%P)$ in the metal is the objective of this exercise, and thus (solving for $(\%P)^2$, objective of this exercise, of the general equation):

$$\frac{\left(x(PO_4^{3-})\right)^2}{\left(x(O^{2-})\right)^3 \cdot (\%O)^5 \cdot (\%P)^2} = 1.473 \times 10^7 \rightarrow (\%P)^2$$

$$= \frac{\left(x(PO_4^{3-})\right)^2}{\left(x(O^{2-})\right)^3 \cdot (\%O)^5 \cdot 1.473 \times 10^7} \rightarrow (\%P)^2$$

$$= \frac{(0.0171)^2}{(0.6901)^3 \cdot (\%O)^5 \cdot 1.473 \times 10^7}$$

$$(6.600)$$

The value of the quantity of oxygen dissolved in the metal is the last of the variables in the last equation that is still unknown. The maximum concentration of oxygen that can be assimilated by the metal is function of the quantity of FeO that is slagged in equilibrium, Exercises 6.18 and 6.19. That is to say, the quantity of oxygen in the metal is function of the activity of the FeO in the slag, Exercise 6.18:

$$Fe\ (melt) + O(dis; a_H) \leftrightarrow FeO(dis; slag) \tag{6.601}$$

With the objective of calculating the standard free energy associated with the reaction (6.601), $\Delta_r G^0(6.601)$, it is necessary to consider the following equilibriums:

$$Fe(l) + \frac{1}{2}O_2(g) \leftrightarrow FeO(l) \tag{6.602}$$

that according to Appendix 1:

$$\Delta_r G^0(\text{reaction } 6.602) = -265.0 + 0.067 \cdot T(K) \left[kJ\,(mol\ FeO)^{-1} \right] \tag{6.603}$$

$$\frac{1}{2}O_2(g) \leftrightarrow O(dis; a_H) \tag{6.604}$$

that, according to the data of Appendix 3:

$$\Delta_r G^0(\text{reaction } 6.604) = -117.0 - 3.14 \cdot T(K) \left[kJ\,(mol\ O)^{-1} \right] \tag{6.605}$$

Consequently,

$$\Delta_r G^0(\text{reaction } 6.601) = \Delta_r G^0(\text{reaction } 6.602) - \Delta_r G^0(\text{reaction } 6.604)$$
$$= -148.0 + 0.068 \cdot T(\text{K}) \left[\text{kJ (mol FeO)}^{-1} \right] \qquad (6.606)$$

The equilibrium constant of the reaction (6.601) at 1700 °C is equal to:

$$\Delta_r G^0(\text{reaction } 6.601) = (-148{,}000 + 68 \cdot 1973) \frac{\text{J}}{\text{mol}}$$

$$= -8.314 \frac{\text{J}}{\text{mol K}} \cdot 1973 \text{ K} \cdot \ln \text{K}_{(\text{reaction}6.601),\text{Fe/FeO}}$$

$$\rightarrow \text{K}_{(\text{reaction}6.601),\text{Fe/FeO}} = 2.3244 = \frac{a(\text{FeO})}{(\%\text{O})}$$
$$(6.607)$$

According to the ionic models, the activity of the FeO in the slag is:

$$a(\text{FeO}) = x_{\text{Fe}^{2+}} \cdot x_{\text{O}^{2-}} = 0.1840 \cdot 0.6901 = 0.1270 \qquad (6.608)$$

To calculate the deviations from the ideal situation, we use a coefficient of activity:

$$a(\text{FeO}) = \left(x_{\text{Fe}^{2+}} \cdot x_{\text{O}^{2-}} \right) \cdot \gamma_{\text{FeO}} \qquad (6.609)$$

According to the IRSID model, the value of the coefficient of activity of the FeO is:

$$\log \gamma_{\text{FeO}} = 5.50 - 1.12 \cdot \text{S} - \frac{5600}{T} + B \cdot x(\text{O}^{2-}) + C \cdot \left(x(\text{O}^{2-}) \right)^2 \qquad (6.610)$$

where B and C are calculated as follows:

$$B = -16.07 + 2.85 \cdot \text{S} + \frac{20{,}800}{T} \qquad (6.611)$$

$$C = 10.97 - 1.62 \cdot \text{S} - \frac{16{,}100}{T} \qquad (6.612)$$

when $S = 0.9316$, $x(\text{O}^{2-}) = 0.6901$, Exercise 6.27, and $T = 1700$ °C (1973 K), the contribution of each one of the terms to the decimal logarithm of the activity coefficient of the FeO is:

$$\log \gamma_{FeO} = 5.50 - 1.12 \cdot 0.9316 - \frac{5600}{1973} + \left(-16.07 + 2.85 \cdot 0.9316 + \frac{20,800}{1973}\right)$$
$$\cdot 0.6901 + \left(10.97 - 1.62 \cdot 0.9316 - \frac{16,100}{1973}\right) \cdot (0.6901)^2$$
$$= 0.2553$$

$$(6.613)$$

Consequently, the activity of the FeO in the slag of Exercise 6.18, according to the IRSID is:

$$a(FeO) = (x_{Fe^{2+}} \cdot x_{O^{2-}}) \cdot \gamma_{FeO} \qquad (6.614)$$

$$\log a(FeO) = \log(x_{Fe^{2+}} \cdot x_{O^{2-}}) + \log \gamma_{FeO} \rightarrow \log a(FeO)$$
$$= \log(0.6901 \cdot 0.1840) + 0.2553 \rightarrow a(FeO) = 0.2286 \qquad (6.615)$$

The concentration of oxygen in the metal in $(\%O)$ is:

$$(\%O) = \frac{a(FeO)}{K_{(reaction6.601),Fe/FeO}} = \frac{0.2286}{2.3244} = 0.098\%(\sim 1000 \text{ ppm}) \qquad (6.616)$$

Finally, the value of the minimum concentration that can reach the phosphorus in the metal will be:

$$(\%P)^2 = \frac{(0.0171)^2}{(0.6901)^3 \cdot (0.098)^5 \cdot 1.473 \times 10^7} = 6.68 \times 10^{-6} \rightarrow (\%P)$$
$$= 2.59 \times 10^{-3}\%(25.9 \text{ ppm}) \qquad (6.617)$$

Commentaries: Habitually, the slags able to catch the phosphorus are generated in the own process (endogenous slags). From the kinetics point of view, the process of endogenous formation of the slag requires from a time, during which, the slagging of the phosphorus does not take place. It is possible to use synthetic slags, which are formed in an exogeneous way (out of the process), that could shorten the time required for the slagging of the phosphorus. From this point of view, the concentrated solar energy might be a tool to be used in the future to produce synthetic slags for the ironmaking process (Fernández-González et al. 2018).

6.6   Alternative Processes to the Blast Furnace

The survival of the technologies based on the smelting reduction, as the blast furnace (that exists since the Fourteenth century), is mainly consequence of the low both energy costs and consumption of reductant gases and the minimal operating and installation costs with respect to the potential competitors.

Regarding the good results with respect to the energy and reductant gases consumptions, in Table 6.14 we collect the comparative data.

The current blast furnace, due to the improvements in the optimization of the reductant mechanism for the smelting of the ferric burden, has very competitive energy and operating costs. Data reflected in the mass and energy balances of Tables 6.7 and 6.8 is closer to the operation line used in the sixties than the one used in the modern blast furnaces.

One of the main drawbacks of the integrated steel plants has been the high installation costs per ton of pig iron. The capital required in 1950 to face such investment came from state organisms. The private sector was not prepared to assume the risks of a low benefits due to high sums of capital with amortization rates excessively long in the time. Recently, the installation costs have been significantly reduced and the private sector has shown interest in the ironmaking market. The investment required for an integrated steel plant with 6×10^6 t year^{-1} of capacity was 4000 million dollars in 1965, while in 1997 the value was reduced to 1800 million dollars.

In Table 6.15, we indicate the installation costs, per ton of product, of a blast furnace (including the sintering and coking plants) compared with the corresponding investment in the Corex, Midrex, Finmet, and iron carbide (Fe$_3$C) processes.

In the battle of competitiveness/costs (production, energy, reductant and installation), the blast furnace was the winner. The main characteristic of the processes that try to compete in the ironmaking business with the blast furnace is, apart from being close to its costs, using different raw materials (least expensive raw

Table 6.14 Consumption of reductant gas, oxygen and electrical energy, in different processes, per ton of product (Lüngen and Steffen 1988)

Reductant	Consumption per ton of product				
	Blast furnace	Corex	Midrex	Finmet	Fe$_3$C (Iron carbide)
Coke, kg	320				
Injection of coal, kg	160				
Natural gas, GJ			10.6	12.5	16.0
Gasification plant, kg		1000			
Oxygen, m^3	30	560			
Electrical energy, kWh	58	75	130	150	230

Table 6.15 Capital required, per ton of product, for the construction of a blast furnace, a Corex plant and the direct reduction plants Midrex, Finmet, and iron carbide (Lüngen and Steffen 1988)

Installation	$ t^{-1}	Production (Mt year^{-1})
Blast furnace (pig iron)	300	3.5
Corex (pig iron)	265	0.8
Midrex (HBI)	165	1.3
Finmet (HBI)	280	2.2
Iron carbide (Fe$_3$C)	200	0.3

materials or recycled products with zero cost) to that required in the blast furnace as the coke, sinter, pellet, and coarse iron ore (<15% of the total ferric burden).

On the other hand, the smelting reduction processes, under development, that aim to compete with the blast furnace, for instance the American Iron and Steel Institute (AISI), tend to manufacture a product the most similar to the steel possible, with carbon percentages <1%, which might mean the disappearance of the refinement installations BOF-BOS (Basic Oxygen Furnace-Steelmaking) of the integrated steel plants.

We analyze in this section in a more detailed way the data of Table 5.1, because the alternatives to the blast furnace in operation (Corex, HYL), those that can be used in the twenty-first century (Romelt, Ausiron, DIOS, CCF, AISI, Tecnored) or those that are under development (Finex, IT Mark 3), are specified. Summarizing, we can conclude that, the technologies that compete in the market of the iron manufacture are four: the blast furnace and the Midrex, HYL, and Corex processes (Figs. 5.3, 5.4, and 6.15). The rest of the alternatives, up to now, are not competitive. However, in the field of the recovery of iron–carbonaceous residues with high volatiles content (Zn, Na, and K), the Rotary Hearth Furnace technology might be an effective instrument for the specific treatment of this kind of residues.

6.6.1 Corex Process

The Corex is a process that uses carbon and oxygen to reduce the ferric burden that comprises coarse fraction iron ore, sinter and pellets. It uses a shaft furnace to reduce the burden, and gasification–smelting reactor as a smelting element and a generator of reductant gases.

At the end of the twentieth century, the Corex plants of South Africa and Korea reached a capacity of production of around 4.0×10^6 tons. In the first fifteen years of the twenty-first century, the capacity of production of the Corex pig iron is significantly greater (>10.0×10^6 t) with new plants built and being built worldwide.

In Fig. 6.15, we schematize a Corex plant connected to a unity of direct reduction that uses the reductant gas generated in the process. The Corex gas has an approximate composition of 35–45% CO, 35% CO$_2$, 15–20% H$_2$, 1% CH$_4$, 3% N$_2$, and others, with heat of combustion of 7.5 MJ Nm^3, while the produced melt has characteristics very similar to that of the blast furnace 4.2–4.8% C, 0.4–0.8% Si, 0.3–1.0% Mn, 0.05–0.10% P, and 0.02–0.06% S.

Other alternative consists in using the gases of the process as a combustible in a plant of electric power generation. Summarizing, we are going to mention the main characteristics of the Corex process:

- It is the first commercial process that is presented as an alternative to the blast furnace for the smelting reduction of the iron ore.
- It does not require coking plant.
- The capacity of production of the Corex unities is still limited if compared with those of the blast furnace. However, the production of its plants, 1.0–2.0 million of tons per year, might be attractive for the new generation steel plants of the twenty-first century.
- It requires accessory plants to use the reductant gas for the competitiveness and efficiency of the Corex unities (the production of gas per ton of Corex pig iron is much greater than the production of gas per ton of pig iron in the blast furnace).

6.6.2   DIOS Process

The development of the Direct Iron Ore Smelting process (DIOS) started in 1988 encouraged by the Japan Iron and Steel Federation (JISF) in cooperation with the Japan Coal Energy Center (JCEC), and the Ministry of Industry and Commerce of Japan.

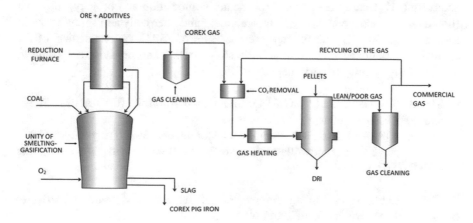

Fig. 6.15 Flow diagram of the Corex process with unity of direct reduction

We show in Fig. 6.16 a flow diagram of the DIOS process that has:

- Two unities, of preheating and prereduction of the iron ore fines (<8 mm), by which 25% of metallization is achieved.
- A unity of final reduction with injection of coal and oxygen.

At the end of the twentieth century, this technology was at a pilot scale, with a capacity of production of 500 t day^{-1}. The coal consumption in the pilot plant was in the range of 730–750 kg of coal per ton of melt. The expectations planed at the end of the century included the construction of the first commercial plant with a capacity of 3000 t day^{-1}, which was not yet built. According to Naito et al. 2015 (pages 30–31), it would be necessary the construction of two smelting unities of 6000 t day^{-1} of capacity to definitely check the industrial competitiveness of the process. It is planned a reduction in the operation and installation costs, with respect to those of the blast furnace, within 35% and 19%, respectively, because neither coke ovens nor sintering plant are required. It is finally considered that an alternative to relaunch the process will be the improvement of the performance in the prereduction unity, for which the possibility of connecting the smelting reduction technology DIOS with the Rotary Hearth Furnace might be taken into account (Fig. 5.6).

6.7 Solved Exercises: Application of the CaO–Al$_2$O$_3$–SiO$_2$ Ternary Diagram and Physical–Chemical Properties of the Slags

Exercise 6.29 Detail the chemical (elemental composition and basicity), physical (liquidus, T_L, and solidus, T_S, temperatures), and thermodynamic (enthalpy of formation at 1500 °C with reference temperature of 25 °C) characteristics, of the following crystallized slag, considering a slow cooling and behavior according to the CaO–Al$_2$O$_3$–SiO$_2$ ternary diagram of Fig. 6.2:

- Slag A: 40% Gehlenite (2CaO·Al$_2$O$_3$·SiO$_2$), 20% Anorthite (CaO·Al$_2$O$_3$·2SiO$_2$), and 40% Pseudowollastonite (CaO·SiO$_2$).
- Slag B: 40% Anorthite, 30% Pseudowollastonite, and 30% Tridymite (SiO$_2$).
- Slag C: 75% Pseudowollastonite, 18% Gehlenite, and 7% Rankinite (3CaO·2SiO$_2$).

The molecular weight of the components of the ternary system, CaO–Al$_2$O$_3$–SiO$_2$, are, respectively: 56, 102, and 60 g/mol.

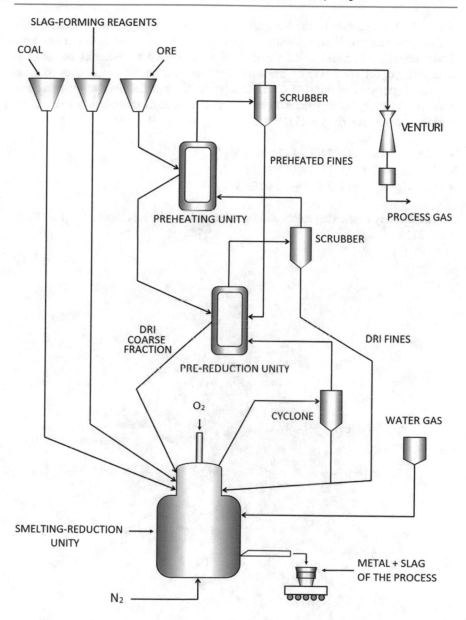

Fig. 6.16 Flow diagram of the DIOS process

When we locate inside of the triangles (Roozebum method) either the Gehlenite–Anorthite–Pseudowollastonite (Slag A, green), the Anorthite–Pseudowollastonite–Tridymite (Slag B, blue) or the Gehlenite–Pseudowollastonite–Rankinite (Slag C, red) and project over one of the sides the percentages of the above-indicated crystalline phases and drawing from these points the parallel lines to the other two sides of the triangle, they cross in one point that defines the pseudoternary composition of the slag (Fig. 6.17):

- Slag A: 40% CaO, 22% Al_2O_3, 38% SiO_2.
- Slag B: 22% CaO, 15% Al_2O_3, 63% SiO_2.
- Slag C: 48% CaO, 7% Al_2O_3, 45% SiO_2.

The basicity of the slag calculated from the index, $I_B(I)$, is (Ballester et al. 2000, page 271):

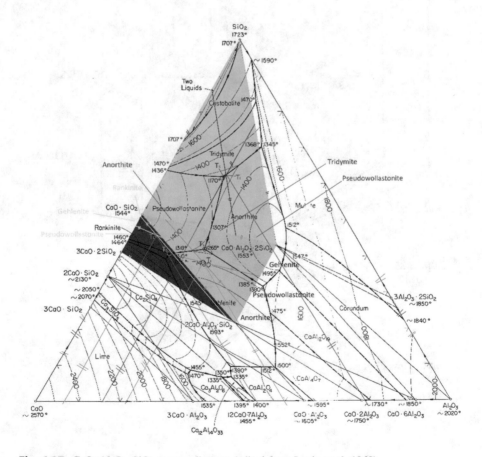

Fig. 6.17 CaO–Al$_2$O$_3$–SiO$_2$ ternary diagram (edited from Levin et al. 1969)

$$I_B(I) = \frac{\%CaO}{\%SiO_2} \tag{6.618}$$

The values of the basicity of the three slags are: 1.05, 0.35, and 1.07 for the slag A, B, and C, respectively.

Equally, the basicity of one slag can be expressed by the optical basicity, Λ_{slag} (Ballester et al. 2000, page 272). Using the values provided by Springorum 1995 (Table 6.10), the results are obtained as follows:

- Considering as reference 100 g of slag, the number of moles of each one of the oxides will be (the percentage mass composition of each one of the oxides is transformed into their corresponding molar fraction):

$$\text{Moles } CaO(n_{CaO}): \frac{40 \text{ g CaO}}{100 \text{ g slag}} \cdot \frac{1 \text{ mol CaO}}{56.08 \text{ g CaO}} = 0.7133 \frac{\text{mol CaO}}{100 \text{ g slag}} \tag{6.619}$$

$$\text{Moles } Al_2O_3(n_{Al_2O_3}): \frac{22 \text{ g Al}_2O_3}{100 \text{ g slag}} \cdot \frac{1 \text{ mol Al}_2O_3}{101.96 \text{ g Al}_2O_3} = 0.2158 \frac{\text{mol Al}_2O_3}{100 \text{ g slag}} \tag{6.620}$$

$$\text{Moles } SiO_2(n_{SiO_2}): \frac{38 \text{ g SiO}_2}{100 \text{ g slag}} \cdot \frac{1 \text{ mol SiO}_2}{60.09 \text{ g SiO}_2} = 0.6324 \frac{\text{mol SiO}_2}{100 \text{ g slag}} \tag{6.621}$$

- Consequently, the number of moles in 100 g of slag is 1.5615 mol. Thus, the composition expressed as molar fraction of each one of the oxides in the slag will be:

$$\text{Molar fraction } CaO(x_{CaO}): \frac{0.7133 \frac{\text{mol CaO}}{100 \text{ g slag}}}{1.5615 \frac{\text{mol slag}}{100 \text{ g slag}}} = 0.4568 \frac{\text{mol CaO}}{\text{mol slag}} \tag{6.622}$$

$$\text{Molar fraction } Al_2O_3(x_{Al_2O_3}): \frac{0.2158 \frac{\text{mol Al}_2O_3}{100 \text{ g slag}}}{1.5615 \frac{\text{mol slag}}{100 \text{ g slag}}} = 0.1382 \frac{\text{mol Al}_2O_3}{\text{mol slag}} \tag{6.623}$$

$$\text{Molar fraction } SiO_2(x_{SiO_2}): \frac{0.6324 \frac{\text{mol SiO}_2}{100 \text{ g slag}}}{1.5615 \frac{\text{mol slag}}{100 \text{ g slag}}} = 0.4050 \frac{\text{mol SiO}_2}{\text{mol slag}} \tag{6.624}$$

- We calculate now the summation resulted from the multiplication of the molar fraction of each oxide by the number of atoms of oxygen in the molecules. The result of the summation is (see equation (6.363)):

$$\sum xO = x_{CaO} \cdot 1 + x_{Al_2O_3} \cdot 3 + x_{SiO_2} \cdot 2 = 1.6814 \qquad (6.625)$$

- We obtain, for each one of the oxides of the slag, the value of $x(M_i)$, consequence of the multiplication of the molar fraction of the oxide by the number of oxygen atoms and dividing the result by the value of the summation obtained in Eq. (6.625):

$$X(M_{CaO}) = \frac{x_{CaO} \cdot 1}{\sum xO} = \frac{0.4568 \cdot 1}{1.6814} = 0.2717 \qquad (6.626)$$

$$X(M_{Al_2O_3}) = \frac{x_{Al_2O_3} \cdot 3}{\sum xO} = \frac{0.1382 \cdot 3}{1.6814} = 0.2466 \qquad (6.627)$$

$$X(M_{SiO_2}) = \frac{x_{SiO_2} \cdot 2}{\sum xO} = \frac{0.4050 \cdot 2}{1.6814} = 0.4817 \qquad (6.628)$$

- Finally, we multiply each one of the values of $x(M_i)$ previously calculated and their corresponding value of the optical basicity:

$$\Lambda_{slag} = \Lambda_{CaO} \cdot X(M_{CaO}) + \Lambda_{Al_2O_3} \cdot X(M_{Al_2O_3}) + \Lambda_{SiO_2} \cdot X(M_{SiO_2})$$
$$= 1 \cdot 0.2717 + 0.605 \cdot 0.2466 + 0.48 \cdot 0.4817 = 0.6521 \qquad (6.629)$$

Using the same procedure, the optical basicities of the slags B and C are 0.5684 and 0.6639, respectively.

The values of the liquidus, T_L, and solidus, T_S, temperatures are determined using Fig. 6.17 and are collected in Table 6.16.

Finally, using the software HSC 5.1, we calculate the enthalpy associated with the slags with the mass and energy balance (Exercise 5.2). In this case, the components of entrance to the system are, at 25 °C, calcium, aluminum, and silicon oxides, while the components of exit, in each case, are the crystalline phases of the enunciation at 1500 °C. The results are:

- Slag A: 103 MJ t^{-1}.
- Slag B: 130 MJ t^{-1}.
- Slag C: 85 MJ t^{-1}.

Table 6.16 Liquidus and solidus temperatures

Slag	T_L (°C)	T_S (°C)	Eutectic
A	1400	1265	Pseudowollastonite–anorthite–gehlenite
B	1200	1170	Pseudowollastonite–anorthite–tridymite
C	1450	1310	Pseudowollastonite–gehlenite–rankinite

The more acid the slag is, the greater the enthalpy of formation is. However, in acid slags, as they have a lower liquidus temperature, operation conditions in the hearth of the furnace might be 100 °C lower. The problem that appears with basic slags is that, when the operation conditions are brought closer to the liquidus temperature, problems in the tapping of the furnaces can appear.

Commentaries: In the smelting reduction process developed by the blast furnace, the process of formation of the slag ends in the hearth. The three considered slags in exercise are the result of different conditions of operation of the furnace, from acid operation (Slag *B*) to basic conditions (Slag *C*).

When inside of the furnace shaft (Fig. 6.3), the temperature reaches 1100 °C and primary slags with high iron content are usually formed (Levin et al. 1969, page 241): the FeO–Al$_2$O$_3$–SiO$_2$ system has eutectic reaction at 1083 °C when 47% FeO, 12% Al$_2$O$_3$, and 41% SiO$_2$. The ferrous content of this primary slag will be transformed in the bosh zone of the furnace; the hearth of the furnace and the slag can chemically evolve toward the pseudoternary compositions indicated in the enunciation, and the iron will be reincorporated to the metal while the aluminum oxide will reintegrate in the melt.

There is software to estimate the liquidus and solidus temperatures in multicomponent systems. In this case, the accurateness reached with the ternary system might be enough, both from the academic point of view and from a first approximation to the industrial problem. For instance, the accurateness in the chemical composition obtained by means of Fig. 6.17 of the CaO–Al$_2$O$_3$–SiO$_2$ ternary system is for the slag *A* 40% CaO, 22% Al$_2$O$_3$, and 38% SiO$_2$, while with the software HSC 5.1 is 39.7% CaO, 22.2% Al$_2$O$_3$, and 38.1% SiO$_2$.

Exercise 6.30 Taking into account the acid slags generated in the blast furnace, using the SiO$_2$–Al$_2$O$_3$–CaO ternary diagram (Fig. 6.18, Roozebum method), calculate:

1. The composition of the slag if at 1307 °C the first solid phases to freeze after melting during the solidification are the pseudowollastonite (W, CaO·SiO$_2$) and the anorthite (A, CaO·Al$_2$O$_3$·2SiO$_2$).
2. The activities of the Al$_2$O$_3$ and SiO$_2$ in the melt at 1307 °C considering that the activity of the calcium oxide in the melt is 5.0×10^{-3}

Use the molecular weights indicated in Exercise 6.29.

In Fig. 6.18, it is possible to locate the chemical composition for which there are only three phases (two solid and one liquid) in equilibrium at the liquidus temperature of 1307 °C: 34% CaO, 18% Al$_2$O$_3$, and 48% SiO$_2$.

On the other hand, and as a consequence of the previously indicated:

$$I_B(l) = \frac{\%\text{CaO}}{\%\text{SiO}_2} = \frac{34}{48} = 0.71 \tag{6.630}$$

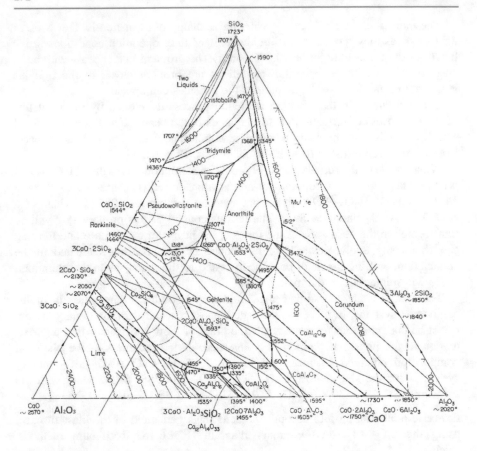

Fig. 6.18 SiO_2–Al_2O_3–CaO ternary diagram (edited from Levin et al. 1969)

Now, we calculate the composition expressed as molar fraction. Considering as reference 100 g of slag, the number of moles of each one of the oxides will be (the percentage mass composition of each one of the oxides is transformed in their corresponding molar fraction):

$$\text{Moles } CaO(n_{CaO}): \frac{34 \text{ g CaO}}{100 \text{ g slag}} \cdot \frac{1 \text{ mol CaO}}{56.08 \text{ g CaO}} = 0.6063 \frac{\text{mol CaO}}{100 \text{ g slag}} \quad (6.631)$$

$$\text{Moles } Al_2O_3(n_{Al_2O_3}): \frac{18 \text{ g Al}_2O_3}{100 \text{ g slag}} \cdot \frac{1 \text{ mol Al}_2O_3}{101.96 \text{ g Al}_2O_3} = 0.1765 \frac{\text{mol Al}_2O_3}{100 \text{ g slag}} \quad (6.632)$$

$$\text{Moles } SiO_2(n_{SiO_2}): \frac{48 \text{ g SiO}_2}{100 \text{ g slag}} \cdot \frac{1 \text{ mol SiO}_2}{60.09 \text{ g SiO}_2} = 0.7988 \frac{\text{mol SiO}_2}{100 \text{ g slag}} \quad (6.633)$$

Consequently, the number of moles in 100 g of slag is 1.5816 mol. Thus, the composition expressed as molar fraction of each one of the oxides in the slag will be:

$$\text{Molar fraction CaO}(x_{\text{CaO}}): \frac{0.6063 \frac{\text{mol CaO}}{100 \text{ g slag}}}{1.5816 \frac{\text{mol slag}}{100 \text{ g slag}}} = 0.3833 \frac{\text{mol CaO}}{\text{mol slag}} \quad (6.634)$$

$$\text{Molar fraction Al}_2\text{O}_3(x_{\text{Al}_2\text{O}_3}): \frac{0.1765 \frac{\text{mol Al}_2\text{O}_3}{100 \text{ g slag}}}{1.5816 \frac{\text{mol slag}}{100 \text{ g slag}}} = 0.1116 \frac{\text{mol Al}_2\text{O}_3}{\text{mol slag}} \quad (6.635)$$

$$\text{Molar fraction SiO}_2(x_{\text{SiO}_2}): \frac{0.7988 \frac{\text{mol SiO}_2}{100 \text{ g slag}}}{1.5816 \frac{\text{mol slag}}{100 \text{ g slag}}} = 0.5051 \frac{\text{mol SiO}_2}{\text{mol slag}} \quad (6.636)$$

Applying the lever rule, the proportion of the two solid phases of the melt that precipitate simultaneously by the binary eutectic transformation is 51% Anorthite (A) and 49% Pseudowollastonite (W).

The data of the ternary represents a thermodynamic equilibrium between phases (between the eutectic melt and the solids Anorthite and Pseudowollastonite). With the data of Appendix 1 for the standard free energy of formation of W and A, from their components (CaO, Al$_2$O$_3$, and SiO$_2$):

$$\Delta_f G^0(T)(W) = -82.9 - 0.003 \cdot T(\text{K}) \left[\text{kJ} \, (\text{mol W})^{-1} \right] \quad (6.637)$$

$$\Delta_f G^0(T)(A) = -55.1 - 0.035 \cdot T(\text{K}) \left[\text{kJ} \, (\text{mol A})^{-1} \right] \quad (6.638)$$

At 1307 °C, the equilibrium constant of the pseudowollastonite is:

$$\Delta_f G^0(T)(W) = -82,900 - 3 \cdot (1307 + 273) \frac{\text{J}}{\text{mol}} = -87,640 \frac{\text{J}}{\text{mol}}$$

$$= -R \cdot T \cdot \ln K_{\text{eq}} \rightarrow K_{\text{eq}} = \exp\left(\frac{87,640}{8.314 \cdot 1580} \right) = 789.72$$

$$(6.639)$$

The equilibrium constant is:

$$K_{eq} = 789.72 = \frac{1}{a_{CaO} \cdot a_{SiO_2}} \rightarrow a_{SiO_2} = \frac{1}{a_{CaO} \cdot 789.72} = \frac{1}{5 \times 10^{-3} \cdot 789.72}$$

$$= 0.253$$

(6.640)

This way:

$$a_{SiO_2} = x_{SiO_2} \cdot \gamma_{SiO_2} \rightarrow \gamma_{SiO_2} = \frac{a_{SiO_2}}{x_{SiO_2}} = \frac{0.253}{0.5051} = 0.5008 \tag{6.641}$$

In the same way, the equilibrium constant for the formation of the Anorthite from the components of the ternary system is:

$$\Delta_f G^0(T)(W) = -55,100 - 35 \cdot (1307 + 273) \frac{J}{mol} = -110,400 \frac{J}{mol}$$

$$= -R \cdot T \cdot \ln K_{eq} \rightarrow K_{eq} = \exp\left(\frac{110,400}{8.314 \cdot 1580}\right) = 4466.27$$

(6.642)

The equilibrium constant is:

$$K_{eq} = 4466.27 = \frac{1}{a_{CaO} \cdot a_{SiO_2}^2 \cdot a_{Al_2O_3}} \rightarrow a_{Al_2O_3} = \frac{1}{a_{CaO} \cdot a_{SiO_2}^2 \cdot 4466.27}$$

$$= \frac{1}{5 \times 10^{-3} \cdot 0.253^2 \cdot 4466.27} = 0.7 \tag{6.643}$$

This way:

$$a_{Al_2O_3} = x_{Al_2O_3} \cdot \gamma_{Al_2O_3} \rightarrow \gamma_{Al_2O_3} = \frac{a_{Al_2O_3}}{x_{Al_2O_3}} = \frac{0.7}{0.1116} = 6.2724 \tag{6.644}$$

Commentaries: According to Springorum 1995 (page 238), the activity of the SiO_2 at 1600 °C, for the same composition of the melt (34% CaO, 18% Al_2O_3, and 48% SiO_2) is equal to 0.1, that is, increasing the temperature means that the capacity of reaction of the silicon oxide in the slag diminishes.

Exercise 6.31 If the basic slags of the blast furnace are analyzed with the SiO_2–Al_2O_3–CaO ternary diagram of Fig. 6.19, calculate:

1. The composition of the melt at 1500 °C in equilibrium with the solid phases Gehlenite ($2CaO \cdot Al_2O_3 \cdot SiO_2$) and dicalcium silicate–Larnite ($2CaO \cdot SiO_2$).
2. The activities of the aluminum and calcium oxides, under the equilibrium conditions of the previous section.

The value of the activity of the silicon oxide at 1500 °C is 0.04.

Using the ternary diagram of Fig. 6.19 with the Roozebum method, the chemical composition of the melt in equilibrium with the solid phases Gehlenite and Larnite is: 49% CaO, 18% Al$_2$O$_3$, and 33% SiO$_2$.
The basicity of the slag is:

$$I_B(I) = \frac{\%CaO}{\%SiO_2} = \frac{49}{33} = 1.48 \qquad (6.645)$$

We calculate the composition expressed as molar fraction. Considering as reference 100 g of slag, the number of moles of each one of the oxides will be (the percentage mass composition of each one of the oxides is transformed in their corresponding molar fraction):

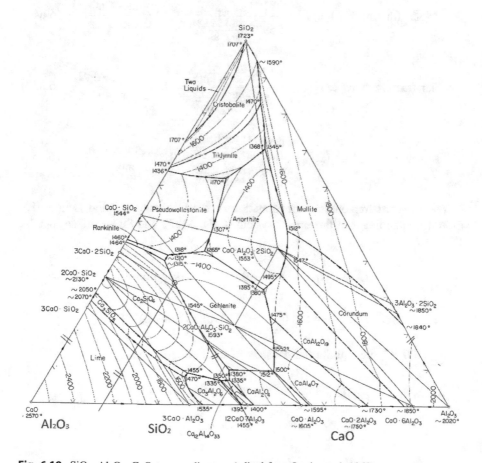

Fig. 6.19 SiO$_2$–Al$_2$O$_3$–CaO ternary diagram (edited from Levin et al. 1969)

$$\text{Moles CaO}(n_{\text{CaO}}) : \frac{49 \text{ g CaO}}{100 \text{ g slag}} \cdot \frac{1 \text{ mol CaO}}{56.08 \text{ g CaO}} = 0.8738 \frac{\text{mol CaO}}{100 \text{ g slag}} \qquad (6.646)$$

$$\text{Moles Al}_2\text{O}_3(n_{\text{Al}_2\text{O}_3}) : \frac{18 \text{ g Al}_2\text{O}_3}{100 \text{ g slag}} \cdot \frac{1 \text{ mol Al}_2\text{O}_3}{101.96 \text{ g Al}_2\text{O}_3} = 0.1765 \frac{\text{mol Al}_2\text{O}_3}{100 \text{ g slag}}$$
$$(6.647)$$

$$\text{Moles SiO}_2(n_{\text{SiO}_2}) : \frac{33 \text{ g SiO}_2}{100 \text{ g slag}} \cdot \frac{1 \text{ mol SiO}_2}{60.09 \text{ g SiO}_2} = 0.5492 \frac{\text{mol SiO}_2}{100 \text{ g slag}} \qquad (6.648)$$

Consequently, the number of moles in 100 g of slag is 1.5995 mol. Thus, the composition expressed as molar fraction of each one of the oxides in the slag will be:

$$\text{Molar fraction CaO}(x_{\text{CaO}}) : \frac{0.8738 \frac{\text{mol CaO}}{100 \text{ g slag}}}{1.5995 \frac{\text{mol slag}}{100 \text{ g slag}}} = 0.5463 \frac{\text{mol CaO}}{\text{mol slag}} \qquad (6.649)$$

$$\text{Molar fraction Al}_2\text{O}_3(x_{\text{Al}_2\text{O}_3}) : \frac{0.1765 \frac{\text{mol Al}_2\text{O}_3}{100 \text{ g slag}}}{1.5995 \frac{\text{mol slag}}{100 \text{ g slag}}} = 0.1103 \frac{\text{mol Al}_2\text{O}_3}{\text{mol slag}} \qquad (6.650)$$

$$\text{Molar fraction SiO}_2(x_{\text{SiO}_2}) : \frac{0.5492 \frac{\text{mol SiO}_2}{100 \text{ g slag}}}{1.5995 \frac{\text{mol slag}}{100 \text{ g slag}}} = 0.3434 \frac{\text{mol SiO}_2}{\text{mol slag}} \qquad (6.651)$$

The standard free energies of formation of the Gehlenite (G) and Larnite (L), from the components of the ternary system (CaO, Al$_2$O$_3$, and SiO$_2$), are (Appendix 1):

$$\Delta_f G^0(T)(G) = -96.0 - 0.05 \cdot T(\text{K}) \left[\text{kJ} \, (\text{mol G})^{-1} \right] \qquad (6.652)$$

$$\Delta_f G^0(T)(L) = -103.0 - 0.02 \cdot T(\text{K}) \left[\text{kJ} \, (\text{mol L})^{-1} \right] \qquad (6.653)$$

At 1500 °C, the equilibrium constant of the larnite is:

$$\Delta_f G^0(T)(L) = -103{,}000 - 20 \cdot (1500 + 273) \frac{\text{J}}{\text{mol}} = -138{,}460 \frac{\text{J}}{\text{mol}}$$
$$= -R \cdot T \cdot \ln K_{\text{eq}} \rightarrow K_{\text{eq}} = \exp\left(\frac{138{,}460}{8.314 \cdot 1773} \right) = 12{,}004.38$$
$$(6.654)$$

The equilibrium constant is:

$$K_{eq} = 12{,}004.38 = \frac{1}{a_{CaO}^2 \cdot a_{SiO_2}} \rightarrow a_{CaO} = \left(\frac{1}{a_{SiO_2} \cdot 12004.38}\right)^{0.5}$$

$$= \left(\frac{1}{0.04 \cdot 12004.38}\right)^{0.5} = 0.0456 \tag{6.655}$$

This way:

$$a_{CaO} = x_{CaO} \cdot \gamma_{CaO} \rightarrow \gamma_{CaO} = \frac{a_{CaO}}{x_{CaO}} = \frac{0.0456}{0.5463} = 0.0835 \tag{6.656}$$

Under the same conditions, the equilibrium constant for the formation of the Gehlenite from the components of the ternary system is:

$$\Delta_f G^0(T)(G) = -96{,}000 - 50 \cdot (1500 + 273)\frac{J}{mol} = -184{,}650\frac{J}{mol}$$

$$= -R \cdot T \cdot \ln K_{eq} \rightarrow K_{eq} = \exp\left(\frac{184{,}650}{8.314 \cdot 1773}\right) = 275{,}549.75 \tag{6.657}$$

The equilibrium constant is:

$$K_{eq} = 275{,}549.75 = \frac{1}{a_{CaO}^2 \cdot a_{SiO_2} \cdot a_{Al_2O_3}} \rightarrow a_{Al_2O_3} = \frac{1}{a_{CaO}^2 \cdot a_{SiO_2} \cdot 275549.75}$$

$$= \frac{1}{0.0456^2 \cdot 0.04 \cdot 275549.75} = 0.0436 \tag{6.658}$$

This way:

$$a_{Al_2O_3} = x_{Al_2O_3} \cdot \gamma_{Al_2O_3} \rightarrow \gamma_{Al_2O_3} = \frac{a_{Al_2O_3}}{x_{Al_2O_3}} = \frac{0.0436}{0.1103} = 0.3946 \tag{6.659}$$

Commentaries: Springorum 1995 (page 238) provides data about the activity of the calcium oxide (0.05) at a temperature slightly higher than that of the exercise, 1550 °C, in agreement with the calculated value.

Exercise 6.32 A blast furnace pseudoternary slag has the following composition in weight (slag C of Exercise 6.29): 48% CaO, 7% Al$_2$O$_3$, and 45% SiO$_2$. Considering the Ionic Model of Manson, calculate the activity of the calcium oxide in the melt at 1500 °C. Consider that the equilibrium constant of the polymerization of the monomer SiO$_4^{2-}$ is 3.5 × 10^{-2}.

The calculation of the calcium oxide activity, assuming the conceptual representation of the slag as a melt of anions and cations, is equal to the product of the equivalent fraction of cations of calcium in the homogeneous mixture of cations, by that of the oxygen anions in the homogeneous mixture of anions:

$$a_{CaO} = N_{e,Ca^{2+}} \cdot N_{e,O^{2-}} \tag{6.660}$$

The Ionic Model of Manson (Ballester et al. 2000, pages 282–283) provides a tool to obtain the molar fraction of oxygen anions in the homogeneous mixture of the slag, using the following equations (see additional information in the commentaries):

$$x(O^{2-}) = \frac{1-a}{1-a+a \cdot K^{-1}} \tag{6.661}$$

where the value of a is obtained by means of the following quadratic equation:

$$A \cdot a^2 + B \cdot a + C = 0 \tag{6.662}$$

where A, B, and C are constants:

$$A = x_{SiO_2} \cdot (K-1) \tag{6.663}$$

$$B = 3 \cdot x_{SiO_2} - 2 \cdot K \cdot x_{SiO_2} - 1 \tag{6.664}$$

$$C = K \cdot x_{SiO_2} \tag{6.665}$$

where x_{SiO_2} and K are, respectively, the molar fraction of the silicon oxide and the constant of polymerization of the silicate in the melt (a slag can be understood as the polymerization of the monomer SiO_4^{2-}). We calculate the composition expressed as molar fraction. Considering as reference 100 g of slag, the number of moles of each one of the oxides will be (the percentage mass composition of each one of the oxides is transformed into their corresponding molar fraction):

$$\text{Moles } CaO(n_{CaO}) : \frac{48 \text{ g CaO}}{100 \text{ g slag}} \cdot \frac{1 \text{ mol CaO}}{56.08 \text{ g CaO}} = 0.8559 \frac{\text{mol CaO}}{100 \text{ g slag}} \tag{6.666}$$

$$\text{Moles } Al_2O_3(n_{Al_2O_3}) : \frac{7 \text{ g Al}_2O_3}{100 \text{ g slag}} \cdot \frac{1 \text{ mol Al}_2O_3}{101.96 \text{ g Al}_2O_3} = 0.0687 \frac{\text{mol Al}_2O_3}{100 \text{ g slag}} \tag{6.667}$$

$$\text{Moles } SiO_2(n_{SiO_2}) : \frac{45 \text{ g SiO}_2}{100 \text{ g slag}} \cdot \frac{1 \text{ mol SiO}_2}{60.09 \text{ g SiO}_2} = 0.7489 \frac{\text{mol SiO}_2}{100 \text{ g slag}} \tag{6.668}$$

Consequently, the number of moles in 100 g of slag is 1.6735 mol. Thus, the composition expressed as molar fraction of each one of the oxides in the slag will be:

$$\text{Molar fraction CaO}(x_{CaO}): \frac{0.8559\frac{\text{mol CaO}}{100\text{ g slag}}}{1.6735\frac{\text{mol slag}}{100\text{ g slag}}} = 0.5114\frac{\text{mol CaO}}{\text{mol slag}} \quad (6.669)$$

$$\text{Molar fraction Al}_2\text{O}_3(x_{Al_2O_3}): \frac{0.0687\frac{\text{mol Al}_2\text{O}_3}{100\text{ g slag}}}{1.6735\frac{\text{mol slag}}{100\text{ g slag}}} = 0.0411\frac{\text{mol Al}_2\text{O}_3}{\text{mol slag}} \quad (6.670)$$

$$\text{Molar fraction SiO}_2(x_{SiO_2}): \frac{0.7489\frac{\text{mol SiO}_2}{100\text{ g slag}}}{1.6735\frac{\text{mol slag}}{100\text{ g slag}}} = 0.4475\frac{\text{mol SiO}_2}{\text{mol slag}} \quad (6.671)$$

In one mol of slag, the number of moles of calcium and aluminum cations in the homogenous mixture of cations is calculated as follows:

$$n(Ca^{2+}) = 0.5114\frac{\text{mol CaO}}{\text{mol slag}} \cdot \frac{\text{mol Ca}^{2+}}{\text{mol CaO}} = 0.5114\frac{\text{mol Ca}^{2+}}{\text{mol slag}} \quad (6.672)$$

$$n(Al^{3+}) = 0.0411\frac{\text{mol Al}_2\text{O}_3}{\text{mol slag}} \cdot \frac{\text{mol Al}^{3+}}{\text{mol Al}_2\text{O}_3} = 0.0411\frac{\text{mol Al}^{2+}}{\text{mol slag}} \quad (6.673)$$

The equivalent number of cations in the homogeneous mixture of cations of the slag will be:

$$n_e(\text{total}) = 2 \cdot n(Ca^{2+}) + 3 \cdot n(Al^{3+}) = 1.1458 \quad (6.674)$$

Thus, the equivalent fraction of calcium cations in the mixture of cations of the slag (the only cations in the slag are the calcium and the aluminum):

$$N_{e,Ca^{2+}} = \frac{2 \cdot n(Ca^{2+})}{n_e(\text{total})} = \frac{2 \cdot 0.5114}{1.1458} = 0.8928 \quad (6.675)$$

The molar fraction of oxygen anions in the homogeneous mixture of anions is calculated, in a first step, by calculating the value of a using the quadratic equation:

$$[x_{SiO_2} \cdot (K-1)] \cdot a^2 + (3 \cdot x_{SiO_2} - 2 \cdot K \cdot x_{SiO_2} - 1) \cdot a + K \cdot x_{SiO_2} = 0 \quad (6.676)$$

$$[0.4475 \cdot (3.5 \times 10^{-2} - 1)] \cdot a^2 + (3 \cdot 0.4475 - 2 \cdot 3.5 \times 10^{-2} \cdot 0.4475 - 1) \cdot a$$
$$+ 3.5 \times 10^{-2} \cdot 0.4475 = 0$$

(6.677)

$$-0.43184 \cdot a^2 + 0.31118 \cdot a + 0.015663 = 0 \qquad (6.678)$$

This way, $a = 0.768$, and consequently, the molar fraction of the anion oxygen is:

$$x(O^{2-}) = \frac{1-a}{1-a+a \cdot K^{-1}} = \frac{1-0.768}{1-0.768+0.768 \cdot (3.5 \times 10^{-2})^{-1}} = 0.01046$$

(6.679)

As the equivalent number of anions in the homogeneous mixture of anions must be equal to that of cations, the equivalent fraction of oxygen anions, $x(O^{2-})$, in the slag is:

$$N_{e,O^{2-}} = \frac{x(O^{2-}) \cdot 2}{n_e(\text{total})} = \frac{2 \cdot 0.01046}{1.1458} = 0.01826 \qquad (6.680)$$

Thus, according to the Model of Manson, the activity of the calcium oxide will be:

$$a_{CaO} = N_{e,Ca^{2+}} \cdot N_{e,O^{2-}} = 0.8928 \cdot 0.01826 = 0.0163 \qquad (6.681)$$

Commentaries: The activity of the calcium oxide proposed by Springorum 1995 (page 238) for a composition in weight percentage of 48% CaO, 7% Al_2O_3, and 45% SiO_2 is 0.0150 at 1550 °C.

In weakly basic slags, $I_B(I) = 1$, the structure of the silica in the melt passes from being a system of discrete anions SiO_4^{4-} to form an ensemble of linear or cyclic links with a variable number of unities (monomers) SiO_4^{4-}.

This Model of Manson, as almost all the ionic formulations, admits the Temkin equation $(a_{Me_2O_n} = N_{e,Me^{n+}} \cdot N_{e,O^{2-}})$ to calculate the activities of the components of the melt. The contribution of the model lies in the form of calculating the equivalent or ionic fraction of oxygen anions in the mixture of anions.

Manson assumes that the structure of a slag, whose basicity index, $I_B(I)$, is around one, is formed by linear chains resulted from the polymerization of the SiO_4^{4-} tetrahedron. Chemically, the polymerization reactions of different linear chains of silicate anions will be:

$$SiO_4^{4-} + SiO_4^{4-} \rightarrow Si_2O_7^{6-} + O^{2-} \qquad (6.682)$$

$$SiO_4^{4-} + Si_2O_7^{6-} \rightarrow Si_3O_{10}^{8-} + O^{2-} \qquad (6.683)$$

$$SiO_4^{4-} + Si_3O_{10}^{8-} \rightarrow Si_4O_{13}^{10-} + O^{2-} \tag{6.684}$$

The structure of a blast furnace slag analyzed from the perspective of the SiO_2–Al_2O_3–CaO ternary diagram (the composition of the pseudoternary slag would be: 35% SiO_2, 15% Al_2O_3, and 50% CaO) would be formed by a mixture of cations (Ca^{2+} and Al^{3+}) and a mixture of anions (O^{2-}, SiO_4^{4-}, $Si_2O_7^{6-}$, $Si_3O_{10}^{8-}$, $Si_4O_{13}^{10-}$, ..., $Si_nO_{3n+1}^{2(n+1)-}$).

The concentration, expressed as molar fraction, x, of the different anions of the mixture, would be calculated by equations as those mentioned below:

$$x_{Si_2O_7^{6-}} = K(I) \cdot \frac{x_{SiO_4^{4-}}}{x_{O^{2-}}} \cdot x_{SiO_4^{4-}} \tag{6.685}$$

$$x_{Si_3O_{10}^{8-}} = K(I) \cdot K(II) \cdot \left(\frac{x_{SiO_4^{4-}}}{x_{O^{2-}}} \right)^2 \cdot x_{SiO_4^{4-}} \tag{6.686}$$

$$x_{Si_4O_{13}^{10-}} = K(I) \cdot K(II) \cdot K(III) \cdot \left(\frac{x_{SiO_4^{4-}}}{x_{O^{2-}}} \right)^3 \cdot x_{SiO_4^{4-}} \tag{6.687}$$

$$x_{Si_nO_{3n+1}^{2(n+1)-}} = K(I) \cdot K(II)...K(n-1) \cdot \left(\frac{x_{SiO_4^{4-}}}{x_{O^{2-}}} \right)^{(n-1)} \cdot x_{SiO_4^{4-}} \tag{6.688}$$

where $K(I) \cdot K(II)...K(n-1)$ are the equilibrium constants of the polymerization reactions of the type shown in Eqs. (6.682) to (6.684). The problem is simplified if it is assumed that all the constants are equal in value; that is to say, there is not any preference in the addition of one SiO_4^{4-} to the linear chain of any silicate.

In this way, to calculate the ionic fraction of oxygen anions, $x_{O^{2-}}$, from the molar fraction of silicon oxide in the slag, x_{SiO_2}, the following equations are used:

$$(K \cdot x_{SiO_2} - x_{SiO_2}) \cdot a^2 + (3 \cdot x_{SiO_2} - 2 \cdot K \cdot x_{SiO_2} - 1) \cdot a + K \cdot x_{SiO_2} = 0 \tag{6.689}$$

$$K \cdot (1 - a) = (K - a \cdot K + a) \cdot x_{O^{2-}} \tag{6.690}$$

where K is the polymerization constant of the SiO_4^{4-} anion and the parameter a is equal to:

$$a = \frac{x_{SiO_4^{4-}}}{x_{O^{2-}}} \cdot K \tag{6.691}$$

Exercise 6.33 Calculate the density, specific heat and viscosity at 1500 °C for the three pseudoternary slags of Exercise 6.29. The partial molar properties of the components of the slags (Ballester et al. 2000, pages 275–276) are the following ones:

- Partial molar volumes at 1500 °C, $\overline{V}_i \cdot 10^6$, in $m^3 \cdot mol^{-1}$: CaO, 20.7; Al_2O_3, $28.31 + 32 \cdot x_{Al_2O_3} - 31.45 \cdot x^2_{Al_2O_3}$; SiO_2, $19.55 + 7.966 \cdot x_{SiO_2}$.
- Partial molar specific heats in molten state, $\bar{c}_{p-i}$, in $J\ mol^{-1}\ K^{-1}$: CaO, 80.8; Al_2O_3, 146.4; SiO_2, 87.0.
- For the calculus of the viscosity, in Pa s, consider an equation of Arrhenius type:

$$\mu = A \cdot T \cdot \exp\left(\frac{B}{T}\right) \tag{6.692}$$

where A and B are constants:

$$ln\, A = -19.81 + 1.73 \cdot x_{CaO} - 33.76 \cdot x_{Al_2O_3} \tag{6.693}$$

$$B = 31{,}140 - 23{,}896 \cdot x_{CaO} + 68{,}833 \cdot x_{Al_2O_3} \tag{6.694}$$

The composition in molar and in weight percentage for the three slags is collected in Table 6.17.

The density of the slag at 1500 °C is estimated from the molar volume of the slag with the partial molar volumes of the three components in $m^3\ mol^{-1}$:

$$V^{slag}_m = \overline{V}_{CaO} \cdot x_{CaO} + \overline{V}_{Al_2O_3} \cdot x_{Al_2O_3} + \overline{V}_{SiO_2} \cdot x_{SiO_2} \tag{6.695}$$

From the molar volume and the average molecular weight of each slag, $\overline{M}^{slag}_m$, in $kg\ mol^{-1}$, the density is calculated in $kg\ m^{-3}$:

$$\rho_{slag} = \frac{1000 \cdot \overline{M}^{slag}_m}{V^{slag}_m} \tag{6.696}$$

which for 1500 °C are collected in Table 6.18.

The specific heat of the slag at 1500 °C is estimated (in general, we assume that the specific heat of the melt does not vary significantly with the temperature) as follows:

$$c^{slag}_p = \bar{c}_{p-CaO} \cdot x_{CaO} + \bar{c}_{p-Al_2O_3} \cdot x_{Al_2O_3} + \bar{c}_{p-SiO_2} \cdot x_{SiO_2} \tag{6.697}$$

Table 6.17 Molar and weight percentages of the three slags

Slag	A		B		C	
	wt %	molar %	wt %	Molar %	wt %	Molar %
CaO	40	45.69	22	24.71	48	44.75
Al_2O_3	22	13.80	15	9.25	7	4.10
SiO_2	38	40.51	63	66.04	45	51.15

Table 6.18 Results of the calculation of the density of the slag

Slag	A	B	C
$\overline{V}_{CaO}\,(m^3\,mol^{-1})$	20.70×10^{-6}	20.70×10^{-6}	20.70×10^{-6}
$\overline{V}_{Al_2O_3}\,(m^3\,mol^{-1})$	32.13×10^{-6}	31.00×10^{-6}	29.57×10^{-6}
$\overline{V}_{SiO_2}\,(m^3\,mol^{-1})$	22.78×10^{-6}	24.81×10^{-6}	23.12×10^{-6}
$V_m^{slag}\,(m^3\,mol^{-1})$	23.12×10^{-6}	24.37×10^{-6}	22.14×10^{-6}
$M_m^{slag}\,(kg\,mol^{-1})$	63.97×10^{-3}	62.90×10^{-3}	59.67×10^{-3}
$\rho_{slag}\,(kg\,m^{-3})$	2767	2581	2695

The results of the molar specific heat at constant pressure, for the three slags, and the viscosity in Pa s (the viscosity of the water at room temperature is 10^{-3} Pa s) are collected in Table 6.19.

Commentaries: Although within the density and the specific heat of the acid and basic slags there are not significant differences, it does not happen the same with the viscosity of the acid slags. For the acid slag *B* of the referred exercise, we mentioned the viscosity at a temperature of 1400 °C. One of the advantages of the acid slags is the possibility of working at lower temperatures than in the case of the basic slags because the liquidus temperature is lower (Exercise 6.29). The disadvantages of the acid slags are mainly connected with the physical chemistry of the metal–slag reactions, with the pouring rate or with the possible erosive wear that they produce in the refractory lining of the runners and tapholes (Verdeja et al. 2014, pages 181–183; Mochón et al. 2012, pages 195–201).

Exercise 6.34 Calculate the solid–gas and solid–liquid surface energies for the slags of Exercise 6.29. Consider, for the values of the solid–gas surface energy, that the partial molar surface energies solid–gas of the components of the slag at 1500 ° C ($\overline{\gamma}_{sg-i}$ in mN m^{-1} or erg cm^{-2}) are 625, 655 and 260 for the CaO, Al$_2$O$_3$ and SiO$_2$, respectively.

To estimate the solid–liquid surface energy of the slags, apart from taking into account the properties calculated in Exercises 6.29 and 6.33, the variation of specific heat (J mol^{-1} K^{-1}) with the temperature of the three solid components are (data from TAPP database):

Table 6.19 Values of the specific heat at constant pressure and viscosity of the three slags

Slag	A	B	B	C
$T(°C)$	1500	1400	1500	1500
$c_p^{slag}\,(J\,mol^{-1} \cdot K^{-1})$	92.4	91.0	91.0	86.3
Constant A	5.23×10^{-11}	2.492×10^{-9}	1.683×10^{10}	1.515×10^{-9}
Constant B	29,718.5	31,140.0	31,602.2	21,736.3
Viscosity (μ) (Pa s)	1.76	505.55	16.44	0.57

$$c_{p-CaO} = 2.439 \cdot 10^{-3} \cdot T - 0.9366 \cdot 10^6 \cdot T^{-2} \qquad (6.698)$$

$$c_{p-Al_2O_3} = 104.547 + 24.174 \cdot 10^{-3} \cdot T - 3.541 \cdot 10^6 \cdot T^{-2} - 4.452 \cdot 10^{-6}$$
$$\cdot T^2 + 118.659 \cdot T^{-0.3}$$

$$(6.699)$$

$$c_{p-SiO_2} = 10.04 \cdot 10^{-3} \cdot T \qquad (6.700)$$

where T is the temperature in K.

The calculation of the solid–gas surface energy of the slags is performed using the following equation:

$$\gamma_{sg-slag} = \overline{\gamma}_{sg-CaO} \cdot x_{CaO} + \overline{\gamma}_{sg-Al_2O_3} \cdot x_{Al_2O_3} + \overline{\gamma}_{sg-SiO_2} \cdot x_{SiO_2} \qquad (6.701)$$

We calculate the molar fraction of the constituents for each slag. Considering as reference 100 g of slag, the number of moles of each one of the oxides will be (the percentage mass composition of each one of the oxides is transformed in their corresponding molar fraction) for the slag A:

$$\text{Moles } CaO(n_{CaO}) : \frac{40 \text{ g CaO}}{100 \text{ g slag}} \cdot \frac{1 \text{ mol CaO}}{56.08 \text{ g CaO}} = 0.7133 \frac{\text{mol CaO}}{100 \text{ g slag}} \qquad (6.702)$$

$$\text{Moles } Al_2O_3(n_{Al_2O_3}) : \frac{22 \text{ g Al}_2O_3}{100 \text{ g slag}} \cdot \frac{1 \text{ mol Al}_2O_3}{101.96 \text{ g Al}_2O_3} = 0.2158 \frac{\text{mol Al}_2O_3}{100 \text{ g slag}}$$

$$(6.703)$$

$$\text{Moles } SiO_2(n_{SiO_2}) : \frac{38 \text{ g SiO}_2}{100 \text{ g slag}} \cdot \frac{1 \text{ mol SiO}_2}{60.09 \text{ g SiO}_2} = 0.6324 \frac{\text{mol SiO}_2}{100 \text{ g slag}} \qquad (6.704)$$

Consequently, the number of moles in 100 g of slag is 1.5615 mol. Thus, the composition expressed as molar fraction of each one of the oxides in the slag A will be:

$$\text{Molar fraction } CaO(x_{CaO}) : \frac{0.7133 \frac{\text{mol CaO}}{100 \text{ g slag}}}{1.5615 \frac{\text{mol slag}}{100 \text{ g slag}}} = 0.4568 \frac{\text{mol CaO}}{\text{mol slag}} \qquad (6.705)$$

$$\text{Molar fraction } Al_2O_3(x_{Al_2O_3}) : \frac{0.2158 \frac{\text{mol Al}_2O_3}{100 \text{ g slag}}}{1.5615 \frac{\text{mol slag}}{100 \text{ g slag}}} = 0.1382 \frac{\text{mol Al}_2O_3}{\text{mol slag}} \qquad (6.706)$$

$$\text{Molar fraction } SiO_2(x_{SiO_2}) : \frac{0.6324\frac{\text{mol } SiO_2}{100 \text{ g slag}}}{1.5615\frac{\text{mol slag}}{100 \text{ g slag}}} = 0.4050\frac{\text{mol } SiO_2}{\text{mol slag}} \qquad (6.707)$$

Considering as reference 100 g of slag, the number of moles of each one of the oxides will be (the percentage mass composition of each one of the oxides is transformed in their corresponding molar fraction) for the slag B:

$$\text{Moles } CaO(n_{CaO}) : \frac{22 \text{ g CaO}}{100 \text{ g slag}} \cdot \frac{1 \text{ mol CaO}}{56.08 \text{ g CaO}} = 0.3923\frac{\text{mol CaO}}{100 \text{ g slag}} \qquad (6.708)$$

$$\text{Moles } Al_2O_3(n_{Al_2O_3}) : \frac{15 \text{ g } Al_2O_3}{100 \text{ g slag}} \cdot \frac{1 \text{ mol } Al_2O_3}{101.96 \text{ g } Al_2O_3} = 0.1471\frac{\text{mol } Al_2O_3}{100 \text{ g slag}}$$
$$(6.709)$$

$$\text{Moles } SiO_2(n_{SiO_2}) : \frac{63 \text{ g } SiO_2}{100 \text{ g slag}} \cdot \frac{1 \text{ mol } SiO_2}{60.09 \text{ g } SiO_2} = 1.0484\frac{\text{mol } SiO_2}{100 \text{ g slag}} \qquad (6.710)$$

Consequently, the number of moles in 100 g of slag is 1.5878 mol. Thus, the composition expressed as molar fraction of each one of the oxides in the slag B will be:

$$\text{Molar fraction } CaO(x_{CaO}) : \frac{0.3923\frac{\text{mol CaO}}{100 \text{ g slag}}}{1.5878\frac{\text{mol slag}}{100 \text{ g slag}}} = 0.2471\frac{\text{mol CaO}}{\text{mol slag}} \qquad (6.711)$$

$$\text{Molar fraction } Al_2O_3(x_{Al_2O_3}) : \frac{0.1471\frac{\text{mol } Al_2O_3}{100 \text{ g slag}}}{1.5878\frac{\text{mol slag}}{100 \text{ g slag}}} = 0.0926\frac{\text{mol } Al_2O_3}{\text{mol slag}} \qquad (6.712)$$

$$\text{Molar fraction } SiO_2(x_{SiO_2}) : \frac{1.0484\frac{\text{mol } SiO_2}{100 \text{ g slag}}}{1.5878\frac{\text{mol slag}}{100 \text{ g slag}}} = 0.6603\frac{\text{mol } SiO_2}{\text{mol slag}} \qquad (6.713)$$

Considering as reference 100 g of slag, the number of moles of each one of the oxides will be (the percentage mass composition of each one of the oxides is transformed in their corresponding molar fraction) for the slag C:

$$\text{Moles } CaO(n_{CaO}) : \frac{48 \text{ g CaO}}{100 \text{ g slag}} \cdot \frac{1 \text{ mol CaO}}{56.08 \text{ g CaO}} = 0.8559\frac{\text{mol CaO}}{100 \text{ g slag}} \qquad (6.714)$$

$$\text{Moles } Al_2O_3(n_{Al_2O_3}) : \frac{7 \text{ g } Al_2O_3}{100 \text{ g slag}} \cdot \frac{1 \text{ mol } Al_2O_3}{101.96 \text{ g } Al_2O_3} = 0.0687 \frac{\text{mol } Al_2O_3}{100 \text{ g slag}}$$

$$(6.715)$$

$$\text{Moles } SiO_2(n_{SiO_2}) : \frac{45 \text{ g } SiO_2}{100 \text{ g slag}} \cdot \frac{1 \text{ mol } SiO_2}{60.09 \text{ g } SiO_2} = 0.7489 \frac{\text{mol } SiO_2}{100 \text{ g slag}} \quad (6.716)$$

Consequently, the number of moles in 100 g of slag is 1.6735 mol. Thus, the composition expressed as molar fraction of each one of the oxides in the slag C will be:

$$\text{Molar fraction } CaO(x_{CaO}) : \frac{0.8559 \frac{\text{mol } CaO}{100 \text{ g slag}}}{1.6735 \frac{\text{mol slag}}{100 \text{ g slag}}} = 0.5114 \frac{\text{mol } CaO}{\text{mol slag}} \quad (6.717)$$

$$\text{Molar fraction } Al_2O_3(x_{Al_2O_3}) : \frac{0.0687 \frac{\text{mol } Al_2O_3}{100 \text{ g slag}}}{1.6735 \frac{\text{mol slag}}{100 \text{ g slag}}} = 0.0411 \frac{\text{mol } Al_2O_3}{\text{mol slag}} \quad (6.718)$$

$$\text{Molar fraction } SiO_2(x_{SiO_2}) : \frac{0.7489 \frac{\text{mol } SiO_2}{100 \text{ g slag}}}{1.6735 \frac{\text{mol slag}}{100 \text{ g slag}}} = 0.4475 \frac{\text{mol } SiO_2}{\text{mol slag}} \quad (6.719)$$

The values of the solid–gas surface energy, in mN m^{-1} (erg cm^{-2}), are the following ones:

$$\gamma_{sg-slag}(\text{slag } A) = 625 \cdot 0.4568 + 655 \cdot 0.1382 + 260 \cdot 0.4050 = 481.32 \frac{\text{erg}}{\text{cm}^2}$$

$$(6.720)$$

$$\gamma_{sg-slag}(\text{slag } B) = 625 \cdot 0.2471 + 655 \cdot 0.0926 + 260 \cdot 0.6603 = 386.77 \frac{\text{erg}}{\text{cm}^2}$$

$$(6.721)$$

$$\gamma_{sg-slag}(\text{slag } C) = 625 \cdot 0.5114 + 655 \cdot 0.0411 + 260 \cdot 0.4475 = 462.90 \frac{\text{erg}}{\text{cm}^2}$$

$$(6.722)$$

To estimate the solid–liquid surface energy of the slags, the following hypothesis should be considered:

(a) The enthalpy associated with the formation of the slag at the eutectic temperature is equal to:

$$L = \left(c_{p-melt} - c_{p-solid}\right) \cdot T_{eutectic} \tag{6.723}$$

Consider also that, while the specific heat of the solids varies with the temperature, in the case of the melts, the specific heat does not vary significantly (Exercise 6.33).

(b) The value of the enthalpy associated with the formation of the slag previously calculated is considered as the value of the volumetric free energy of the process of clustering and growth of the new solid phase in the melt (L, latent heat of melting).

(c) At dimensional scale (nanometric, atomic) of 10^{-10} m (Å), the magnitude of the solid–liquid surface energy ΔG_S, in J cm^{-2}, can be estimated from the volumetric energy ΔG_V, in J cm^{-3}, associated with the formation of one cluster of few dozens of atoms:

$$\Delta G_V \times 10^{-8} = \Delta G_S \tag{6.724}$$

The radius of one cluster is $\simeq 10$ nm ($\simeq 100$ atoms–1000 atoms) as calculated in (Pero-Sanz et al. 2017, Chap. 2). In this way:

$$r^* = \frac{2 \cdot G_{LS}}{\Delta G_V} \simeq 10\,nm \rightarrow 10^{-8}m = \frac{2 \cdot G_{LS}}{\Delta G_V} \rightarrow 10^{-8} \cdot \Delta G_V = 2 \cdot \Delta G_S$$
$$\rightarrow 10^{-8} \cdot \Delta G_V \simeq \Delta G_S \tag{6.725}$$

(d) As a function of both the quantity of eutectic liquid (for a composition of the slag different from the eutectic one) and the temperature of it, it will be necessary to correct the values of the surface free energy of the eutectic slag at 1500 °C. For it, we should consider, for any type of slag, that the diminishing of the surface energy with the temperature is 0.15 erg cm^{-2} for each degree Celsius or Kelvin.

In Fig. 6.20, we represent the eutectic point of each slag, and the method to be used to determine the quantity of liquid at the eutectic temperature.

The proportion of CaO, SiO₂, and Al₂O₃ in the eutectics of each one of the three slags is:

- Slag A: 38% CaO, 42% SiO₂, and 20% Al₂O₃.
- Slag B: 24% CaO, 61% SiO₂, and 15% Al₂O₃.
- Slag C: 47% CaO, 41% SiO₂, and 12% Al₂O₃.

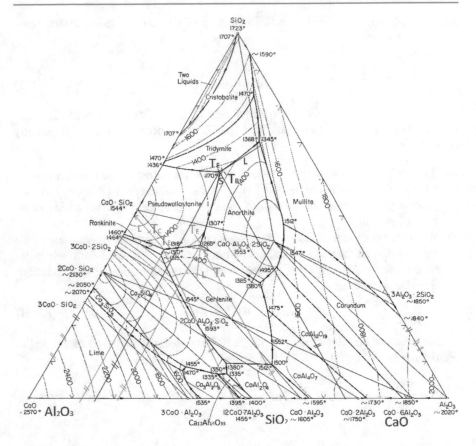

Fig. 6.20 Al_2O_3–CaO-SiO_2 ternary diagram, determination of the eutectic liquid, and composition of the slag (edited from Levin et al. 1969) (wt%)

And the molar fractions, calculated as in the previous exercises, are:

- Slag A: x_{CaO}, 0.4309; x_{SiO_2}, 0.4445; and, $x_{Al_2O_3}$, 0.1245.
- Slag B: x_{CaO}, 0.2692; x_{SiO_2}, 0.6385; and, $x_{Al_2O_3}$, 0.0924.
- Slag C: x_{CaO}, 0.5117; x_{SiO_2}, 0.4166; and, $x_{Al_2O_3}$, 0.0717.

The molar specific heat for the CaO, SiO_2, and Al_2O_3 are (at the eutectic temperature):

- Slag A:

$$c_{p-CaO} = 2.439 \cdot 10^{-3} \cdot (1265 + 273) - 0.9366 \cdot 10^6 \cdot (1265 + 273)^{-2}$$
$$= 3.4 \frac{J}{mol\ K}$$

(6.726)

$$c_{p-Al_2O_3} = 104.547 + 24.174 \cdot 10^{-3} \cdot (1265 + 273) - 3.541 \cdot 10^6$$
$$\cdot (1265 + 273)^{-2} - 4.452 \cdot 10^{-6} \cdot (1265 + 273)^2 \tag{6.727}$$
$$+ 118.659 \cdot (1265 + 273)^{-0.3} = 142.8 \frac{J}{mol\ K}$$

$$c_{p-SiO_2} = 10.04 \cdot 10^{-3} \cdot (1265 + 273) = 15.4 \frac{J}{mol\ K} \tag{6.728}$$

- Slag B:

$$c_{p-CaO} = 2.439 \cdot 10^{-3} \cdot (1170 + 273) - 0.9366 \cdot 10^6 \cdot (1170 + 273)^{-2}$$
$$= 80.8 \frac{J}{mol\ K} \tag{6.729}$$

$$c_{p-Al_2O_3} = 104.547 + 24.174 \cdot 10^{-3} \cdot (1170 + 273) - 3.541 \cdot 10^6$$
$$\cdot (1170 + 273)^{-2} - 4.452 \cdot 10^{-6} \cdot (1170 + 273)^2 \tag{6.730}$$
$$+ 118.659 \cdot (1170 + 273)^{-0.3} = 146.4 \frac{J}{mol\ K}$$

$$c_{p-SiO_2} = 10.04 \cdot 10^{-3} \cdot (1170 + 273) = 87.0 \frac{J}{mol\ K} \tag{6.731}$$

- Slag C:

$$c_{p-CaO} = 2.439 \cdot 10^{-3} \cdot (1310 + 273) - 0.9366 \cdot 10^6 \cdot (1310 + 273)^{-2}$$
$$= 35 \frac{J}{mol\ K} \tag{6.732}$$

$$c_{p-Al_2O_3} = 104.547 + 24.174 \cdot 10^{-3} \cdot (1310 + 273) - 3.541 \cdot 10^6$$
$$\cdot (1310 + 273)^{-2} - 4.452 \cdot 10^{-6} \cdot (1310 + 273)^2 \tag{6.733}$$
$$+ 118.659 \cdot (1310 + 273)^{-0.3} = 143.3 \frac{J}{mol\ K}$$

$$c_{p-SiO_2} = 10.04 \cdot 10^{-3} \cdot (1310 + 273) = 15.9 \frac{J}{mol\ K} \tag{6.734}$$

And the molar specific heat of the eutectic solid is (at the eutectic temperature):

- Slag *A*:

$$
\begin{aligned}
c_{p-\text{slag}-\text{eu}} &= x_{\text{CaO}} \cdot c_{p-\text{CaO}} + x_{\text{SiO}_2} \cdot c_{p-\text{SiO}_2} + x_{\text{Al}_2\text{O}_3} \cdot c_{p-\text{Al}_2\text{O}_3} \\
&= 0.4309 \cdot 3.4 + 0.4445 \cdot 15.4 + 0.1245 \cdot 142.8 \\
&= 26.1 \frac{\text{J}}{\text{mol K}} \cdot \frac{\text{mol slag}}{(0.4309 \cdot 56 + 0.4445 \cdot 60 + 0.1245 \cdot 102)\text{g}} \\
&= 0.4109 \frac{\text{J}}{\text{g K}}
\end{aligned} \tag{6.735}
$$

- Slag *B*:

$$
\begin{aligned}
c_{p-\text{slag}-\text{eu}} &= x_{\text{CaO}} \cdot c_{p-\text{CaO}} + x_{\text{SiO}_2} \cdot c_{p-\text{SiO}_2} + x_{\text{Al}_2\text{O}_3} \cdot c_{p-\text{Al}_2\text{O}_3} \\
&= 0.2692 \cdot 3.1 + 0.6385 \cdot 14.5 + 0.0924 \cdot 141.9 \\
&= 23.2 \frac{\text{J}}{\text{mol K}} \cdot \frac{\text{mol slag}}{(0.2692 \cdot 56 + 0.6385 \cdot 60 + 0.0924 \cdot 102)\text{g}} \\
&= 0.3691 \frac{\text{J}}{\text{g K}}
\end{aligned} \tag{6.736}
$$

- Slag *C*

$$
\begin{aligned}
c_{p-\text{slag}-\text{eu}} &= x_{\text{CaO}} \cdot c_{p-\text{CaO}} + x_{\text{SiO}_2} \cdot c_{p-\text{SiO}_2} + x_{\text{Al}_2\text{O}_3} \cdot c_{p-\text{Al}_2\text{O}_3} \\
&= 0.5117 \cdot 3.5 + 0.4166 \cdot 15.9 + 0.0717 \cdot 143.3 \\
&= 18.7 \frac{\text{J}}{\text{mol K}} \cdot \frac{\text{mol slag}}{(0.5117 \cdot 56 + 0.4166 \cdot 60 + 0.0717 \cdot 102)\text{g}} \\
&= 0.3064 \frac{\text{J}}{\text{g K}}
\end{aligned} \tag{6.737}
$$

The molar specific heat for the CaO, SiO_2, and Al_2O_3 is (at the 1500 °C, liquid slag, Exercise 6.33):

$$
c_{p-\text{CaO}} = 80.8 \frac{\text{J}}{\text{mol K}} \tag{6.738}
$$

$$
c_{p-\text{SiO}_2} = 146.4 \frac{\text{J}}{\text{mol K}} \tag{6.739}
$$

$$
c_{p-\text{Al}_2\text{O}_3} = 87.0 \frac{\text{J}}{\text{mol K}} \tag{6.740}
$$

And the molar specific heat of the eutectic solid is (at the 1500 °C):

- Slag A:

$$
\begin{aligned}
c_{p-slag-1500°C} &= x_{CaO} \cdot c_{p-CaO} + x_{SiO_2} \cdot c_{p-SiO_2} + x_{Al_2O_3} \cdot c_{p-Al_2O_3} \\
&\quad - 0.4309 \cdot 80.8 + 0.4445 \cdot 87.0 + 0.1245 \cdot 146.4 \\
&= 91.7 \frac{J}{mol\ K} \cdot \frac{mol\ slag}{(0.4309 \cdot 56 + 0.4445 \cdot 60 + 0.1245 \cdot 102)g} \\
&= 1.4443 \frac{J}{g\ K}
\end{aligned}
$$

(6.741)

- Slag B:

$$
\begin{aligned}
c_{p\ slag-1500°C} &= x_{CaO} \cdot c_{p-CaO} + x_{SiO_2} \cdot c_{p-SiO_2} + x_{Al_2O_3} \cdot c_{p-Al_2O_3} \\
&= 0.2692 \cdot 80.8 + 0.6385 \cdot 87.0 + 0.0924 \cdot 146.4 \\
&= 90.8 \frac{J}{mol\ K} \cdot \frac{mol\ slag}{(0.2692 \cdot 56 + 0.6385 \cdot 60 + 0.0924 \cdot 102)g} \\
&- 1.4461 \frac{J}{g\ K}
\end{aligned}
$$

(6.742)

- Slag C

$$
\begin{aligned}
c_{p-slag-1500\ °C} &= x_{CaO} \cdot c_{p-CaO} + x_{SiO_2} \cdot c_{p-SiO_2} + x_{Al_2O_3} \cdot c_{p-Al_2O_3} \\
&= 0.5117 \cdot 80.8 + 0.4166 \cdot 87.0 + 0.0717 \cdot 146.4 \\
&= 88.1 \frac{J}{mol\ K} \cdot \frac{mol\ slag}{(0.5117 \cdot 56 + 0.4166 \cdot 60 + 0.0717 \cdot 102)g} \\
&= 1.4449 \frac{J}{g\ K}
\end{aligned}
$$

(6.743)

The density of the eutectic solid is calculated as follows (TAPP database):

$$
\begin{aligned}
\rho_{eu-sol} &= \left[2193 \cdot \left(\frac{\%SiO_2}{100} \right) + [4027 \cdot 0.111 \cdot (T_E + 273)] \cdot \left(\frac{\%Al_2O_3}{100} \right) \right. \\
&\quad \left. + [3400 - 0.153 \cdot (T_E + 273)] \cdot \left(\frac{\%CaO}{100} \right) \right] \frac{kg}{m^3} \cdot \frac{1\ m^3}{10^6 cm^3} \\
&\quad \cdot \frac{10^3 g}{1\ kg}
\end{aligned}
$$

(6.744)

Thus, the values for each slag are:

- Slag A: 2.895 g cm^{-3}.
- Slag B: 2.681 g cm^{-3}.
- Slag C: 1.946 g cm^{-3}.

This way, the values of ΔG_V and ΔG_S are calculated as follows:

$$\Delta G_V(\text{slag A}) = \left(c_{p-\text{slag}-1500\,°C} - c_{p-\text{slag}-\text{eu}}\right) \cdot (T_E + 273) \cdot \rho_{\text{eu}-\text{sol}}$$

$$= (1.4443 - 0.4109)\left(\frac{J}{g \cdot K}\right) \cdot (1265 + 273)(K) \qquad (6.745)$$

$$\cdot\, 2.895 \left(\frac{g}{cm^3}\right) = 4601.22 \frac{J}{cm^3}$$

$$\Delta G_S(\text{slag A}) = \Delta G_V(\text{slag A}) \times 10^{-8} = 4601.22 \frac{J}{cm^3} \times 10^{-8}$$

$$= 4.6012 \times 10^{-5} \frac{J}{cm^2} \cdot \frac{erg}{10^{-7}J} = 460.12 \frac{erg}{cm^2} \qquad (6.746)$$

The value ΔG_S must be corrected with the percentage of eutectic liquid

$$\gamma_{\text{SL}}(\text{slag A}) = \Delta G_S(\text{slag A}) \cdot \%\text{Eutectic liquid} = 460.12 \frac{erg}{cm^2} \cdot 0.46$$

$$= 211.66 \frac{erg}{cm^2} \qquad (6.747)$$

The surface energy diminishes 0.15 erg cm^{-2} each degree Celsius of temperature:

$$\gamma_{\text{SL}}(\text{slag A}) = 211.66 \frac{erg}{cm^2} - 0.15 \frac{erg}{cm^2} \cdot (1500 - 1265)$$

$$= 176.41 \frac{erg}{cm^2} \times \frac{10^{-7}J}{erg} \times \frac{10^4\,cm^2}{1\,m^2} = 0.176 \frac{J}{m^2} \qquad (6.748)$$

In the case of the slag B:

$$\Delta G_V(\text{slag B}) = \left(c_{p-\text{slag}-1500°C} - c_{p-\text{slag}-\text{eu}}\right) \cdot (T_E + 273) \cdot \rho_{\text{eu}-\text{sol}}$$

$$= (1.4461 - 0.3691)\left(\frac{J}{gK}\right) \cdot (1170 + 273)(K) \qquad (6.749)$$

$$\cdot\, 2.681 \left(\frac{g}{cm^3}\right) = 4166.57 \frac{J}{cm^3}$$

$$\Delta G_S(\text{slag B}) = \Delta G_V(\text{slag B}) \times 10^{-8} = 4166.57 \frac{\text{J}}{\text{cm}^3} \times 10^{-8}$$

$$= 4.1666 \times 10^{-5} \frac{\text{J}}{\text{cm}^2} \cdot \frac{\text{erg}}{10^{-7}\,\text{J}} = 416.66 \frac{\text{erg}}{\text{cm}^2} \tag{6.750}$$

The value ΔG_S must be corrected with the percentage of eutectic liquid

$$\gamma_{SL}(\text{slag B}) = \Delta G_S(\text{slag B}) \cdot \%\text{Eutectic liquid} = 416.66 \frac{\text{erg}}{\text{cm}^2} \cdot 0.92 = 383.32 \frac{\text{erg}}{\text{cm}^2} \tag{6.751}$$

The surface energy diminishes 0.15 erg cm^{-2} each degree Celsius of temperature:

$$\gamma_{SL} = 416.66 \frac{\text{erg}}{\text{cm}^2} - 0.15 \frac{\text{erg}}{\text{cm}^2} \cdot (1500 - 1170) = 333.82 \frac{\text{erg}}{\text{cm}^2} \times \frac{10^{-7}\text{J}}{\text{erg}} \times \frac{10^4 \text{cm}^2}{1\,\text{m}^2}$$

$$= 0.334 \frac{\text{J}}{\text{m}^2} \tag{6.752}$$

In the case of the slag C:

$$\Delta G_V(\text{slag C}) = \left(c_{\text{p–slag–1500°C}} - c_{\text{p–slag–eu}}\right) \cdot (T_E + 273) \cdot \rho_{\text{eu–sol}}$$

$$= (1.4449 - 0.3064) \left(\frac{\text{J}}{\text{g} \cdot \text{K}}\right) \cdot (1310 + 273)(\text{K}) \tag{6.753}$$

$$\cdot 1.946 \left(\frac{\text{g}}{\text{cm}^3}\right) = 3507.17 \frac{\text{J}}{\text{cm}^3}$$

$$\Delta G_S(\text{slag C}) = \Delta G_V(\text{slag C}) \times 10^{-8} = 3507.17 \frac{\text{J}}{\text{cm}^3} \times 10^{-8}$$

$$= 3.5072 \times 10^{-5} \frac{\text{J}}{\text{cm}^2} \cdot \frac{\text{erg}}{10^{-7}\text{J}} = 350.72 \frac{\text{erg}}{\text{cm}^2} \tag{6.754}$$

The value ΔG_S must be corrected with the percentage of eutectic liquid:

$$\gamma_{SL}(\text{slag C}) = \Delta G_S(\text{slag C}) \cdot \%\text{Eutectic liquid} = 350.72 \frac{\text{erg}}{\text{cm}^2} \cdot 0.31 = 108.72 \frac{\text{erg}}{\text{cm}^2} \tag{6.755}$$

The surface energy diminishes 0.15 erg cm^{-2} each degree Celsius of temperature:

$$\gamma_{SL} = 108.72 \frac{erg}{cm^2} - 0.15 \frac{erg}{cm^2} \cdot (1500 - 1310) = 80.22 \frac{erg}{cm^2} \times \frac{10^{-7}J}{erg} \times \frac{10^4 cm^2}{1\ m^2}$$

$$= 0.08\ \frac{J}{m^2}$$

(6.756)

The γ_{SL} above-indicated is calculated at the reference temperature of 1500 °C.

Commentaries: It is important to know the surface energies of the slags in contact with other metallic or ceramic phases. The contact slag–metal allows analyzing the operations of separation of these components and the surface energies of the interactions with ceramic oxides for the study of the corrosion of the refractories in the furnace.

References

Babich A, Senk D, Gudenau HW, Mavrommatis TT (2008) Ironmaking. RWTH Aachen University, Department of Ferrous Metallurgy, Aachen, Germany

Ballester A, VerdejaLF, Sancho J (2000) Metalurgia extractiva, vol I. Síntesis, Madrid, Spain

Coudurier L, Hopkins DW, Wilkomirsky I (1985) Fundamentals of metallurgical processes. Pergamon Press, Oxford

Fernández-González D, Ruiz-Bustinza I, González-Gasca C, Piñuela-Noval J, Mochón-Castaños J, Sancho-Gorostiaga J, Verdeja LF (2018) Concentrated solar energy applications in materials science and metallurgy. Solar Energy 170:520–540

Flinn RA (1963) Fundamentals of metal casting. Addison-Wesley Publishing Company Inc., Reading, Massachusetts, USA

Gaye H (1982) Equilibria and kinetics in steelmaking, Journées d'Etude LBE. First LBE Workshop, Avignon, 10–14 October, p 4

Kawai Y, Shiraishi Y (1988) Handbook of physico-chemical properties of high temperatures. The Iron and Steel Institute of Japan—Committee of Japan Society for Promotion of Science, Tokyo, Japan

Levin EM, Robbins CR, McMurdie HF, Reser MK (1969) Phase diagrams for ceramists. American Ceramic Society, Colombus, Ohio

Lüngen HB, Steffen R (1988) Comparison of production cost for hot metal and sponge iron. Coquemaking Intl 1(1):28–36

Massalski TB, Okamoto H, Subramanian PR, Kacprzac L (1990) Binary alloy phase diagrams. ASM International, Materials Park, Ohio

McCloy JS, Schweiger MJ, Rodríguez CP, Vienna JD (2011) Nepheline crystallization in nuclear waste glasses. Progress toward acceptance of high-alumina formulations. Int J Appl Glass Sci 2 (3):201–214

Mochón J, Quintana MJ, Ruiz-Bustinza I, González R, Marinas E, Barbés MA, Verdeja LF (2012) Protection mechanisms for blast furnace crucible using titanium oxides. Metall Mater Eng 18 (3):195–201

Naito M, Takeda K, Matsui Y (2015) Ironmaking technology for the last 100 years: Development to advanced technologies from introduction of technological know-how, and evolution to next-generation process. ISIJ Int 55(1):7–35

Pero-Sanz JA, Quintana MJ, Verdeja LF (2017) Solidification and solid-state transformations of metals and alloys, 1st edn. Elsevier, Cambridge, Massachusetts

Pero-Sanz JA, Fernández-González D, Verdeja LF (2018) Physical metallurgy of cast irons. Springer, Cham

Pero-Sanz JA, Fernández-González D, Verdeja LF (2019) Structural materials: properties and selection. Springer, Cham

Rosenqvist T (1983) Principles of extractive metallurgy. McGraw-Hill, New York

Sigworth GH, Elliot JM (1974) Thermodynamic of dilute iron alloys. Metals Sci 8:298–310

Springorum D (1995) Slag Atlas. Verein Deutscher Eisenhüttenleute (VDeh), Düsseldorf, Germany

The Japan Society for the Promotion of Science. The 19th Committee on Steelmaking (1988): steelmaking data sourcebook. Gordon and Breach Science Publishers, New York, USA

Verdeja LF, Sancho JP, Ballester A, González R (2014) Refractory and ceramic materials. Síntesis, Madrid, Spain

Strengths and Uncertainties of the Iron Metallurgy

7

7.1 Introduction

It was shown in the six previous chapters that the production of crude steel is carried out using one of the following alternatives:

- The smelting reduction of the raw materials in a shaft furnace (blast furnace, Corex process), Chap. 6.
- Reduction with a gas of the iron oxide raw materials to produce Direct Reduction Iron (DRI), Chap. 5.

If it is assumed that 65% of the steel production is obtained from iron ore, and considering a production of crude steel of approximately 1810 million tons every year (2018, data from the World Steel Association), the production of crude steel, using iron oxides as raw material, is around 1177 million tons.

However, from the 1177 million tons of crude steel nowadays produced from iron ore, it is necessary to make the following considerations:

- That 1060 million tons of crude steel come from the reduction of the iron ore in the blast furnace to produce pig iron.
- That 95 million tons of crude steel have their origin in the direct reduction processes using reductant gases.
- Finally, that 22 million tons of crude steel are produced from Corex melt, which is a liquid with physical–chemical characteristics very similar to those of the blast furnace pig iron, although as opposed to the process in the blast furnace, ironmaking coke is not used as reductant but rather coals that do not have the suitable characteristics for being subjected to the coking process.

On the other hand, it is necessary to mention that only a small quantity of the iron ore (coarse iron ore grains with suitable chemical characteristics and grain size greater than 20 mm) is used in the blast furnace. The typical burden, furnace charge

J. I. Verdeja González et al., *Operations and Basic Processes in Ironmaking*, Topics in Mining, Metallurgy and Materials Engineering, https://doi.org/10.1007/978-3-030-54606-9_7

of iron-bearing materials, of the blast furnace is usually pretreated, either by pelletizing of iron ore fines (grain size smaller than 100 microns) or by sintering of iron ore with grain size of around 5 mm (Fernández-González et al. 2017a, b, c, d, 2018a). The coarse iron ore fraction charged into the blast furnace does not represent more than 10%. The same way, the ferric burden employed in the commercial processes to obtain the DRI is usually pretreated by means of operations and basic processes of agglomeration.

Within the raw materials used to perform the smelting reduction of the hematite, Fe_2O_3, the ironmaking coke is the most important. Although the physical and chemical specifications of the ironmaking coke are related with the operating parameters of the blast furnace, the utilization of other carbon-based reductants is habitually grouped under the generic word "ironmaking coke" (Schwartz 2018). The ironmaking coke or the metallurgical coke is used without distinction in the production of ferroalloys and cast irons in the cupola furnace (see Pero-Sanz et al. 2018, pages 313–330). The differences between ironmaking and metallurgical cokes regarding the physical and chemical characteristics are usually minimal (Habashi 2002). The most significant differences among these cokes are related with the physical characteristics of the coke: granulometry, specific surface area, and porosity.

Finally, the following question could be set out: What will be the future (next decades) to produce crude steel using iron oxides as raw materials? The answer to this question is closely related with the evolution of the following variables:

- Apart from considering a growth of the world population, to achieve an increase of the crude steel production, it is necessary that the developing continents, countries or regions could incorporate to the first stages of the industrial development, although being compatible with the environmental equilibrium.
- Considering that the working hypothesis until now considered, which sets that the production of crude steel represents around 65% of the production demand of steel every year, can be maintained over the time.

However, an increase of the world demand of steel could negatively affect the production of crude steel as this demand might be covered with the recycling–reuse of steel. On the other hand, the improvement in the quality of the finished products and the extension of the product life cycle could negatively affect the supply of scrap (recycled steel) and, thus, the increase of the crude steel production might be favored (Pero-Sanz et al. 2019).

7.2   Aspects That Favor the Production of Iron

The *Book of Deuteronomy* (Chap. 8, verses 9–11) associates the material prosperity of the people to have "a land where bread will not be scarce and you will lack nothing; a land where the rocks are iron and you can dig copper out of the hills.

When you have eaten and are satisfied, praise the Lord your God for the good land he has given you" (Biblia de Navarra 2008).

Nowadays, without going to such extremes, it is not reasonable to disregard the idea of living in a land that could provide abundant harvest and with mountains plenty of high-quality hematite. However, other reasons can be provided to explain that the iron should be still considered as an essential element for the development and prosperity of the countries. The facts that could be indicated are the following ones:

- The iron is the fourth element most abundant in the earth crust. The availability and abundance of iron ore resources in the planet are significant.
- The presence of military or social conflicts in different regions and countries seems to have influence neither in the supply nor in the prices of the raw materials (nowadays).
- As opposed to other metallic elements, the iron mineral resources currently mined are characterized by having a high iron grade and low presence of undesired elements: sulfur, phosphorus, and alkalis. In general, in the nonferrous metallurgies, there are less and less content in the valuable metals in the mined mineral resources (Habashi 1996). On the contrary, the content in the harmful elements is greater and greater, and this question penalties the extraction of the valuable element.

The "development" of the iron metallurgy that is possible and environmentally sustainable is closely related with the continuous improvement of the production processes and operations, as well as with the quality of the finished products, in this case pig iron and Direct Reduction Iron. The only option to survive in this "implacably competitive" world is living-working in an environment of continuous improvement of processes and products, offering to the market materials with a better performance at lower prices. That is to say, reaching the concept known as "virtuous circle of the competitivity" that consists in supplying to the market products with better properties at lower prices.

Other of the aspects that, in the last decades has been being optimized in the ensemble of operations and basic industrial processes was the energy efficiency. In the particular situation of the iron and steel production, the data of the energy consumption for all the operation processes of the iron and steelmaking industry indicates that 25.00×10^{18} J are consumed every year (25.00 EJ year^{-1}); see Cavaliere (2019, pages 6–12). To make an estimation of the thermal energy exclusively associated with the production of iron, it is necessary to use the data provided in Exercise 6.15, where it is indicated that the energy that it is necessary to supply to equilibrate the energy balance in the case of producing one mol of iron would be 168.48 kJ. Besides, to make such estimation, the following assumptions were made:

- That the production of steel derived from the obtaining of iron from the ore is equivalent to 65% of the steel consumption for one year worldwide. The quantity of steel produced–consumed in 2018 was 1810 million of tons.
- That the mass unit of iron produced is equivalent to the mass unit finally obtained (neglecting alloy elements).
- That the energy costs of the blast furnace route are equivalent to those of the DRI production. It is well known that the energy costs are greater in the case of the blast furnace route than in the DRI product, but this assumption was made because the errors associated with the second assumption are avoided: identification of the mass unit of iron with that of the produced steel.

Consequently, the estimation of the energy consumption in the production of iron would be:

$$\frac{168.46\,\text{kJ}}{\text{mol Fe}} \cdot \frac{\text{mol Fe}}{55.85\,\text{g Fe}} \cdot \frac{1\,\text{g Fe}}{1\,\text{g steel}} \cdot \frac{10^6\,\text{g steel}}{1\,\text{t steel}} \cdot \frac{1810 \times 10^6\,\text{t steel} \cdot 0.65}{\text{year}} \tag{7.1}$$

$$3.55 \times 10^{15}\,\text{kJ year}^{-1} = 3.55 \times 10^{18}\,\text{J year}^{-1} = 3.55\,\text{EJ year}^{-1} \tag{7.2}$$

If this last value, 3.55×10^{18} J year^{-1}, is compared with the value provided by Cavaliere (2019), 25.00×10^{18} J year^{-1}, it is possible to see that only 14.2% of the energy consumed in the iron and steelmaking processes is used in the ironmaking.

7.3 Aspects That Influence Negatively in the Production of Iron

The old and traditional image associated with the ironmaking industry (the first blast furnaces started their operation in Central Europe around the middle of the XVth century) contrasts with the current application of the modern technologies to the traditional operations and processes of the ironmaking: blockchain, additive manufacturing, artificial intelligence or big data, that are supported in the concept of 4.0 Industry (Bolton 2019; McDonald 2019). In January 2019, the World Economic Forum (WEF) announced that the steel factory of Ijmuiden from Tata Steel Europe (close to Amsterdam, the Netherlands) would be included in the section of "lighthouses," a distinction that is given to those installations that are leader in technology of the 4.0 Industry. On the other hand, the recovery–decontamination of the ground affected by either obsolete installations or gaseous emissions that carried out organic molecules with benzene rings (associated with sintering plants or coke batteries) are being treated with modern techniques linked to biotechnology and photocatalysis (Diaz 2012).

The topic of CO_2 emissions occupies nowadays great number of the debates and discussions that are being raised and that are related with those industrial sectors that require a great quantity of energy to offer to the market the finished products.

Considering the data provided by Cavaliere (2019, pages 13–22), the production of iron and steel (the complete iron and steelmaking process) contributes with a CO_2 emission of 2.00 Gt every year. If we assume as working hypothesis:

- that the steel production via blast furnace route requires 400 kilograms of carbon per ton of pig iron,
- that the steel production from direct reduction iron requires the same quantity of reductant that the blast furnace route,
- that the steel production, derived from the iron production from the ore is equivalent to the 65% of the steel consumed during one year in the world. This quantity was fixed, as in the case of the estimation of the energy consumption, in 1810 million of tons every year.

Consequently, the CO_2 emissions associated with the iron production would be:

$$\frac{1177\,t\,steel}{year} \cdot \frac{0.400\,t\,C}{t\,pig\,iron} \cdot \frac{1.11\,t\,pig\,iron}{t\,steel} \cdot \frac{1\,mol\,C}{12.01\,g\,C} \cdot \frac{1\,mol\,CO_2}{1\,mol\,C} \cdot \frac{44.01\,g\,CO_2}{1\,mol\,CO_2}$$
$$\cdot \frac{10^6\,g\,C}{1\,t\,C} \cdot \frac{1\,t\,CO_2}{10^6\,g\,CO_2} = 1.91\,Gt\,CO_2\,year^{-1}$$

$$(7.3)$$

If the carbon dioxide emissions of the iron and steelmaking process were estimated in 2.20 Gt of CO_2, 95% of the CO_2 emissions are related with the production of primary iron from the iron ore.

There is not a single solution to try to reduce the CO_2 emissions in the production of iron. The alternatives that are being considered nowadays can be summarized in the following points:

1. CO_2 capture from dilute gas stream and subsequent storage of the carbon dioxide from a gas with content greater than 90%.
2. Possible replacement of the carbon, as single reductant reagent, with hydrogen or electric power (aqueous electrolysis).
3. Progress introduction of the renewable energies as the biomass or the concentrated solar thermal energy in operations and processes of the iron and steelmaking (Fernández-González et al. 2018b).

The CO_2 capture from a gaseous stream (wet cleaning of the gases) using an aqueous solution of lime (milk of lime) is a process industrially known to capture the CO_2, existing in the gas, as $CaCO_3$. The question that, nowadays, could displace the CO_2 capture toward processes that do no generate CO_2, might be the production of activated quicklime, CaO, required to carry out the capture of carbon dioxide by means of the limestone, $CaCO_3$, calcination in a solar furnace; see Fernández--González et al. (2018c).

On the other hand, the biomass is starting to be used in the production of iron by means of the replacement of the ironmaking coke by charcoal, as for instance in ArcelorMittal Brazil (at least two small blast furnaces at Juiz de Fora in Minas Gerais use charcoal as "bio-reductant") where the company ArcelorMittal Bio-Florestas produces charcoal from renewable eucalyptus forests (see project *Use of Charcoal from Renewable Biomass Plantations as Reducing Agent in Pig Iron Mill of ArcelorMittal Juiz de Fora, Brazil*). The current challenge would be determining what would be the rate of replacement of the ironmaking coke by biomass that could be reached without increasing the concentrations of CO_2 in the atmosphere.

Finally, we want to present some considerations about the implications of the carbon dioxide emissions over the environment and the production of materials:

(a) First, it is necessary to consider and reiterate that the presence of CO_2 in the atmosphere is necessary for the growth of plants and the generation of biomass.

(b) The CO_2 concentration in the atmosphere is around 414 ppm (January 2020). It is estimated that during the Jurassic period, characterized by abundant vegetation, the CO_2 concentration could even reach 2000 ppm.

(c) During the period comprised between 1940 and 1975, there was a slight cooling of the planet although the CO_2 concentration in the atmosphere had increased during that period.

(d) It is possible that the increase of the CO_2 concentration in the atmosphere could be related with the increase of other compounds-macromolecules or organic radicals (available, for instance, in the washing water of the coke–oven gas that can be reduced by means of remediation bioprocesses) or of nonmetallic oxides that, by itself or indirectly, can strengthen the influence of the CO_2 in the atmosphere over its nominal concentration. Equally, the presence of CO_2 might potentiate–activate the chemical activity of other chemical species dissolved in the atmosphere and being the real responsible of the "climate changes."

(e) The alternation of wet periods with other drought periods is collected even in the Bible. For instance, in Chaps. 17 and 18 of the First Book of Kings it is indicated that during the period of the prophet Elias (874 BC), there was a long drought in all Palestine (Biblia de Navarra 2008). It does not seem that in that period the values of CO_2 concentration in the atmosphere were greater than the current ones.

(f) The massive manufacture of different qualities of carbon fibers is closely related with the coke production: reductant material, up to now, par excellence in the metallurgical industry. A drastic reduction in the coke production could be harmful and seriously alter the manufacture of carbon fiber composites, every day more and more used in the transportation industry (ground, air, and maritime) and goods.

References

Biblia de Navarra (2008) Ediciones Universitarias de Navarra S. A. (ENUNSA). Pamplona, Navarra, Spain

Bolton J (2019) Green steel and future for the sector in UK. Mater World Mag 27(8):21

Cavaliere P (2019) Clean ironmaking and steelmaking process. Springer, Cham, Switzerland

Díaz M (2012) Ingeniería de Bioprocesos. Paraninfo, Madrid, Spain

Fernández-González D, Ruiz-Bustinza I, Mochón J, González-Gasca C, Verdeja LF (2017a) Iron ore sintering: raw materials and granulation. Miner Process Extr Metall Rev 38(1):36–46

Fernández-González D, Ruiz-Bustinza I, Mochón J, González-Gasca C, Verdeja LF (2017b) Iron ore sintering: process. Miner Process Extr Metall Rev 38(4):215–227

Fernández-González D, Ruiz-Bustinza I, Mochón J, González-Gasca C, Verdeja LF (2017c) Iron ore sintering: quality indices. Miner Process Extr Metall Rev 38(4):254–264

Fernández-González D, Ruiz-Bustinza I, Mochón J, González-Gasca C, Verdeja LF (2017d) Iron ore sintering: environment, automatic and control techniques. Miner Process Extr Metall Rev 38(4):238–249

Fernández-González D, Piñuela-Noval J, Verdeja LF (2018a) Iron ore agglomeration technologies. In: Shatokha V (ed) Iron ores and iron oxide materials. IntechOpen, London, United Kingdom, pp 61–80 (chapter 4)

Fernández-González D, Prazuch J, Ruiz-Bustinza I, González-Gasca C, Piñuela-Noval J, Verdeja LF (2018b) Iron metallurgy via concentrated solar energy. Metals 8(11):art. 873

Fernández-González D, Ruiz-Bustinza I, González-Gasca C, Piñuela-Noval J, Mochón-Castaños J, Sancho-Gorostiaga J, Verdeja LF (2018c) Concentrated solar energy applications in materials science and metallurgy. Sol Energy 170(8):520–540

Habashi F (1996) Pollution problems in the mineral and metallurgical industries. Metallurgie Extractive Quebec, Quebec, Canada

Habashi F (2002) Textbook of pyrometallurgy. Metallurgie Extractive Quebec, Quebec, Canada

McDonald Ch (2019) Steel innovations. Mater World Mag 27(5):52–55

Pero-Sanz JA, Fernández-González D, Verdeja LF (2018) Physical metallurgy of cast irons. Springer, Cham, Switzerland

Pero-Sanz JA, Fernández-González D, Verdeja LF (2019) Structural materials: properties and selection. Springer, Cham, Switzerland

Schwartz M (2018) Coking coal needed. Mater World Mag 26(10):52–54

Appendix 1

Standard free energies for some iron and steelmaking reactions, $\Delta_r G^0$.

$\Delta_r G^0 (\text{kJ mol}^{-1}) = A + B \cdot T(K)$	Linear correlation coefficient				Ref. (*)
	$-A$	B	$(\pm)$ kJ	°C	
Aluminum					
$2Al(s) + 3/2O_2(g) = Al_2O_3(s)$	1672.0	0.313	6.9	25–659	1, 2, 3
$2Al(l) + 3/2O_2(g) = Al_2O_3(s)$	1680.0	0.324	4.5	659–1700	1, 2, 3, 4
$Al(s) + 1/2N_2(g) = AlN(s)$	319.0	0.101	6.6	25–659	1, 2, 3
$Al(l) + 1/2N_2(g) = AlN(s)$	328.0	0.112	19.8	659–1700	1, 2, 3, 4
$4Al(s) + 3C(s) = Al_4C_3(s)$	213.0	0.042	4.1	25–659	1, 2, 3
$4Al(l) + 3C(s) = Al_4C_3(s)$	283.0	0.107	4.5	659–1700	1, 2, 3
$2Al(l) + 1/2O_2(g) = Al_2O(g)$	193.0	−0.046	0.2	1227–1727	2, 3, 4
$Al(l) + 1/2O_2(g) = AlO(g)$	−42.5	−0.059	0.1	1227–1727	2, 3
Boron					
$2B(s) + 3/2O_2(g) = B_2O_3(s)$	1272.0	0.264	0.8	25–450	1, 2, 3
$2B(s) + 3/2O_2(g) = B_2O_3(l)$	1225.0	0.208	3.5	450–1700	1, 2, 3
$B(s) + 1/2N_2(g) = BN(s)$	252.0	0.088	1.0	25–900	1, 2, 3
$4B(s) + C(s) = B_4C(s)$	60.6	0.003	6.2	25–900	1, 2, 3
$B(s) + 1/2N_2(g) = BN(g)$	−648.0	−0.112	-	1227–1727	4
$B(s) + 1/2O_2(g) = BO(g)$	29.8	−0.084	34.8	1227–1727	2, 3, 4
Barium					
$Ba(s) + 1/2O_2(g) = BaO(s)$	554.0	0.094	11.1	25–704	1, 2, 3
$Ba(l) + 1/2O_2(g) = BaO(s)$	561.0	0.102	8.1	704–1638	1, 2, 3, 4
$3Ba(s) + N_2(g) = Ba_3N_2(s)$	352.0	0.234	8.9	25–704	1, 3
$BaO(s) + SiO_2(s) = BaSiO_3(s)$	135.0	0.004	22.4	25–1300	1, 3
Beryllium					
$Be(s) + 1/2O_2(g) = BeO(s)$	604.0	0.098	8.3	25–1283	1, 2, 3
$Be(l) + 1/2O_2(g) = BeO(s)$	605.0	0.100	8.4	1283–1700	1, 2, 3, 4
$3Be(s) + N_2(g) = Be_3N_2(s)$	578.0	0.175	14.9	25–700	1, 2, 3
Carbon					
$C(s) + 2H_2(g) = CH_4(g)$	95.6	0.113	1.6	25–2000	1, 3, 4
$C(s) + 1/2O_2(g) = CO(g)$	118.0	−0.084	3.7	25 − 2000	1,2,3,4

(continued)

© The Editor(s) (if applicable) and The Author(s), under exclusive license to Springer   305
Nature Switzerland AG 2020
J. I. Verdeja González et al., *Operations and Basic Processes in Ironmaking*,
Topics in Mining, Metallurgy and Materials Engineering,
https://doi.org/10.1007/978-3-030-54606-9

(continued)

$\Delta_r G^0\left(\text{kJ mol}^{-1}\right) = A + B \cdot T(K)$	Linear correlation coefficient				Ref. (*)
	$-A$	B	$(\pm)$ kJ	°C	
$C(s) + O_2(g) = CO_2(g)$	395.0	0.001	5.1	25–2000	1, 2, 3, 4
$C(s) + 1/2S_2(g) = CS(g)$	−176.0	−0.067	10.7	1600–1800	1, 2, 3, 4
$CO(g) + 1/2O_2(g) = CO_2(g)$	277.0	0.085	–	–	
$C(s) + S_2(g) = CS_2(g)$	−26.5	−0.050	43.4	25–1300	1, 2, 3
$CO(g) + 1/2S_2(g) = COS(g)$	132.0	0.065	98.2	25–1200	1, 2, 3
$2C(s) + H_2(g) = C_2H_2(g)$	−217.0	−0.049	–	1227–1727	4
Calcium					
$Ca(s) + 1/2O_2(g) = CaO(s)$	633.0	0.103	1.5	25–850	1, 2, 3
$Ca(l) + 1/2O_2(g) = CaO(s)$	640.0	0.109	2.9	850–1487	1, 2, 3, 4
$Ca(g) + 1/2O_2(g) = CaO(s)$	788.0	0.193	1.6	1487–1700	1, 2, 3, 4
$Ca(s) + 1/2S_2(g) = CaS(s)$	517.0	0.072	27.7	25–850	1, 2, 3
$Ca(l) + 1/2S_2(g) = CaS(s)$	549.0	0.106	3.5	850–1487	1, 2, 3, 4
$Ca(g) + 1/2S_2(g) = CaS(s)$	696.0	0.189	3.1	1487–1700	2, 4
$3Ca(s) + N_2(g) = Ca_3N_2(s)$	437.0	0.155	31.6	25–850	1, 3
$Ca(s)a + 2C(s) = CaC_2(s)$	57.0	−0.024	0.3	25–400	1, 3
$Ca(s)b + 2C(s) = CaC_2(s)$	49.4	−0.035	0.5	400–850	1, 3
$Ca(l) + 2C(s) = CaC_2(s)$	60.6	−0.025	1.2	850–1487	1, 3, 4
$Ca(g) + 2C(s) = CaC_2(s)$	212.0	0.058	13.3	1487–1900	1, 3, 4
$Ca(s) + Si(s) = CaSi(s)$	150.0	0.002	–	25–850	1
$Ca(l) + Si(s) = CaSi(s)$	107.0	−0.028	–	850–1444	1
$2Ca(l) + Si(s) = Ca_2Si(s)$	178.0	−0.019	–	850–1444	1
$3CaO(s) + Al_2O_3(s) = Ca_3Al_2O_6(s)$	18.9	−0.028	11.1	25–1550	1, 3, 5
$12CaO(s) + 7Al_2O_3(s) = Ca_{12}Al_{14}O_{33}(s)$	72.8	−0.210	–	25–1500	1
$CaO(s) + Al_2O_3(s) = CaAl_2O_4(s)$	20.4	0.017	4.2	25–1600	1, 3, 5
$CaO(s) + CO_2(g) = CaCO_3(s)$	168.0	0.143	–	25–880	1
$2CaO(s) + Fe_2O_3(s) = Ca_2Fe_2O_5(s)$	38.3	−0.010	–	600–1435	1
$2CaO(s) + Fe_2O_3(s) = Ca_2Fe_2O_5(l)$	−31.5	−0.051	–	1435–1600	1
$4CaO(s) + P_2(g) + 5/2O_2(g) = Ca_4P_2O_9(s)$	2348.0	0.600	–	1300–1600	1
$3CaO(s) + P_2(g) + 5/2O_2(g) = Ca_3P_2O_8(s)$	2306.0	0.600	–	1300–1600	1
$2CaO(s) + SiO_2(s) = Ca_2SiO_4(s)$	103.0	−0.020	18.7	25–400	1, 5
$CaO(s) + SiO_2(s) = CaSiO_3(s)\alpha$	88.7	0.001	–	25–1210	1
$CaO(s) + SiO_2(s) = CaSiO_3(s)\beta$	82.9	−0.003	–	1210–1543	1
$CaO(s) + SiO_2(s) = CaSiO_3(s)$	76.7	−0.003	–	25–727	5
$3CaO(s) + 2SiO_2(s) = Ca_3Si_2O_7(s)$	202.0	−0.013	–	25–727	5
$3CaO(s) + SiO_2(s) = Ca_3SiO_5(s)$	115.0	−0.023	–	25–727	5
$CaO(s) + 2Al_2O_3(s) = CaAl_4O_7(s)$	16.8	−0.034	3.8	25–1765	3, 5
$CaO(s) + 6Al_2O_3(s) = CaAl_{12}O_{19}(s)$	22.5	−0.032	–	25–727	5
	96.0	−0.050	31.5	25–1590	3,5

(continued)

(continued)

$\Delta_r G^0 \left(\text{kJ mol}^{-1}\right) = A + B \cdot T(K)$	Linear correlation coefficient				Ref. (*)
	$-A$	B	$(\pm)$ kJ	°C	
2CaO (s) + SiO$_2$(s) + Al$_2$O$_3$(s) = Ca$_2$SiAl$_2$O$_7$(s)					
CaO (s) + 2SiO$_2$(s) + Al$_2$O$_3$(s) = CaSi$_2$Al$_2$O$_8$(s)	55.1	−0.035	35.4	25–1553	3, 5
CaO(s) + MgO (s) + 2SiO$_2$(s) = CaMgSi$_2$O$_6$(s)	37.6	−0.015	29.6	25–1392	3
CaO(s) + MgO (s) + SiO$_2$(s) = CaMgSiO$_4$(s)	99.2	−0.009	14.3	25–1227	3, 5
2CaO(s) + MgO (s) + 2SiO$_2$(s) = Ca$_2$MgSi$_2$O$_7$(s)	157.0	−0.030	24.8	25–1454	3, 5
3CaO(s) + MgO (s) + 2SiO$_2$(s) = Ca$_3$MgSi$_2$O$_8$(s)	204.0	−0.032	–	–	5
Cerium					
2Ce(s) + 3/2O$_2$(g) = Ce$_2$O$_3$(s)	1805.0	0.309	2.0	25–804	1, 3
2Ce(l) + 3/2O$_2$(g) = Ce$_2$O$_3$(s)	1810.0	0.325	68.4	804–1700	1, 3, 4
Ce(s) + O$_2$(g) = CeO$_2$(s)	1053.0	0.207	35.5	25–804	1, 3
Ce(l) + O$_2$(g) = CeO$_2$(s)	1046.0	0.211	52.8	804–1700	1, 3, 4
Cobalt					
Co(s) + 1/2O$_2$(g) = CoO(s)	235.0	0.074	1.4	25–1495	1, 2, 3
3CoO(s) + 1/2O$_2$(g) = Co$_3$O$_4$(s)	198.0	0.158	10.3	25–1000	1, 2, 3
9Co(s) + 4S$_2$(g) = Co$_9$S$_8$(s)	1316.0	0.634	18.7	25–778	1, 3
2Co(s) + C(s) = Co$_2$C(s)	−16.5	−0.009	-	25–900	1
Chromium					
2Cr(s) + 3/2O$_2$(g) = Cr$_2$O$_3$(s)	1129.0	0.264	11.6	25–1898	1, 2, 3
2Cr(s) + 1/2N$_2$(g) = Cr$_2$N(s)	117.0	0.067	11.4	25–1898	1, 2, 3
Cr(s) + 1/2N$_2$(g) = CrN(s)	113.0	0.075	4.8	25–1898	1, 2, 3
23Cr(s) + 6C(s) = Cr$_{23}$C$_6$(s)	352.0	−0.045	55.1	25–1400	1, 2, 3
7Cr(s) + 3C(s) = Cr$_7$C$_3$(s)	164.0	−0.026	10.5	25–1200	1, 2, 3
3Cr(s) + 2C(s) = Cr$_3$C$_2$(s)	84.7	−0.011	5.4	25–1700	1, 2, 3
Copper					
2Cu(s) + 1/2O$_2$(g) = Cu$_2$O(s)	168.0	0.074	2.9	25–1084	1, 2, 3
2Cu(l) + 1/2O$_2$(g) = Cu$_2$O(s)	182.0	0.082	1.9	1084–1230	1, 2, 3
1/2Cu$_2$O(s) + 1/4O$_2$(g) = CuO(s)	72.1	0.054	3.5	25–1000	1, 2, 3
2Cu(l) + 1/2O$_2$(g) = Cu$_2$O(l)	129.0	0.049	6.0	1229–1727	2, 3, 4
Cu(l) + 1/2O$_2$(g) = CuO(l)	154.0	0.088	1.6	1447–1727	2, 4
Iron					
Fe(s) + 1/2O$_2$(g) = FeO(s)-α	265.0	0.065	6.0	25–1371	1, 2, 3
0.95Fe(s) + 1/2O$_2$(g) = Fe$_{0.95}$O(s)	261.3	0.065	–	25–1371	–
Fe(s) + 1/2O$_2$(g) = Wustita(s)-β	31.2	0.019	–	25–1371	5
Fe(l) + 1/2O$_2$(g) = FeO(l)	244.0	0.050	4.4	1537–1700	1, 2

(continued)

(continued)

$\Delta_r G^0 \left(\text{kJ mol}^{-1}\right) = A + B \cdot T(K)$	Linear correlation coefficient				Ref. (*)
	$-A$	B	$(\pm)$ kJ	°C	
$3Fe(s) + 2O_2(g) = Fe_3O_4(s)$	1105.0	0.328	14.0	25–560	1, 2, 3
$3FeO(s) + 1/2O_2(g) = Fe_3O_4(s)$	296.0	0.112	14.6	560–1371	1, 2, 3
$2/3Fe_3O_4(s) + 1/6O_2(g) = Fe_2O_3(s)$	80.2	0.045	2.0	25–1400	1, 2, 3
$Fe(s) + 1/2S_2(g) = FeS(s)\alpha$	164.0	0.080	6.0	25–140	1, 2, 3
$Fe(s) + 1/2S_2(g) = FeS(s)\beta$	138.0	0.038	12.3	140–906	1, 2, 3
$FeS(s)b + 1/2S_2(g) = FeS_2(s)$	135.0	0.134	23.9	300–800	1, 2, 3
$4Fe(s) + 1/2N_2(g) = Fe_4N(s)$	7.7	0.044	1.9	25–600	1, 3
$3Fe(s) + 1/2P_2(g) = Fe_3P(s)$	224.0	0.066	5.6	25–1170	1, 3
$3Fe(s) + C(s) = Fe_3C(s)$	−26.5	−0.023	0.7	25–190	1, 3
$3Fe(s) + C(s) = Fe_3C(s)$	−28.1	−0.027	0.5	190–840	1, 3
$3Fe(s) + C(s) = Fe_3C(s)$	−9.8	−0.010	–	840–1537	1, 3
$FeO(s) + Al_2O_3(s) = FeAl_2O_4(s)$	47.1	0.017	15.3	25–1371	1, 3
$FeO(s) + Cr_2O_3(s) = FeCr_2O_4(s)$	33.2	−0.006	18.9	35–1371	1, 3
$2FeO(s) + SiO_2(s) = Fe_2SiO_4(s)$	55.0	0.028	0.1	25–1217	1, 3, 5
$2FeO(s) + SiO_2(s) = Fe_2SiO_4(l)$	−62.0	−0.048	–	1217–1371	1
$2FeO(l) + SiO_2(s) = Fe_2SiO_4(l)$	−14.4	−0.019	–	1371–1700	1
$FeO(s) + TiO_2(s) = FeTiO_3(s)$	11.2	−0.003	7.4	900–1371	1, 3
Hydrogen					
$H_2(g) + 1/2O_2(g) = H_2O(g)$	249.0	0.057	1.0	25–1700	1, 2, 3, 4
$H_2(g) + 1/2S_2(g) = H_2S(g)$	87.3	0.045	1.4	25–1500	1, 2, 3
$3/2H_2(g) + 1/2N_2(g) = NH_3(g)$	49.3	0.109	0.5	25–700	1, 3
Lanthanum					
$2La(s) + 3/2O_2(g) = La_2O_3(s)$	1823.0	0.283	33.1	25–700	1, 3
$La(s) + 1/2N_2(g) = LaN(s)$	299.0	0.097	3.3	25–700	1, 3
Magnesium					
$Mg(s) + 1/2O_2(g) = MgO(s)$	600.0	0.105	1.6	25–650	1, 2, 3
$Mg(l) + 1/2O_2(g) = MgO(s)$	609.0	0.115	2.5	650–1120	1, 2, 3
$Mg(g) + 1/2O_2(g) = MgO(s)$	732.0	0.205	5.6	1120–1700	1, 2, 3, 4
$Mg(s) + 1/2S_2(g) = MgS(s)$	391.0	0.070	28.5	25–650	1, 2, 3
$Mg(l) + 1/2S_2(g) = MgS(s)$	419.0	0.103	1.6	650–1120	1, 2, 3
$Mg(g) + 1/2S_2(g) = MgS(s)$	548.0	0.196	3.2	1120–1700	1, 2, 3, 4
$3Mg(s) + N_2(g) = Mg_3N_2(s)$	460.0	0.199	3.0	25–650	1, 2, 3
$3Mg(l) + N_2(g) = Mg_3N_2(s)$	487.0	0.228	3.9	650–1120	1, 2, 3
$MgO(s) + CO_2(g) = MgCO_3(s)$	106.0	0.158	14.1	25–700	1, 2, 3
$3MgO(s) + P_2(g) + 5/2O_2(g) = Mg_3P_2O_8(s)$	2047.0	0.576	12.2	1000–1250	1, 2, 3
$2MgO(s) + SiO_2(s) = Mg_2SiO_4(s)$	60.1	−0.002	7.1	650–1120	1, 2, 3, 5
$MgO(s) + SiO_2(s) = MgSiO_3(s)$	37.8	0.003	2.8	25–1300	1, 2, 3, 5

(continued)

(continued)

$\Delta_r G^0 \left(\text{kJ mol}^{-1} \right) = A + B \cdot T(K)$	Linear correlation coefficient				Ref. (*)
	$-A$	B	$(\pm)$ kJ	°C	
$Mg(g) + 2C(s) = MgC_2(s)$	30.9	0.072	37.9	122–1727	2, 4
$3Mg(g) + N_2(g) = Mg_3N_2(l)$	865.0	0.507	2.1	1227–1727	2, 4
$MgO(s) + Al_2O_3(s) = MgAl_2O_4(s)$	17.0	−0.015	7.7	25–1727	3, 5
Manganese					
$Mn(s) + 1/2O_2(g) = MnO(s)$	383.0	0.073	1.0	25–1244	1, 3
$Mn(l) + 1/2O_2(g) = MnO(s)$	402.0	0.086	3.2	1244–1700	1, 3, 4
$Mn(s) + 1/2S_2(g) = MnS(s)$	274.0	0.066	4.3	25–1244	1, 3
$Mn(l) + 1/2S_2(g) = MnS(s)$	291.0	0.078	9.0	1244–1530	1, 3, 4
$Mn(l) + 1/2S_2(g) = MnS(l)$	265.0	0.064	9.3	1530–1700	1, 3, 4
$3Mn(s) + C(s) = Mn_3C(s)$	14.0	−0.001	0.2	25–740	1, 3
$MnO(s) + SiO_2(s) = MnSiO_3(s)$	25.4	0.004	5.7	25–1300	1, 3, 5
$3Mn(l) + 2O_2(g) = Mn_3O_4(\beta)$	1411.0	0.366	7.0	1243–1560	3, 4
$7Mn(l) + 3C(s) = Mn_7C_3(s)$	65.8	−0.058	–	1243–1727	4
$2MnO(s) + SiO_2(s) = Mn_2SiO_4(s)$	55.6	0.016	9.0	25–1345	3, 5
$MnO(s) + Al_2O_3(s) = MnAl_2O_4(s)$	45.2	0.010	2.0	25–1727	3, 5
Molybdenum					
$Mo(s) + O_2(g) = MoO_2(s)$	584.0	0.177	1.4	25–1000	1, 3
$MoO_2(s) + 1/2O_2(g) = MoO_3(s)$	158.0	0.074	1.6	25–1000	1, 2, 3
$2Mo(s) + 3/2S_2(g) = Mo_2S_3(s)$	560.0	0.242	6.7	850–1200	1, 2, 3
$2Mo(s) + 1/2N_2(g) = Mo_2N(s)$	82.2	0.090	0.1	25–1000	1, 3
$2Mo(s) + C(s) = Mo_2C(s)$	36.8	0.004	17.3	25–1000	1, 3
$Mo(s) + C(s) = MoC(s)$	−40.4	−0.058	–	1227–1727	4
$Mo(s) + 3/2O_2(g) = MoO_3(g)$	392.0	0.057	87.9	1280–1727	2,3,4
$Mo(s) + S_2(g) = MoS_2(s)$	379.0	0.181	41.4	1227–1727	2, 3, 4
Niobium					
$2Nb(s) + 2O_2(g) = Nb_2O_4(s)$	1560.0	0.336	–	1227–1727	4
$2Nb(s) + 5/2O_2(g) = Nb_2O_5(s)$	1860.0	0.404	5.8	1227–1453	2, 3, 4
$2Nb(s) + 5/2O_2(g) = Nb_2O_5(l)$	1741.0	0.337	6.4	1453–1727	2, 3, 4
Nickel					
$Ni(s) + 1/2O_2(g) = NiO(s)$	236.0	0.088	1.7	25–1452	1, 3
$Ni(l) + 1/2O_2(g) = NiO(s)$	249.0	0.093	0.5	1452–1900	1, 3, 4
$Ni(s) + 1/2S_2(g) = NiS(s)$	135.0	0.057	24.2	300–580	1, 2, 3
$3Ni(s) + C(s) = Ni_3C(s)$	−63.5	−0.055	–	25,700	1
Phosphorus					
$1/2P_2(g) + 1/2O_2(g) = PO(g)$	104.0	−0.001	11.3	1227–1727	2, 4
$2P_2(g) + 5O_2(g) = P_4O_{10}(g)$	3125.0	0.980	38.9	1227–1727	2, 3, 4
Platinum					
$Pt(s) + 1/2S_2(g) = PtS(s)$	16.0	−0.005	0.3	1227–1727	3
Sulphur					

(continued)

(continued)

$\Delta_r G^0 \left(\text{kJ mol}^{-1} \right) = A + B \cdot T(K)$	Linear correlation coefficient				Ref. (*)
	$-A$	B	$(\pm)$ kJ	°C	
$1/2S_2(g) = S(g)$	−213.0	−0.057	1.3	25–700	1, 2, 3
$1/2S_2(g) + 1/2O_2(g) = SO(g)$	61.7	−0.006	3.7	25–1700	1, 2, 3, 4
$1/2S_2(g) + O_2(g) = SO_2(g)$	361.0	0.072	1.3	25–1700	1, 2, 3, 4
$1/2S_2(g) + 3/2O_2(g) = SO_3(g)$	455.0	0.151	10.7	25–1500	1, 2, 3
Silicon					
$Si(s) + 1/2O_2(g) = SiO(g)$	102.0	−0.072	6.4	25–1410	1, 2, 3
$Si(l) + 1/2O_2(g) = SiO(g)$	157.0	−0.047	5.5	1410–1700	1, 2, 3, 4
$Si(s) + O_2(g) = SiO_2(s)\text{cristob.}$	903.0	0.175	3.6	400–1410	1, 2, 3
$Si(l) + O_2(g) = SiO_2(s)\text{cristob.}$	945.0	0.198	5.7	1410–1700	1, 2, 3, 4
$3Si(s) + 2N_2(g) = Si_3N_4(s)$	738.0	0.326	14.2	25–1410	1, 2, 3
$3Si(l) + 2N_2(g) = Si_3N_4(s)$	874.0	0.406	3.0	1410–1700	1, 2, 3, 4
$Si(s) + C(s) = SiC(s)\beta$	66.3	0.006	11.2	25–1410	1, 2, 3
$Si(l) + C(s) = SiC(s)\beta$	113.0	0.036	11.1	1410–1700	1, 2, 3, 4
$2SiO_2(s) + 3Al_2O_3(s) = Si_2Al_6O_{13}(s)$	−22.0	−0.032	5.3	25–1850	3, 5
Tin					
$Sn(l) + O_2(g) = SnO_2(s)$	584.0	0.213	1.7	500–700	1, 3
$Sn(l) + 1/2O_2(g) = SnO(s)$	268.0	0.089	13.1	1227–1727	3, 4
$Sn(l) + 1/2O_2(g) = SnO(g)$	14.0	−0.045	11.5	1227–1727	3, 4
$Sn(l) + O_2(g) = SnO_2(l)$	511.0	0.168	–	1227–1727	4
Tantalum					
$2Ta(s) + 5/2O_2(g) = Ta_2O_5(s)$	1991.0	0.395	5.0	1227–1727	2, 3
$2Ta(s) + C(s) = Ta_2C(s)$	179.0	0.007	35.9	1227–1727	3, 4
$Ta(s) + C(s) = TaC(s)$	157.0	0.006	6.5	1227–1727	3, 4
$Ta(s) + 1/2N_2(g) = TaN(s)$	235.0	0.078	2.5	1227–1727	3, 4
Thorium					
$Th(s) + O_2(g) = ThO_2(s)$	1226.0	0.185	3.5	25–1500	1, 3
$3Th(s) + 2N_2(g) = Th_3N_4(s)$	1298.0	0.364	11.0	25–1700	1, 3
$Th(s) + 2C(s) = ThC_2(s)$	156.0	0.001	29.0	25–2000	1, 3
Titanium					
$Ti(s) + 1/2O_2(g) = TiO(s)$	530.0	0.092	18.3	300–1700	1, 2, 3
$2TiO(s) + 1/2O_2(g) = Ti_2O_3(s)$	448.0	0.087	32.0	25–1700	1, 2, 3
$3/2Ti_2O_3(s) + 1/4O_2(g) = Ti_3O_5(s)$	176.0	0.032	7.3	400–1700	1, 2, 3
$1/3Ti_3O_5(s) + 1/6O_2(g) = TiO_2(s)$	116.0	0.031	15.5	25–1850	1, 2, 3
$Ti(s)\alpha + 1/2N_2(g) = TiN(s)$	337.0	0.095	1.6	25–882	1, 2, 3
$Ti(s)\beta + 1/2N_2(g) = TiN(s)$	336.0	0.094	1.3	882–1200	1, 2, 3
$Ti(s)\alpha + C(s) = TiC(s)$	183.0	0.011	0.9	25–882	1, 2, 3
$Ti(s)\beta + C(s) = TiC(s)$	187.0	0.014	1.3	882–1700	1, 2, 3
$Ti(s) + 1/2O_2(g) = TiO(g)$	−32.4	−0.079	1.3	1227–1667	3, 4
Vanadium					

(continued)

$\Delta_r G^0 (\text{kJ mol}^{-1}) = A + B \cdot T(K)$	Linear correlation coefficient				Ref. (*)
	$-A$	B	$(\pm)$ kJ	°C	
$V(s) + 1/2O_2(g) = VO(s)$	425.0	0.081	11.0	600–1500	1, 2, 3
$2VO(s) + 1/2O_2(g) = V_2O_3(s)$	393.0	0.065	58.2	550–1112	1, 2, 3
$V(s) + C(s) = VC(s)$	102.0	0.009	1.7	900–1100	1, 3
$2V(s) + 2O_2(g) = V_2O_4(s)$	1380.0	0.294	2.7	1227–1545	2, 4
$2V(s) + 2O_2(g) = V_2O_4(l)$	1256.0	0.225	1.9	1545–1727	2, 4
$2V(s) + 5/2O_2(g) = V_2O_5(l)$	1442.0	0.320	9.1	1227–1727	2, 3, 4
$V(s) + 1/2N_2(g) = VN(s)$	195.0	0.079	28.8	1227–1727	2, 3, 4
Wolfram					
$W(s) + C(s) = WC(s)$	38.8	0.002	1.3	25–1700	1, 3
$W(s) + O_2(g) = WO_2(s)$	567.0	0.164	2.3	1227–1727	2, 3, 4
$W(s) + 3/2O_2(g) = WO_3(s)$	815.0	0.232	3.8	1227–1470	2, 3, 4
$W(s) + 3/2O_2(g) = WO_3(l)$	736.0	0.186	4.0	1470–1727	2, 3, 4
Zirconium					
$Zr(s)\alpha + O_2(g) = ZrO_2(s)$	1092.0	0.191	11.3	25–870	1, 2, 3
$Zr(s)\beta + O_2(g) = ZrO_2(s)$	1076.0	0.177	11.9	1205–1865	1, 2, 3, 4
$Zr(s)\alpha + 1/2N_2(g) = ZrN(s)$	365.0	0.094	2.6	25–862	1, 2, 3
$Zr(s)\beta + 1/2N_2(g) = ZrN(s)$	365.0	0.093	3.9	862–1200	1, 2, 3
$Zr(s) + C(s) = ZrC(s)$	196.0	0.010	11.6	25–1900	1, 2, 3

(*) In the elaboration of this table we took as reference the work of the Association of Iron and Steel Engineers (AISE) previously included in the database of Janaf, Knacke, Pehlke and Gaye (according to Verdeja, L. F., Alfonso, A. and Suárez, M.: Energías libres de formación de compuestos siderúrgicos. Revista de Minas, No. 15-16, 109-112, 1997). The numbers of the last column are contributions to the values of A and B

1. Lankford, W. T.: The making, shaping and treating of steel. Association of Iron and Steel Engineers, 10ª edición. Pittsburgh, EEUU. 1985
2. Chase, M. W. Jr., Davies, C. A., Downey, J. R. Jr., Frurip, D. J., McDonald, R. A., Syverud, A. N.: JANAF Thermochemical Tables. American Chemical Society y American Institute of Physics for the National Bureau Standards, 3ª edition, New York, USA. 1986
3. Knacke, O., Kubaschewski, O. and Hesselmann, K.: Thermochemical properties of inorganic substances. Springer-Verlag. 2ª edition, Berlin, Germany. 1991
4. Pehlke, R. D.: Unit processes of extractive metallurgy. Elsevier, 2ª edición. New York, USA. 1975
5. Gaye, H. y Welfringer, J.: Modelling of the thermodynamic properties of complex metallurgical slags. Proceedings Second International Symposium on Metallurgical Slags and Fluxes. Editors Fine, U. A. and Gaskell, D. R., Metallurgical Society of AIME, Warrendale, USA. 1984

Appendix 2

Standard enthalpies of several iron and steelmaking reactions, $\Delta_r H^0$.

$\Delta_r H^0 \left(\text{kJ mol}^{-1} \right) = A + B \cdot T(K)$	Linear correlation coefficient				Ref. (*)
	$-A$	B	$(\pm)$ kJ	°C	
Aluminum					
$2Al(s) + 3/2O_2(g) = Al_2O_3(s)$	1676.0	0.0016	0.6	25–659	1, 2
$2Al(l) + 3/2O_2(g) = Al_2O_3(s)$	1710.0	0.0149	0.2	659–1700	1, 2
$Al(s) + 1/2N_2(g) = AlN(s)$	318.0	−0.0026	0.4	25–659	1, 2
$Al(l) + 1/2N_2(g) = AlN(s)$	331.0	0.0014	03	659–1700	1, 2
$4Al(s) + 3C(s) = Al_4C_3(s)$	212.0	−0.0005	3.5	25–659	1, 2
$4Al(l) + 3C(s) = Al_4C_3(s)$	257.0	−0.0058	7.7	659–1700	1, 2
$2Al(l) + 1/2O_2(g) = Al_2O(g)$	157.0	−0.0203	0.1	1227–1727	1, 2
$Al(l) + 1/2O_2(g) = AlO(g)$	−55.0	−0.0074	0.1	1227–1727	1, 2
Boron					
$2B(s) + 3/2O_2(g) = B_2O_3(s)$	1273.0	0.0020	0.4	25–450	1, 2
$2B(s) + 3/2O_2(g) = B_2O_3(l)$	1251.0	0.0172	0.2	450–1700	1, 2
$B(s) + 1/2N_2(g) = BN(s)$	251.0	−0.0012	0.8	25–900	1, 2
$4B(s) + C(s) = B_4C(s)$	62.7	0.0003	0.2	25–900	1, 2
$B(s) + 1/2O_2(g) = BO(g)$	−8.0	−0.0108	0.1	1227–1727	1, 2
Barium					
$Ba(s) + 1/2O_2(g) = BaO(s)$	547.0	−0.0016	0.5	25–704	1, 2
$Ba(l) + 1/2O_2(g) = BaO(s)$	564.0	0.0040	0.6	704–1638	1, 2
Beryllium					
$Be(s) + 1/2O_2(g) = BeO(s)$	608.0	−0.0003	0.2	25–1283	1, 2
$Be(l) + 1/2O_2(g) = BeO(s)$	627.0	0.0069	3.5	1283–1700	1, 2
$3Be(s) + N_2(g) = Be_3N_2(s)$	599.0	0.0311	10.5	25–700	1, 2
Carbon					
$C(s) + 1/2O_2(g) = CO(g)$	104.0	−0.0075	0.2	25–2000	1, 2
$C(s) + O_2(g) = CO_2(g)$	393.0	−0.0017	1.0	25–2000	1, 2
$C(s) + 1/2S_2(g) = CS(g)$	−224.0	−0.0074	0.1	1600–1800	1, 2

(continued)

© The Editor(s) (if applicable) and The Author(s), under exclusive license to Springer
Nature Switzerland AG 2020
J. I. Verdeja González et al., *Operations and Basic Processes in Ironmaking*,
Topics in Mining, Metallurgy and Materials Engineering,
https://doi.org/10.1007/978-3-030-54606-9

(continued)

$\Delta_r H^0 \left(\text{kJ mol}^{-1}\right) = A + B \cdot T(K)$	Linear correlation coefficient				Ref. (*)
	$-A$	B	$(\pm)$ kJ	°C	
$CO(g) + 1/2S_2(g) = COS(g)$	142.0	−0.0196	9.2	25–1200	1, 2
Calcium					
$Ca(s) + 1/2O_2(g) = CaO(s)$	636.0	0.0043	0.1	25–850	1, 2
$Ca(l) + 1/2O_2(g) = CaO(s)$	647.0	0.0030	0.1	850–1487	1, 2
$Ca(g) + 1/2O_2(g) = CaO(s)$	823.0	0.0184	0.1	1487–1700	1, 2
$Ca(s) + 1/2S_2(g) = CaS(s)$	503.0	−0.0068	32.2	25–850	1, 2
$Ca(l) + 1/2S_2(g) = CaS(s)$	548.0	0.0016	0.1	850–1487	1, 2
$Ca(g) + 1/2S_2(g) = CaS(s)$	724.0	0.0168	0.1	1487–1700	1, 2
Cobalt					
$Co(s) + 1/2O_2(g) = CoO(s)$	241.0	0.0110	0.3	25–1495	1, 2
$3CoO(s) + 1/2O_2(g) = Co_3O_4(s)$	192.0	−0.0298	5.0	25–1000	1, 2
Chromium					
$2Cr(s) + 3/2O_2(g) = Cr_2O_3(s)\beta$	1143.0	0.0180	3	25–1898	1, 2
$2Cr(s) + 1/2N_2(g) = Cr_2N(s)$	128.0	0.0071	0.6	25–1898	1, 2
$Cr(s) + 1/2N_2(g) = CrN(s)$	120.0	0.0088	0.2	25–1898	1, 2
$23Cr(s) + 6C(s) = Cr_{23}C_6(s)$	342.0	0.0458	2	25–1400	1, 2
$7Cr(s) + 3C(s) = Cr_7C_3(s)$	167.0	0.0209	0.4	25–1200	1, 2
$3Cr(s) + 2C(s) = Cr_3C_2(s)$	89.9	0.0151	0.3	25–1700	1, 2
Copper					
$2Cu(l) + 1/2O_2(g) = Cu_2O(s)$	10.5	−0.1230	9	1084–1230	1, 2
$1/2Cu_2O(s) + 1/4O_2(g) = CuO(s)$	72.4	0.0057	0.1	25–1000	1, 2
$2Cu(l) + 1/2O_2(g) = Cu_2O(l)$	149.0	0.0158	0.2	1229–1727	1, 2
Iron					
$Fe(s) + 1/2O_2(g) = FeO(s)\text{-}\alpha$	271.0	0.0080	3	25–1371	1, 2
$3Fe(s) + 2O_2(g) = Fe_3O_4(s)$	1129.0	0.0338	3.6	25–560	1, 2
$3FeO(s) + 1/2O_2(g) = Fe_3O_4(s)$	292.0	0.0035	4.4	560–1371	1, 2
$2/3Fe_3O_4(s) + 1/6O_2(g) = Fe_2O_3(s)$	78.6	−0.0011	1	25–1400	1, 2
$Fe(s) + 1/2S_2(g) = FeS(s)\alpha$	175.0	0.0188	4	25–140	1, 2
$Fe(s) + 1/2S_2(g) = FeS(s)\beta$	170.0	0.0149	4.6	140–906	1, 2
$FeS(s)b + 1/2S_2(g) = FeS_2(s)$	135.0	−0.0066	3	300–800	1, 2
$4Fe(s) + 1/2N_2(g) = Fe_4N(s)$	10.7	−0.0006	0.4	25–600	1, 2
Hydrogen					
$H_2(g) + 1/2O_2(g) = H_2O(g)$	245.0	−0.0034	0.2	25–1700	1, 2
$H_2(g) + 1/2S_2(g) = H_2S(g)$	82.0	−0.0097	0.2	25–1500	1, 2
$3/2H_2(g) + 1/2N_2(g) = NH_3(g)$	40.8	−0.0176	0.3	25–700	1, 2
Magnesium					
$Mg(s) + 1/2O_2(g) = MgO(s)$	600.0	0.0024	2	25–650	1, 2
$Mg(l) + 1/2O_2(g) = MgO(s)$	607.0	0.0007	2.4	650–1120	1, 2
$Mg(g) + 1/2O_2(g) = MgO(s)$	755.0	0.0157	3	1120–1700	1, 2

(continued)

(continued)

$\Delta_r H^0 \left(\text{kJ mol}^{-1} \right) = A + B \cdot T(K)$	Linear correlation coefficient				Ref. (*)
	$-A$	B	$(\pm)$ kJ	°C	
$Mg(s) + 1/2S_2(g) = MgS(s)$	376.0	−0.0074	32.2	25–650	1, 2
$Mg(l) + 1/2S_2(g) = MgS(s)$	607.0	0.0007	2.4	650–1120	1, 2
$Mg(g) + 1/2S_2(g) = MgS(s)$	563.0	0.0140	05	1120–1700	1, 2
$3Mg(s) + N_2(g) = Mg_3N_2(s)$	461.0	−0.0012	0.5	25–650	1, 2
$3Mg(l) + N_2(g) = Mg_3N_2(s)$	477.0	−0.0108	1.3	650–1120	1, 2
$MgO(s) + CO_2(g) = MgCO_3(s)$	112.0	0.0057	6.4	25–700	1, 2
$3MgO(s) + P_2(g) + 5/2O_2(g) = Mg_3P_2O_8(s)$	2155.0	0.0902	0.3	1000–1250	1, 2
$2MgO(s) + SiO_2(s) = Mg_2SiO_4(s)$	66.0	−0.0027	4	650–1120	1, 2
$MgO(s) + SiO_2(s) = MgSiO_3(s)$	37.2	−0.0032	1.9	25–1300	1, 2
$MgO(s) + Al_2O_3(s) = MgAl_2O_4(s)$	31.5	0.0068	20.7	25–1727	1, 2
Molybdenum					
$MoO_2(s) + 1/2O_2(g) = MoO_3(s)$	159.0	0.0043	0.1	25–1000	1, 2
$2Mo(s) + 3/2S_2(g) = Mo_2S_3(s)$	597.0	0.1430	195	850–1200	1, 2
$Mo(s) + 3/2O_2(g) = MoO_3(g)$	337.0	−0.0080	0.3	1280–1727	1, 2
$Mo(s) + S_2(g) = MoS_2(s)$	407.0	0.0112	0.1	1227–1727	1, 2
Niobium					
$2Nb(s) + 5/2O_2(g) = Nb_2O_5(s)$	1909.0	0.0300	1.2	1227–1453	1, 2
$2Nb(s) + 5/2O_2(g) = Nb_2O_5(l)$	1898.0	0.0821	1.3	1453–1727	1, 2
Nickel					
$Ni(s) + 1/2S_2(g) = NiS(s)$	98.7	−0.0356	38.4	300–580	1, 2
Phosphorus					
$2P_2(g) + 5O_2(g) = P_4O_{10}(g)$	3228.0	0.0618	0.8	1227–1727	1, 2
Sulphur					
$1/2S_2(g) = S(g)$	−211.0	0.0058	0.1	25–700	1, 2
$1/2S_2(g) + 1/2O_2(g) = SO(g)$	57.8	−0.0008	0.4	25–1700	1, 2
$1/2S_2(g) + O_2(g) = SO_2(g)$	363.0	0.0009	0.1	25–1700	1, 2
$1/2S_2(g) + 3/2O_2(g) = SO_3(g)$	459.0	−0.0025	0.5	25–1500	1, 2
Silicon					
$Si(s) + 1/2O_2(g) = SiO(g)$	97.6	−0.0060	1.1	25–1410	1, 2
$Si(l) + 1/2O_2(g) = SiO(g)$	147.0	−0.0084	1.1	1410–1700	1, 2
$Si(s) + O_2(g) = SiO_2(s)$cristob.	912.0	0.0082	0.7	400–1410	1, 2
$Si(l) + O_2(g) = SiO_2(s)$cristob.	962.0	0.0093	1	1410–1700	1, 2
$3Si(s) + 2N_2(g) = Si_3N_4(s)$	741.0	−0.0121	0.4	25–1410	1, 2
$3Si(l) + 2N_2(g) = Si_3N_4(s)$	934.0	0.0296	3	1410–1700	1, 2
$Si(s) + C(s) = SiC(s)\beta$	72.6	0.0006	0.9	25–1410	1, 2

(continued)

(continued)

$\Delta_r H^0 \left(kJ \, mol^{-1} \right) = A + B \cdot T(K)$	Linear correlation coefficient				Ref. (*)
	$-A$	B	$(\pm)$ kJ	°C	
$Si(l) + C(s) = SiC(s)\beta$	125.0	0.0017	0.8	1410–1700	1, 2
Tantalum					
$2Ta(s) + 5/2O_2(g) = Ta_2O_5(s)$	2068.0	0.0424	0.2	1227–1727	1, 2
$Ta(s) + C(s) = TaC(s)$	146.0	0.0027	0.2	1227–1727	1, 2
Titanium					
$Ti(s) + 1/2O_2(g) = TiO(s)$	546.0	0.0082	0.2	300–1700	1, 2
$2TiO(s) + 1/2O_2(g) = Ti_2O_3(s)$	441.0	0.0180	1	25–1700	1, 2
$3/2Ti_2O_3(s) + 1/4O_2(g) = Ti_3O_5(s)$	165.0	−0.0176	5.9	400–1700	1, 2
$1/3Ti_3O_5(s) + 1/6O_2(g) = TiO_2(s)$	120.0	−0.0094	5.4	25–1850	1, 2
$Ti(s)\alpha + 1/2N_2(g) = TiN(s)$	339.0	0.0027	0.4	25–882	1, 2
$Ti(s)\beta + 1/2N_2(g) = TiN(s)$	345.0	0.0068	1.4	882–1200	1, 2
$Ti(s)\alpha + C(s) = TiC(s)$	185.0	0.0022	0.2	25–882	1, 2
$Ti(s)\beta + C(s) = TiC(s)$	186.0	0.0004	1.4	882–1700	1, 2
Vanadium					
$V(s) + 1/2O_2(g) = VO(s)$	438.0	0.0136	0.4	600–1500	1, 2
$2VO(s) + 1/2O_2(g) = V_2O_3(s)$	350.0	−0.0048	3	550–1112	1, 2
$2V(s) + 5/2O_2(g) = V_2O_5(l)$	1482.0	0.0237	1.1	1227–1727	1, 2
$V(s) + 1/2N_2(g) = VN(s)$	219.0	0.0062	0.2	1227–1727	1, 2
Wolfram					
$W(s) + O_2(g) = WO_2(s)$	606.0	0.0227	0.7	1227–1727	1, 2
$W(s) + 3/2O_2(g) = WO_3(s)$	852.0	0.0221	0.4	1227–1470	1, 2
$W(s) + 3/2O_2(g) = WO_3(l)$	817.0	0.0443	0.4	1470–1727	1, 2
Zirconium					
$Zr(s)\alpha + O_2(g) = ZrO_2(s)$	1101.0	0.0080	1.7	25–870	1, 2
$Zr(s)\beta + O_2(g) = ZrO_2(s)$	1095.0	0.0068	1.8	1205–1865	1, 2
$Zr(s)\alpha + 1/2N_2(g) = ZrN(s)$	368.0	0.0042	1.6	25–862	1, 2
$Zr(s)\beta + 1/2N_2(g) = ZrN(s)$	375.0	0.0078	2.4	862–1200	1, 2
$Zr(s) + C(s) = ZrC(s)$	203.0	0.0035	5.9	25–1900	1, 2

(*) The numbers indicated in the last column of the table are the contributions that were considered to calculate the constants A and B of the linear regression line to obtain the standard enthalpies of the above mentioned reactions

1. Chase, M. W. Jr., Davies, C. A., Downey, J. R. Jr., Frurip, D. J., McDonald, R. A., Syverud, A. N.: JANAF Thermochemical Tables. American Chemical Society y American Institute of Physics for the National Bureau Standards, 3ª edition, New York, USA. 1986
2. Knacke, O., Kubaschewski, O. and Hesselmann, K.: Thermochemical properties of inorganic substances. Springer-Verlag. 2ª edition, Berlin, Germany. 1991

Appendix 3

Free energies associated with the change of standard state: from Raoult's activity one to Henry's activity one (*).

$\Delta_r G^0 (\text{J mol}^{-1}) = A + B \cdot T(K)$	Linear correlation coefficient			Ref
	A	B	$(\pm)$ J	
Ag(l)	82,400	−43.8	36	1, 2
Al(l)	−67,100	−23.7	4857	1,2
B(s)	−65,300	−21.5	17	1, 2
C(s)	19,900	−41.1	565	1, 2
Ca(g)	−39,455	49.37	−	1
Co(l)	1004	−38.74	−	1
Cr(l)	0	−37.7	2	1, 2
Cr(s)	19,200	−46.9	23	1, 2
Cu(l)	33,500	−39.4	16	1, 2
$1/2H_2(g)$	36,500	30.5	12	1, 2
Mn(l)	4084	−38.16	−	1
Mo(l)	0	42.8	−	1
Mo(s)	27,614	−52.38	−	1
$1/2N_2(g)$	6757	22	698	1, 2
Nb(l)	−	−42.68	−	1
Nb(s)	23,000	−52.3	6	1, 2
Ni(l)	−20,900	−31.1	47	1, 2
$1/2O_2(g)$	−117,000	−3.14	500	1, 2
$1/2P_2(g)$	−140,000	−6.92	7782	1, 2
Pb(l)	212,547	−106.27	−	1
$1/2S_2(g)$	−130,000	21	609	1, 2
Si(l)	−132,000	−17.2	1	1, 2
Sn(l)	15,983	−44.43	−	1
Ti(l)	−46,024	−37.03	−	1
Ti(s)	−13,690	−44.98	−	1
U(l)	−56,100	−50.3	113	1, 2
V(l)	−42,300	−36.0	39	1, 2

(continued)

© The Editor(s) (if applicable) and The Author(s), under exclusive license to Springer Nature Switzerland AG 2020
J. I. Verdeja González et al., *Operations and Basic Processes in Ironmaking*,
Topics in Mining, Metallurgy and Materials Engineering,
https://doi.org/10.1007/978-3-030-54606-9

(continued)

$\Delta_r G^0 \left(\text{J mol}^{-1} \right) = A + B \cdot T(K)$	Linear correlation coefficient			Ref
	A	B	$(\pm)$ J	
V(s)	−20,700	−45.6	11	1, 2
W(l)	–	48.1	11	1, 2
W(s)	31,380	−63.6	-	1
Zr(l)	−51,000	−42.4	26	1, 2
Zr(s)	−34,700	−50	13	1, 2

*According to: (1) Sigworth, G. K., and Elliot, J. F., 1994, The thermodynamic of liquid dilute iron alloy, *Metal Science*, 8, pp. 298–310; (2) The Japanese Society for the Promotion of Science. The 19th Committee on Steelmaking: Steelmaking data sourcebook. Gordon and Breach Science, New York, USA, 1988

Printed in the United States
by Baker & Taylor Publisher Services